# City Adrift

# City Adrift

## NEW ORLEANS BEFORE AND AFTER KATRINA

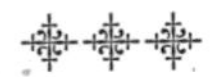

Jenni Bergal
Sara Shipley Hiles
Frank Koughan
John McQuaid
Jim Morris
Katy Reckdahl
Curtis Wilkie

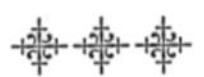

Foreword by Dan Rather

A CENTER FOR PUBLIC INTEGRITY INVESTIGATION

LOUISIANA STATE UNIVERSITY PRESS
BATON ROUGE

Published by Louisiana State University Press

Manufactured in the United States of America
First printing

Designer: Laura Roubique Gleason
Typeface: Whitman
Printer and binder: Edwards Brothers, Inc.

Library of Congress Cataloging-in-Publication Data

City adrift : New Orleans before and after Katrina / Jenni Bergal ... [et al.] ; foreword by Dan Rather ; a Center for Public Integrity Investigation.
p. cm.
Includes bibliographical references and index.
ISBN-13: 978-0-8071-3284-5 (cloth : alk. paper)
1. Hurricane Katrina, 2005. 2. New Orleans (La.)—History—21st century. 3. Disaster victims—Services for—Louisiana—New Orleans. 4. United States. Federal Emergency Management Agency. 5. Disaster relief—Louisiana—New Orleans. 6. Crisis management in government—Louisiana—New Orleans. I. Bergal, Jenni. II. Center for Public Integrity.
HV636 2005 .N4 C58 2007
976.3'35064—dc22

2007005276

The paper in this book meets the guidelines for permanence and durability of the Committee on Production Guidelines for Book Longevity of the Council on Library Resources. ♾

*For the people of New Orleans,*
*who are struggling to rebuild their city and their lives*

## THE TEAM

EDITORIAL DIRECTOR
Diane Fancher

PROJECT MANAGER
Jenni Bergal

RESEARCHER
Shelia Jackson

COPY EDITORS
Tonia E. Moore, Marcia Kramer

FACT-CHECKERS
Peter Smith, Robin Palmer, Alex Cohen

ADDITIONAL REPORTING
John Perry

# Contents

# Acknowledgments

Within days of Hurricane Katrina's assault on New Orleans, the Center for Public Integrity recognized the tragic severity of the government's lack of preparation and the even more woeful response. We decided to create Katrina Watch, a daily online roundup of news and government contract information. Then we launched a major journalistic investigation into what happened in the battered city, resulting in this book.

At the Center, we believe strongly in the power of investigative reporting to educate the wider society. Only by digging into and understanding what happened before, during and after this disaster, could we as a nation learn from our mistakes and perhaps prevent far worse from happening in the future.

For help on this massive project, the Center turned to renowned investigative journalist Bill Moyers. Over his distinguished career at CBS and PBS, Moyers has been honored with more than 30 Emmys as well as the Gold Baton, broadcast journalism's highest award, bestowed in the name of Alfred I. duPont by Columbia University. Moyers is currently president of the Schumann Center for Media and Democracy, an independent foundation that supports programs in effective governance and the environment.

Moyers had also been struck by the government's dismal response and was concerned that the poor and disadvantaged would suffer most from Katrina's wrath, which was indeed the case. He has been quoted as saying, "the rich are getting richer, which arguably wouldn't matter if the rising tide lifted all boats. But the inequality gap is the widest it's been since 1929; the middle class is besieged and the working poor are barely keeping their heads above water." In New Orleans, that water overwhelmed them.

We are enormously grateful to Bill Moyers and the Schumann Center for providing the financial support to thoroughly investigate this disaster. I believe this book helps hold government and emergency agencies accountable—at the local, state and federal levels. In doing so, the Center found that Katrina and its devastating aftermath were much more the result of cumulative, incompetent, insensitive acts of man than an act of nature.

*City Adrift* was completed through the dedication, talent and diligent effort of many Center for Public Integrity staffers along with several others. Then–Executive Director Roberta Baskin got this project rolling. Interim Executive Director Wendell "Sonny" Rawls kept it moving. Editorial Director Diane Fancher guided the project to fruition. Jenni Bergal was the project manager who conceived the idea of the book and kept the writers on track. Shelia Jackson was the invaluable researcher and assistant. Tonia Moore copy-edited much of the book, with help from Marcia Kramer. Robin Palmer, Peter Smith and Alex Cohen provided rigorous fact-checking. Jyoti Sauna conceptualized the cover design. John Perry helped with reporting. In addition to acknowledging the great work of our authors, we appreciate the vital contributions of Leah Rush, Brad Glanzrock, Helena Bengtsson and intern Erika Kaneko. And a special thank you to veteran newsman Dan Rather for his thoughtful and powerful foreword.

William E. Buzenberg
Executive Director
The Center for Public Integrity

# Foreword

On August 29, 2005, a hurricane named Katrina barreled out of the Gulf of Mexico, making landfall on the Louisiana coast between Grand Isle and the mouth of the Mississippi River. Over the days that followed, America saw the slow unfolding of a national tragedy.

Katrina had visited a worst-case scenario upon low-lying New Orleans, its storm surge breaching the levees that have long been that city's sole bulwark against the waters of the gulf. But what brought shock, dismay, and a mounting sense of anger to the Americans who witnessed the events from afar was not the natural disaster wrought by Katrina. It was the human disaster that followed.

Entire neighborhoods in one of America's oldest and most beloved cities sat under many feet of water, with thousands needing help. And for days that help did not come. The images that came out of New Orleans that week were gut-wrenching and heartbreaking, and they challenged directly some of our country's most cherished notions about itself: about our readiness and ingenuity in the face of trouble, about the quality of our leaders, and about the equality of our society.

A lot has been written and said about what happened in New Orleans in the more than a year since Katrina struck. The botched relief effort has been the subject of congressional investigations and of countless editorials and news analyses. In national debates as far-ranging as those on race relations, terrorism preparedness and politics, Katrina has become a touchstone and a symbol.

But in New Orleans, Katrina's aftermath was, and is still, all too real. As we were reminded briefly on the one-year anniversary of the storm—before our national attention turned again to other things—whole stretches of that storied city remain in ruin. Thousands of its onetime citizens, a "Katrina diaspora," are scattered throughout the wider region and beyond. And those who have returned face bureaucratic confusion that, while no longer a threat to life and limb, does call into question the future of a once-vibrant city so rich in history.

Americans know that federal, state and local government all failed New Orleans in its time of need. But between the confusion of real-time events and the inevitable finger-pointing and political spin that came afterward, I'm not sure we've ever been granted a full understanding and accounting of just what went wrong, with whom, and why.

In this book, seven authors who reported on-site provide just such an understanding. The picture that emerges is one in which, despite decades of warnings and studies, every institution and system put into place to protect us—from emergency preparedness to health care—instead collapsed when the storm slammed into New Orleans. In these pages is a concise portrait of a civic breakdown, made all the more frightening by the realization that the civilization in question was and is our own.

There are scenes in the pages that follow that are every bit as harrowing as things this writer has seen on reporting trips to the Third World. The chapter on health care, in particular, seems sprung from the verses of Dante, with its extremes of suffering and privation. And throughout these reports, the word and the theme that emerges time and again is chaos—chaos in the streets and hospitals of New Orleans, yes, but also in the decision-making and communication processes that were supposed to avert and ameliorate such catastrophe.

In the long sagas of New Orleans' levees and the deterioration of the wetlands that protect the city, we see the chaos that can attend a democratic process in which more prosaic environmental concerns and long-range civic planning have no ready champions or constituency against short-term budgetary pressures.

And in the accounts of how government agencies and non-governmental organizations responded—or did not respond—to New Orleans' week of hell, we can discern the chaos that rushes in to fill a near-total vacuum of leadership. The people of New Orleans could have been helped, and well-meaning, dedicated people tried to help; but far too often the best intentions were stymied by red tape, bureaucratic inefficiency and the apparent inability of anyone who was in a position to be of aid to see the big picture.

There is blame to be found here, and there is plenty of it to go around.

But more importantly, there is also a good-faith and thoroughly objective effort to make some sense of this chaos, if only in hindsight. This is the greatest potential value of this book, because we need to absorb the lessons of Katrina, and we can only do that if we have a clear appreciation of just what those lessons are.

Of all the things said here by those who experienced Katrina first-hand, perhaps the most succinct and telling summation of the man-made disaster that came in the wake of the natural one was offered by a 76-year-old woman named Pearl Ellis. Her words provide the epigraph for the first chapter. "It could have been prevented," said Mrs. Ellis. "A lot of lives could have been saved."

Will we as a nation learn from what happened in New Orleans, or will similar words be spoken again after we are called upon to face the next calamity?

Dan Rather
New York

# City Adrift

ONE

# The Storm

Jenni Bergal

*"It could have been prevented. A lot of lives would have been saved."*
—Pearl Ellis, Hurricane Katrina victim

Pearl Ellis remembers the storm approaching New Orleans. She woke her husband, and yelled that the water had come inside the house and that they had to get out.

He draped the baby around his neck and the other eight children linked hands as the family fled in waist-deep water to the local school, where they were rescued by boat and brought to City Hall.

"I thought the whole world was underwater," she recalls.

Many residents of her Lower Ninth Ward neighborhood escaped to attics and rooftops and hoped to be rescued. Dozens didn't make it. Bodies were found floating in the streets. Thousands of houses were severely damaged by the flood, which burst into the area after nearby levees were overtopped and breached during the storm.

When Pearl Ellis returned, she found her home under nearly 4 feet of water. Mud caked the walls. The furniture was covered with muck. Nothing much was salvageable, other than some whiskey her husband had stored in a bar stool.

"Everything was ruined. You had to start from scratch," she says.[1]

The year was 1965. The storm was Hurricane Betsy.

President Lyndon B. Johnson rushed to New Orleans and offered federal assistance to the city. Congress scrambled to authorize a massive hurricane protection project. The head of the Louisiana National Guard later conceded that confusion and communications breakdowns had impeded the rescue and clean-up efforts.[2]

"We have just suffered a disaster that is comparable to an earthquake. Something we couldn't foresee, couldn't anticipate," Louisiana's then-governor, John J. McKeithen, announced at a news conference. "I think this terrible tragedy will start the building of a tremendous hurricane protection of this area," he said.[3]

Pearl Ellis will never forget Hurricane Betsy. She proudly describes how her working-class neighbors came together and cleaned, gutted and rebuilt their homes, just as she and her husband Jessie had done.

That was then.

Today, the 76-year-old widow, whom everyone calls "Miz Pearl," stands in front of her small, white, one-story brick house in tears. Once again, her home has been wrecked.

Forty years after Betsy ravaged the Lower Ninth Ward, the community had been devastated by a storm again. This time, it was August 29, 2005, and the hurricane was Katrina.

"Everything we owned was in this house. When we came back, there was nothing left, nothing but the brick walls. All I had was the memories," says Ellis, a small, animated woman with dark hair tied neatly into a bun, who wears a silver St. Christopher medal around her neck.

But the "Lower Nine," as locals call it, was only one of many parts of New Orleans seriously damaged this time. Less than a day after Katrina struck, about 80 percent of the city was flooded, some areas by as much as 20 feet of water.[4] Lower income neighborhoods with modest homes and tidy yards were swamped with floodwater that left them uninhabitable. So were upscale waterfront neighborhoods with huge homes boasting gourmet kitchens and swimming pools.

Hundreds of thousands of New Orleans residents were displaced by Katrina. More than 1,400 Louisianans died, more than half of them from New Orleans.[5]

Tens of thousands fled to the Superdome and the Ernest N. Morial Convention Center downtown, where they sweltered in filthy, unsafe conditions, waiting for rescuers who took days to come. Others gathered on the interstate because it was high ground. Some suffered from chronic medical conditions such as diabetes and heart disease, but there was no medication to be had.

More than 100,000 houses in New Orleans sustained major damage or were destroyed.[6] The storm caused tens of billions of dollars in property and infrastructure damage in Louisiana and was the most costly catastrophe in U.S. history.[7]

These were the scenes filmed by television crews that shocked viewers around the world.

Post-Katrina, many residents of "the Big Easy" say they've lost confidence in their government, from the local politicians to the bureaucrats in Washington, D.C. They say that those officials should have been better prepared, should have made the engineering changes that would have protected their city and should have learned lessons from Hurricane Betsy and other storms.

New Orleans wasn't devastated by an act of God. It was devastated by the inaction of man.

The strong winds from Katrina didn't decimate New Orleans. This wasn't a Category 5 storm like Hurricane Andrew, which flattened parts of south Miami-Dade County in Florida with winds up to 165 mph more than a decade earlier.[8] Official records show that while Katrina was a strong Category 3 hurricane when it slammed into the Gulf Coast near the Mississippi-Louisiana border, by the time it blew over the city of New Orleans, the winds had weakened considerably. The National Hurricane Center says gauges in the area measured sustained winds ranging from 70 mph to 78 mph, and that most of the city experienced a Category 1 or Category 2 hurricane.[9]

It was the storm surge that caused the levee and canal breaches and failures, dumping more than 100 billion gallons of water into the streets.[10] The pumping stations that could have helped alleviate the problems stopped working because of power outages and flooded equipment.

Katrina extended 460 miles[11] and brought death and destruction to many communities along the Mississippi and Louisiana Gulf Coast. But this book focuses on New Orleans because Katrina was the deadliest natural disaster to strike a major U.S. city in a century, since more than 8,000 people perished in a 1900 hurricane that hit Galveston, Texas.[12] This book also targets New Orleans because for years experts had warned that the bowl-shaped city surrounded by water could be knocked out by "the Big One."

"It remains difficult to understand how government could respond so ineffectively to a disaster that was anticipated for years, and for which specific dire warnings had been issued for days. This crisis was not only predictable, it was predicted," concluded a report by the U.S. House Select Bipartisan Committee to Investigate the Preparation for and Response to Hurricane Katrina.[13]

As early as 1976, a report by the U.S. General Accounting Office (now called the Government Accountability Office) had made it clear: "The greatest natural threat to the New Orleans area is posed by flooding from hurricane-induced sea surges, waves, and rainfall."[14] Numerous studies before Katrina cautioned that storm protection plans weren't moving fast enough, that the levees might not hold in a strong hurricane, that the U.S. Army Corps of Engineers had used outdated data in its engineering plans to build the levees and floodwalls and that the wetlands buffering the area from storms were disappearing.

In 1982, for example, a GAO review examined the massive $924 million hurricane protection project Congress had authorized after Betsy. The plan was for the corps to build a series of levees and barriers that would protect the New Orleans area from flooding after a hurricane. According to the GAO report, the completion date had been bumped from 1978 to 2008, and only about half the project had been finished. "Seventeen years after project approval, residents of the New Orleans area are still without the hurricane protection anticipated when the project was initiated," the GAO wrote.[15]

After hurricanes Hugo in 1989 and Andrew in 1992, government auditors also slammed the Federal Emergency Management Agency for its poorly coordinated response and recovery efforts.

Over the years, articles and books also had exposed the deeply entrenched cronyism and corruption that permeated city and state government, including the local levee boards responsible for overseeing the structures built to protect the area from flooding.

The major local newspaper, *The Times-Picayune*, documented numerous flaws in the levee system and how coastal erosion was exposing the city to storm surges. In a five-part series published in 2002, the New Orleans newspaper predicted what would happen if a storm surge flooded the city: tens of thousands of people with no transportation would be trapped on rooftops and in attics, struggling for their lives; thousands more would be stranded at the Superdome; and in the aftermath, the city would be all but destroyed, with many of its displaced residents living indefinitely in FEMA trailers.

Academic experts also had issued reports warning about looming problems, from the levee system's inability to protect the city from a major hurricane to potential evacuation nightmares.

Consider a 2003 survey by a Louisiana State University research team, which found that only 43 percent of those who identified themselves as in poor health would evacuate the city during a major hurricane. The survey revealed that residents with higher incomes were more likely to say they would evacuate, and that more white residents than black residents said they would leave. One of the researchers described "a kind of New Orleans culture of staying through hurricanes."[16]

Emergency preparedness officials maintain that they were well aware of those potential problems.

In 2004, after nearly five years of delays, FEMA funded a massive hurricane preparedness exercise for New Orleans. The scenario turned out to be eerily similar to Katrina, supposing widespread flooding and hundreds of thousands of people displaced. Many of the same breakdowns in communications, evacuation and health care that occurred in the simulation would come to pass when Katrina struck New Orleans.[17]

The real storm rendered the communications systems—local, state and federal—practically inoperable. The buses that were supposed to evacuate thousands of people never came. Most hospitals lost power and had made no arrangements to evacuate patients. The nation's disaster medical system, which deploys teams to assist in such circumstances, was plagued by bureaucratic problems that stymied its effectiveness. The social services network that was to have been there to pick up the pieces was nonexistent in many areas.

Despite all the warnings and studies, every system—from emergency preparedness to health care to post-hurricane relief—seemed to crumble when the storm barreled into New Orleans. This book will examine how and why that happened. The chapters were written by seven authors, each of whom investigated a specific aspect of the storm and its effect on the lives of hundreds of thousands of people like Pearl Ellis.

She and many other New Orleanians put the blame squarely on the government. She lost several elderly friends who died in their homes because of the storm. She almost lost her son, Solomon, then 46, who spent three days stranded in her attic without food or water before he finally chopped a hole in the roof and escaped to a neighbor's house, where he was rescued by a Good Samaritan in a boat.

"It could have been prevented. A lot of lives would have been saved," says Pearl Ellis.

The congressional committees that investigated the government's preparation for and response to Katrina agreed with her. "Before the storm, government planning was incomplete and preparation was often ineffective, inadequate, or both. Afterward, government responses were often tentative, bureaucratic, or inert. These failures resulted in unnecessary suffering," concluded a U.S. Senate committee that investigated Katrina and its aftermath.[18]

Nine months after the hurricane, much of the Lower Nine was still a ghost town. Houses had been washed away, with nothing but concrete steps remaining. Tall weeds were creeping across the overturned cars and trucks that had ended up in front yards or underneath houses. Half of the area still had no electricity or potable water. FEMA trailers were yet to be seen.

Pearl Ellis, who lived in her house at the corner of Flood and Derbigny streets for more than 55 years, says she won't be moving back. She will remain in "the country," in the rural area about two hours northwest of New Orleans where she evacuated before Katrina hit. That's where she grew up and where she lives now, in the house her father built.

While many in her neighborhood are committed to returning and rebuilding, Ellis, like others, says she's had enough.

"I don't have no desire to come back. I'm too old," she says, shaking her head and pointing to what little remains of the home that once was filled with antiques, family photos and china figurines.

"I told my kids they can do as they like with the house. They were born and raised there. They want to rebuild their family home. I don't have that desire. . . . It doesn't feel like home anymore."

TWO

# The Environment

Sara Shipley Hiles

*"The bottom line here is, when people decided to live in this fragile area, they disrupted the natural system. . . . What's going to happen is the ocean's going to take it over again."*

—Roy Dokka, Louisiana State University geologist

Drive down south of New Orleans about 45 minutes, hang a left at the lift bridge crossing Bayou La Loutre and you'll find yourself at Shell Beach, La., ground zero of one of the worst natural disasters in American history.

Hurricane Katrina's fury nearly wiped the close-knit bayou community of Shell Beach off the map. The storm's powerful winds splintered homes as if they were built out of Popsicle sticks. The mighty storm surge wiped some buildings clean away, leaving only concrete slabs. A few steel beams are all that's left of a marina beloved by generations of New Orleans fishermen.

Oak trees whose graceful branches once dipped to the ground lie upended, their root-balls exposed to the air. The trees that remain standing are festooned with debris—an office chair, a quilt, seafood-packing bags flapping like flags in the wind.

The storm ranked as the most costly hurricane ever to hit the United States, and Shell Beach, like the rest of the New Orleans area, is struggling to recuperate. Yet there is evidence of a much deeper problem going on here—a slow-motion environmental catastrophe that makes the New Orleans area more vulnerable to hurricanes every year.

The entire Mississippi River Delta, including the city of New Orleans, is slowly disappearing into the Gulf of Mexico. The process is natural, but human actions have made it far worse. Levees, oil and gas extraction, and shipping canals—all products of man's attempt to control nature—are partly to blame.

First, the land itself is sinking. The soft soil of the Mississippi Delta is gradually settling into the earth. For New Orleans, that means houses, office buildings and even flood-protection levees get lower every year.

To make matters worse, the Big Easy is losing its only natural shield against hurricanes. Since the 1930s, Louisiana has lost about 1,900 square miles of coastal land.[1] More marshland is disappearing there than in any other state.[2] As a result, the open ocean and its perilous storms draw closer to New Orleans every day.

These problems aren't so apparent in New Orleans, where tourists can party on Bourbon Street oblivious to the fact that most of the city sits below sea level, just 55 miles from the Gulf of Mexico. But Shell Beach offers a preview of what New Orleans can expect 50 years from now. Scientists have predicted that under current conditions Louisiana could lose an additional 700 square miles of coastal land by 2050.[3] If that happens, "some of these levees would be subject to open gulf conditions," said Greg Miller, a project manager with the U.S. Army Corps of Engineers.[4]

Signs of the sea's inward march are evident at Shell Beach. Back from the roadway, dead cypress trees stand like skeletons in swamps choked by saltwater intrusion. Home sites that used to be on dry land now lie under several feet of water. And across from the remains of the Shell Beach marina, a man-made shipping channel has replaced thousands of acres of wetlands that once provided a buffer from storms.[5]

"We think we can control nature," scoffed Roy Dokka, a Louisiana State University geologist who studies subsidence, land's gradual sinking to a lower level. "The bottom line here is, when people decided to live in this fragile area, they disrupted the natural system. . . . What's going to happen is the ocean's going to take it over again."[6]

Louisiana's disappearing coast is no secret, but decades of study and more than half a billion dollars spent on restoration projects have not reversed the trend. By the latest tally, the state is still losing a football field's worth of land every 90 minutes.[7]

Kerry St. Pé, a conservationist who has dedicated his career to restoring the fragile coast, has seen it happen firsthand. He grew up in Port Sulphur in Plaquemines Parish, a narrow strip of land that hugs the

Mississippi as it travels to the gulf. As a child, St. Pé played among live oak trees that covered low ridges jutting out of the surrounding marsh. Today, those trees are dead, overrun by salt water. "They've essentially drowned," said St. Pé, who is program director of the Barataria-Terrebonne National Estuary Program.[8]

The sinking of the delta is a natural process put on steroids by human activity. For some 6,000 years, before European settlement, the Mississippi River acted like a massive earthmover,[9] carrying nearly 400 million tons of sediment each year from the Appalachian Mountains and the Great Plains of the Midwest to the Louisiana coast.[10] Moving back and forth across the delta like a garden hose gone crazy, the river built the delta one piece at a time. The sediment compacted gradually, occasionally being renewed when a flood or a new river channel delivered a fresh veneer of soil.

That earth-building process slowed after the U.S. Army Corps of Engineers undertook a flurry of levee-building beginning in the 19th century that effectively contained the river in an earthen straitjacket. The massive levees kept cities and farms dry, but they also sent the river's land-building sediment streaming over the edge of the continental shelf into the deep water of the Gulf of Mexico.[11]

Without nourishment from the river, the delta began to die. Engineers continued to raise levees, cutting off more of the marsh from its freshwater supply. Continued development drained swampland, leading to more soil compaction.

The oil and gas industry cut channels through the marsh to install more than 9,000 miles of pipeline.[12] The Corps of Engineers built navigation canals to promote shipping. Salt water from the gulf flowed up the network of canals, killing cypress swamps and marsh grasses that held the earth together.

"A lot of this was done fairly innocently. I doubt you could find anybody in the '50s concerned with coastal land loss," said Al Naomi, a longtime project manager with the Corps of Engineers. "They thought wetlands were wastelands. They were [considered] mosquito breeders that needed to be drained."[13]

Underneath it all, shifting of tectonic plates and pressure put on the Earth's crust by the heavy sediment load have caused more sinking, according to Dokka's research.[14] The effect was that a thousand years of natural land-building had been wiped out in a single century.[15]

By the 1970s, some scientists noticed the change and began to sound the alarm. Oceanographer Paul Kemp moved from New York to Louisiana in 1975 to study the Mississippi Delta. "I fell in love with it," he recalled. "One of the things I realized quickly was it was disappearing, it was being destroyed. . . . A lot of people said, 'Yes, it's a shame' . . . but it didn't translate into a call to action."[16]

The next few decades saw the formation of conservation groups and public plans for change, but the efforts amounted to barely a drop in the bucket of land loss. In 1990, Congress passed the first major effort to address Louisiana's coastal land loss. The Coastal Wetlands Planning, Protection and Restoration Act, also known as the "Breaux Act" for one of its sponsors, Democratic senator John Breaux of Louisiana, receives roughly $50 million in federal funding each year for quick-turnaround projects such as the planting of marsh grass to shore up eroding soil. In the past 15 years, the program has approved projects designed to build or protect more than 66,000 acres of wetlands.[17] It sounds like a lot until you realize that's just 5 percent of the wetlands lost in the past century.

"It really wasn't creating enough land," said Denise Reed, a coastal geomorphologist at the University of New Orleans who worked on the project. "We realized this was much more difficult, much more challenging, than we thought."[18]

In the mid-1990s, with evidence of land loss mounting, the state invited engineers, scientists, politicians and the public to create a wide-ranging plan. The resulting Coast 2050 plan, completed in 1998, sketched out a solution with a price tag of $14 billion. It was the costliest environmental restoration project ever proposed, nearly double the $8 billion project to restore the Florida Everglades.

In 2004, the Bush administration balked at the magnitude of the program and directed the corps to downsize it.[19] The revised Louisiana Coastal Area restoration plan sought $1.9 billion over 10 years—spending that had yet to be approved more than a year after Katrina. If completed, the plan's five restoration projects would not stop land loss, but would merely reduce it from 10.3 square miles a year to 8.6 square miles, according to the National Academy of Sciences.[20] "Clearly, execution of the LCA Study alone will not achieve its stated goal to 'reverse the current trend of degradation of the coastal ecosystem,'" a panel of the academy wrote in a 2006 review.[21]

The man-made problems are so huge that it may be impossible to entirely compensate for them. For example, consider the environmental impact of the Mississippi River–Gulf Outlet, a 76-mile, deep-draft shipping canal that runs past Shell Beach. The Army Corps of Engineers built the canal in the 1960s as a shortcut between the Port of New Orleans and the gulf, but big ships rarely used it. Less than 3 percent of the port's cargo uses the canal—not even one ship per day—yet the corps spends an average of $16 million a year dredging it, according to *The Times-Picayune*.[22]

Building the canal destroyed more than 27,000 acres of wetlands, said John Lopez, director of the Coastal Sustainability Program for the Lake Pontchartrain Basin Foundation. Additionally, the canal injects salt water—and in the case of Katrina, storm surge—straight into the heart of New Orleans. The canal ultimately empties into Lake Pontchartrain, the 630-square-mile water body that borders New Orleans on the north. There, the heavy salt water sinks to the bottom and creates a "dead zone" that snuffs out life in up to a sixth of the lake.[23]

Naomi, the corps project manager, agreed that the MR-GO—locally called "Mr. Go"—"has a tremendous environmental impact." He added that engineers predicted its destructive effects when it was built. "It's pretty much given us what we expected," he said.

Corps officials didn't agree with critics who called the canal a "hurricane highway" and blamed it for catastrophic flooding after Katrina. But in December 2006, the agency recommended that the waterway be closed, citing economic reasons. A final decision on how and when to plug the outlet isn't expected until at least the end of 2007.

Louisiana's elected officials have called for a massive infusion of federal money to fix the coastal erosion problem, but critics say the state has a poor record when it comes to spending tax dollars. A 2006 report by Environmental Defense, Taxpayers for Common Sense and other groups says Katrina's massive flooding occurred in Louisiana "despite the fact that Congress, over the past five years, has spent more on water projects there—$1.9 billion—than in any other state."[24]

While debates rage and communities struggle with recovery, the land continues to sink and global warming–influenced sea levels keep rising. According to data gathered for *The Times-Picayune* in 2002, the combination of sinking land and rising sea level had put the Mississippi Delta, on average, 2 feet lower than it was 60 years earlier.[25]

Well before Katrina hit, many scientists, public officials and emergency planners knew that a hurricane could inundate New Orleans, killing as many as 100,000 people as water filled the city like a bathtub.[26] University of New Orleans geologist Shea Penland told *National Geographic*, in a 2004 magazine story that eerily predicted a devastating hurricane, "It's not if it will happen. It's when."[27]

By the time Katrina washed ashore in August 2005, relatively little had been accomplished to stem wetland loss or address land subsidence in urban areas, leaving the New Orleans region more exposed to storms than ever. It was one disaster that would soon lead to another. Katrina didn't just kill more than 1,000 people and flood nearly 200,000 homes in the New Orleans area; the storm created a whole new set of environmental problems, from contamination and trash to increased erosion.

---

Lynnell Rovaris has a record of the precise moment that Katrina changed her life. Leaning into a closet full of objects salvaged from the storm, she pulled out an electronic kitchen clock whose hands froze in place when the floodwaters consumed her New Orleans community of Press Park. "See? The clock stopped at 3:10 a.m.," she said.

Rovaris, 36, had been sleeping on the couch when her boyfriend woke her. "He said, 'Get up—there's water on the floor,'" she said. Groggy and confused, she sat on the steps leading to the second floor of her townhouse and watched the water rise from ankle-deep to near ceiling height in 25 minutes. A tough city girl who had grown up in New Orleans, she had not thought Katrina was enough of a threat to evacuate. "I've never seen so much water in my life. It looked like a river," she said.[28]

For three days, Rovaris and her boyfriend lived on the second floor of her home without food or water until police rescuers came by in a boat. The pair escaped by using a sheet tied to a bedpost to lower themselves out the window. Eventually, she made it to the convention center, where she was loaded onto a bus to Arkansas and later caught a train to Chicago.

Rovaris returned in December 2005, determined to fix up the government-subsidized housing she had been buying on a $346-a-month rent-to-own plan. She later moved her two children and her sister into a tiny trailer provided by FEMA and began fixing the place herself. Petite with short, black hair, she darted about the gutted building, pointing to new doors she hoped to install on her own.

But the future is unclear for Rovaris and her neighbors, whose houses are built on a Superfund hazardous waste cleanup site. Before Katrina, the U.S. Environmental Protection Agency had declared the area safe for residents, but now, environmental concerns are back. In January 2006, a judge declared the area "unreasonably dangerous" and said the neighborhood should be abandoned.[29]

The neighborhood sits on top of the Agriculture Street Landfill, an old city dump that was developed for subsidized housing in the 1970s. After years of complaints among residents of mysterious cancers and other illnesses, the federal government declared it a Superfund site in 1994. The EPA later excavated 2 feet of dirt from residents' yards, laid down a plastic mesh and put clean fill on top. The top layer of soil was clean—until Katrina hit.

The hurricane flooded 80 percent of New Orleans, the swirling water sweeping up dirt, pesticides, bacteria, chemical waste and other debris, and distributing them throughout the city in a layer of brown muck. Environmental activists warned of a "toxic gumbo" that could render the entire area a hazardous waste dump.

The Louisiana Department of Environmental Quality and the EPA tested the water in New Orleans and downplayed the level of contamination. "You didn't want to drink it, but I wouldn't classify it as a toxic soup either," said Tom Harris, an administrator with the DEQ's Environmental Technology Division in Baton Rouge who was involved in the testing. "My interpretation was the levels were pretty similar to what you'd normally see in urban runoff."[30]

In addition to water, the agencies tested sediment and initially found 43 spots in the region[31] that tested high for one of several toxic materials: arsenic, a metal associated with cancer of the bladder, lung and skin;

lead, a neurological poison; and benzo(a)pyrene, a petroleum by-product that can cause chromosomal damage and cancer.[32]

After additional tests at those sites, however, officials said Katrina caused no serious long-term contamination. Much of the bacteria and oil residue found after the storm degraded on its own. The only place where the agencies still had concerns was in the Agriculture Street Landfill neighborhood. Four samples there tested above the acceptable risk level for benzo(a)pyrene.[33]

EPA officials said they were waiting for the Housing Authority of New Orleans to decide whether to demolish or renovate the neighborhood's public housing before taking further action. The agency still intends to remove the site from the Superfund list, said Sam Coleman, director of EPA's regional Superfund division. "Environmentally, there is nothing wrong with this area," he said.[34]

Wilma Subra, a Louisiana environmental chemist who took her own samples on behalf of environmental groups, said the agencies were too quick to dismiss the test results. "It's very serious," Subra said of the chemicals in the sediment, which she said can easily turn to dust and become airborne. "Exposure to all these things will result in increased rates of cancer, miscarriages and birth defects. Those will be the things that are hardest to associate back to the hurricane, because we don't have a good tracking method."[35]

Eight months after the flood, Subra gave a tour of the city's ravaged neighborhoods. She was still frustrated with the state and federal response to the disaster. For example, she said the EPA should have held arsenic test results to its own stricter screening standards instead of using the DEQ's standard, which allows arsenic levels to be 30 times higher.

"I've done Superfund sites since before there were Superfund sites," said Subra, who has acted as a community advisor at nearly a dozen cleanups, including the one at Ag Street. "If this was a Superfund site, we'd be cleaning it up."

State officials say their standard is well below the level where a person would be expected to suffer health problems, even after years of exposure.

For Subra, Katrina showed how environmental mistakes tend to compound themselves. For the people living atop the Ag Street dump, things were bad enough before the storm; now, the site is covered in contaminated sediment and its homes are filled with unhealthful mold and mildew, she said. "I would prefer that this area not be redeveloped, but I didn't think they should have been living here anyway," she said.

It's still unclear whether the neighborhood will be rebuilt. In January 2006, an Orleans Parish judge ordered the city, the housing authority and other parties to pay residents for emotional distress and diminished property values.[36] A settlement could cost up to $300 million, said Suzette Bagneris, an attorney for the residents who filed the suit.[37] The ruling has been appealed. Meanwhile, Bagneris thinks the residents should be offered buyouts. A spokesman for the housing authority said that Katrina destroyed the 56 rental units in the neighborhood, and these may be demolished. The 165 homeowners in the area will have to decide on their own what to do.[38]

Neighborhood leader Elodia Blanco doesn't plan to come back. Post-Katrina, she moved across Lake Pontchartrain to a small town called Tickfaw, which she calls a haven from the crime and chemicals of Ag Street.[39] "No matter what, I will fight the rest of my life to let people know that is a community you should not live in. It's just not healthy," said Blanco, 55.

---

Among Katrina's other major environmental fallouts was the spilling of more than 7 million gallons of oil into the waterways of southeast Louisiana, according to the U.S. Coast Guard.[40] In comparison, the Exxon Valdez spill in 1989 in Alaska's Prince William Sound—the largest oil spill in U.S. history—released 11 million gallons.[41]

A single incident at the Murphy Oil Meraux Refinery in St. Bernard Parish caused by the hurricane unleashed 1 million gallons of oil, coating 1,700 adjacent homes[42] and the surrounding canals, parks and schools with a brown film. According to Coast Guard records, only some of the oil from the spills was cleaned up.[43]

Another massive problem: disposing of waste and debris from the storm—including downed trees, flooded cars, malodorous refrigerators, soggy drywall, moldy carpeting, asbestos roofing tiles and household chemicals. Officials estimated that Katrina created 22 million tons of debris.[44]

To speed the cleanup, the DEQ allowed construction and demolition waste to be taken to two controversial dumpsites in the swampland of eastern New Orleans. The Old Gentilly Landfill, an old, unlined city dump that had been capped with clay, reopened first.[45] Officials later approved dumping at the Chef Menteur Landfill,[46] a new facility that sits near the Bayou Sauvage National Wildlife Refuge and a Vietnamese community. Neighbors protested against the landfill, and New Orleans Mayor C. Ray Nagin closed it a few months later.[47] The landfill's operators fought back in court and were hoping to get it reopened.

Environmentalists and neighbors have filed legal challenges to the use of both dumps, alleging that officials relaxed their own environmental rules and ignored community concerns. "The DEQ is redefining [construction and demolition waste] to include hurricane debris. That includes a lot of things that are essentially household waste," said Adam Babich, director of the Tulane Environmental Law Clinic, who represents the Louisiana Environmental Action Network in a case involving Old Gentilly.[48] "The more you put things in that decompose . . . the bigger risk you're creating that you'll have landfill problems with gas, groundwater contamination, et cetera."

State officials say every possible measure—from curbside checks to pickers and spotters at the landfill—is being taken to ensure that unapproved waste doesn't enter the landfills. Opponents, however, fear that such items as pesticides, bleach and rotting food will inevitably find their way into the dumps.

State regulations stipulate that landfills containing regular household waste have a synthetic liner and system to collect liquid that percolates through the garbage. Although construction and demolition landfills—such as Old Gentilly and Chef Menteur—are required to have only a 2-foot liner of compacted clay, state officials insist that the two landfills are safe for use.

Chuck Brown, a DEQ assistant secretary who oversees the Office of Environmental Services, called the hurricane debris "innocuous" and said the state has a responsibility to open additional dumps to facilitate the city's recovery. "You can't rebuild until you clean up," he said. "We've got 18 million cubic yards of debris left. We've got to get this done."[49]

The need for expediency concerns Babich, the environmental attorney. He worries that a number of environmental laws are being overlooked in Katrina's wake, from rules dealing with everything from oil refineries and lead paint to asbestos roof shingles and landfills. Babich also fears that minority groups are suffering even more after the storm. "It's typical for low-income communities to bear a disproportionate burden of environmental impacts, and here we seem to be exacerbating that even more," he said.

Brown denied that the DEQ engaged in any kind of environmental racism. "We're not picking on any one community," he said. "And Hurricane Katrina didn't discriminate against anybody. It destroyed every neighborhood: black, white, Vietnamese, Mexican."

---

Besides ravaging the man-made landscape, Katrina took a toll on the natural environment.

Satellite images from the U.S. Geological Survey's National Wetlands Research Center showed that Katrina's raging winds and pounding waves devastated the fragile coast.[50] Scientists estimated that the storm destroyed more than 100 square miles of wetlands, including chopping up the Chandeleur Islands, an important barrier island chain.

"We lost more than all we gained" under the Coastal Wetlands Planning, Protection and Restoration Act, the state-federal restoration effort, said Lopez of the Lake Pontchartrain Basin Foundation. (Lopez's home in Slidell also was wiped away by the storm.)

Scientists think that parts of the Louisiana coast will bounce back. What's unclear is how much.

"Hurricanes can be regarded as totally natural," said biologist Michael Poirrier, sitting in his laboratory at the University of New Orleans. "The question is, if we have a stressed ecosystem, how does it recover from a disturbance?"[51]

Poirrier studies sea grasses and clams in Lake Pontchartrain. Both organisms are ecologically important: The grasses provide habitat and food, stabilize sediments, buffer wave energy, produce oxygen and process pollutants. The clams in Lake Pontchartrain filter a volume of water equivalent to the whole lake every 34 days and they create a million tons of shells a year that help stabilize shorelines.[52]

In surveys after the storm, Poirrier found that "Katrina just tore up the bottom." He found a lot of dead clams and little sea grass. In one of his favorite spots, a duck-hunting lease in the marsh south of the city, "I could not find a blade of submersed aquatic vegetation anywhere," he said.

Poirrier jumped in a truck to offer a firsthand look at the damaged Fritchie Marsh, near his home in Slidell. "This area got chewed up," he said, pointing out toilet seats, wrecked boats and other debris piled in the marsh. The once-lush green carpet of grass had been shredded.

---

It's hard to say exactly how much damage Katrina would have caused if Louisiana still had its protective barrier of coastal land. As early as the 1960s, scientists found that wetlands reduced hurricane storm surges by 1 foot for each 2.7 miles of wetland the storm crossed. Later studies found that the number varied, depending on the storm and terrain.

According to America's WETLAND, a public education campaign created by the state of Louisiana to promote coastal restoration, more wetlands would have made all the difference. The campaign's Web site says that "about 80 miles of restored coastal marsh below New Orleans would have prevented most of the flooding from Hurricanes Katrina and Rita."[53]

The Corps of Engineers is working on a simulation that will run a Katrina-type storm over the 1956 coastline. "We think that simulation will give us a good sense of what a healthy ecosystem would look like," said Miller, the corps project manager. "We want to caution everyone, though: Katrina was a monster storm. It went over a lot of land, and it still had a 30-foot storm surge in Mississippi."

But Mark Davis, executive director of the Coalition to Restore Coastal Louisiana, is convinced that Katrina would have been less deadly if the state, the federal government and Louisiana residents had heeded the long-ago warnings of environmental ruin.

The cost of Katrina, Davis said, makes $14 billion in coastal restoration look cheap. In the first five months after the storm, the federal government spent $85 billion, in part to clean up debris and house the displaced, he added. "Not one nickel of that went to preventing this from happening again."[54]

THREE

# The Levees

John McQuaid

> *"Our experience in southeastern Louisiana has been sobering for us, because this is the first time that the Corps of Engineers has had to stand up and say, 'We had a catastrophic failure in one of our projects.'"*
>
> —Lt. Gen. Carl Strock, U.S. Army Corps of Engineers chief

The breaches in the floodwalls protecting New Orleans are exhibit A in one of the biggest engineering blunders in American history. Design errors, flawed decision-making, cost-cutting and political infighting resulted in the deadly flooding that destroyed much of the city.

It wasn't as if the U.S. Army Corps of Engineers had been ignoring the problem. In July 1985, on a test site about 80 miles southwest of New Orleans, scientists with the corps embedded a 200-foot-long steel wall in the earth. They erected an enclosure around their experimental barrier and flooded it to a depth of 4 feet. Then they measured the weight of the water against the steel to gauge how far the wall leaned or otherwise shifted in the soft Mississippi River Delta muck. At intervals lasting a week or two, they raised the water level to 6, 7, then 8 feet.[1]

The scientists wanted to determine whether the steel wall could withstand severe flood conditions. At the time, the corps was planning a major expansion of the hurricane levee system around New Orleans and its suburbs. Plans called for ringing inhabited areas with a combination of high earthen levees and floodwalls made of sheet pile—flat, interlocking steel columns—topped with concrete.

The test had two main purposes—one scientific, the other financial. There wasn't much hard data back then on how the proposed floodwalls (called "I-walls" for their simple vertical profile) would handle a hurricane storm surge. And corps engineers wondered if their standards governing the depth of the steel foundations—standards that could add millions to the installation cost—were too stringent. "Over the next few years, construction of many miles of these I-type floodwalls is proposed at an estimated cost of over $100,000,000. The cost of these walls is obviously highly dependent on the sheet pile penetration required for stability," noted a 1988 report outlining the experiment and the reasons behind it.[2]

After analyzing the data, the scientists concluded that they could indeed relax the standards—sheet pile would not have to be driven as deep as had once been required.

That would later prove to be one of the most spectacularly off-base conclusions in the history of American engineering.

The test also revealed a problem—little-noticed at the time—that would play a pivotal role in Katrina's flooding of New Orleans. The experimental data suggested that as rising water pushed against a wall, gaps could open up between the steel foundation and the earth in which it was supposedly firmly anchored and that water pouring into such a gap could destabilize the entire structure.

The original report made no mention of this vulnerability. Scientists examining the results for another study years later, however, did take note: "As the water level rises, the increased loading may produce separation of the soil from the pile on the flooded side (i.e., a 'tension crack' develops behind the wall). Intrusion of free water into the tension crack produces additional hydrostatic pressures on the wall side of the crack and equal and opposite pressures on the soil side of the crack," they wrote in a paper published in 1997, based on work done in the late 1980s.[3]

But the designers of the New Orleans levee system either didn't know about this well-documented problem or they ignored it. It was exactly such "tension cracks" that precipitated breaches in New Orleans floodwalls on August 29, 2005, and catapulted Hurricane Katrina from a disaster to a catastrophe.

Katrina's storm surge overwhelmed the city's levee system and flooded dozens of neighborhoods on the east side of town and adjacent St. Bernard Parish—some in the space of minutes. The water wasn't high enough to overflow I-walls further to the west in two city drainage channels—the 17th Street and London Avenue canals. But as the floodwaters rose in the canals, the telltale gaps opened up.[4] Wall sections soon collapsed in three places, and there was no stopping the water from inundating central New Orleans.

After the disaster, scientists at the corps' Engineer Research and Development Center in Vicksburg, Miss., ran another floodwall experiment. This time, they built scale models of the New Orleans I-walls in 3-foot-long Lucite boxes and replicated soil conditions at the breaches. They spun them in a giant centrifuge to simulate real-world conditions. A camera recorded what happened: As each mini-wall leaned slightly, a gap developed, and the wall's earthen base split in two like a pat of butter sliced by a knife. Then the model floodwalls breached. Investigators concluded that the gaps helped precipitate all three canal breaches, as well as a fourth one in the nearby Inner Harbor Navigation Canal.[5]

On June 1, 2006, nine months after Katrina struck, the corps chief, Lt. Gen. Carl Strock, admitted that design errors committed by the agency's engineers caused the floodwalls to collapse. "Our experience in southeastern Louisiana has been sobering for us, because this is the first time that the Corps of Engineers has had to stand up and say, 'We had a catastrophic failure in one of our projects,'" said Strock, who in August 2006 announced plans to retire.

Team Louisiana, a panel put together by the state, traced 88 percent of the flooding of the central New Orleans "bowl" to the design errors;[6] the Interagency Performance Evaluation Task Force (known as IPET), the federal team investigating the levees, put it at 70 percent.[7] An analysis by Knight Ridder News Service showed that at least 588 bodies were recovered from areas flooded by the breaches in the drainage canals, compared with 286 in other areas, including the Lower Ninth Ward, eastern New Orleans and St. Bernard Parish.[8]

The susceptibility to being breached under pressure was the most glaring flaw in the design of the complex system that was devised to protect New Orleans from flooding, but it was not the only one. Problems ranged from errors in designs of individual structures to the system's basic architecture. The levees, walls and floodgates were supposed to act in concert to repel flooding. Yet, according to the IPET report, they were "a system in name only."[9]

The disastrous performance of the levee system that most thought safe has left New Orleans residents profoundly shaken, uncertain and angry. "The levee puts a lot of people in fear—not the hurricane. That's why a lot of people haven't come back. They're worried about the seal of approval on the levees. . . . They're worried about levees that don't stand the test of time," said the Rev. Bruce Davenport, 54, pastor of the St. John #5 Faith Church near the London Avenue Canal.

On August 29, the wind ripped off the church's roof. Water rose up 8 feet, inundating the building's first story but not the chapel itself on the second floor. Davenport remained in the city during the storm and refused transport out afterward. He helped rescue neighbors and tend to the dozens of people who remained across the street in the sprawling St. Bernard Housing Project (since tagged for demolition). He pulled several bodies out of project apartments and tied them to light or utility poles. One, the body of an old man, floated eerily for weeks a half-block from the church. "We tied him when it was here," Davenport said, pointing to a water line about 4 feet off the ground, "and they came and got him when it was here." He bent over and touched a point a foot off the grass.[10]

To the west, in the upscale Lakeview neighborhood, Kevin Lair's house on Bellaire Drive sat adjacent to the 17th Street Canal floodwall. "We bought it along the levee because it was real secluded," Lair said. "We would walk the dogs on the path along the levee. We had a pool and a hot tub. People would ask us all the time, 'Don't you think this levee's going to break?' I never did." Others on the block complained of mysterious water leaks that could have indicated trouble, but Lair said his property was always dry.[11]

The family had heeded warnings of the impending storm and evacuated prior to August 29. That morning, floodwater initially broke through a few hundred feet south of the house. Then the opening widened, and soon the flood roared through the Lairs' backyard, blasting through the rear door and windows. The torrent ripped the wall clean off one side of the house and carried it away. Then, for several days, the house sat in the equivalent of a flowing river as water from the breach gradually filled up the city. When the flow subsided, that river became a lake. And after the water level went down a few weeks later, everything that remained on the ground floor of the house—shoeboxes, broken plates and dishes, tables and chairs, old magazines—was covered with a fine, silty mud.

Since then, Lair, 49, who runs a real estate brokerage, has changed his whole perspective on how government works. "I used to be a world-class, white Republican: never question anything [the government does]," he said. "Now, I've done a 180. Question everything. Don't believe anything and everything they tell you. This is something to be outraged about."

Lair got even angrier when, he said, local corps officials gave him the runaround for months after the storm, refusing to accept responsibility for the destruction of his house. In June 2006, after Strock's admission that design flaws caused the levees' failure, the corps agreed to buy out houses along the breach and a contractor demolished them in August. The corps plans to erect a project to permanently repair the floodwall on the site.

---

How could a key infrastructure project intended to insulate a major American city from calamity—in a nation of engineering know-how and ample resources—fail so spectacularly?

Many of the failings, such as the designers' apparent ignorance about floodwall gaps, stem from what critics call the Corps of Engineers' hidebound culture, in which things are done slowly, in a certain traditional way. The Independent Levee Investigation Team, made up of engineers from the University of California at Berkeley and other academic institutions, concluded that this culture slowed the adoption of new technologies and rendered the agency unable to resist political pressures for cheaper, less safe structures. "This is an organizational failure, pure and simple," said Robert Bea, a professor of engineering at Berkeley and a leader of the independent team. "The corps operates with lethal arrogance—'My way or the highway: I already have my mind made up, don't bother me with the facts.'"[12]

Then there was politics. The nation, it seems, always had other bigger priorities than safeguarding New Orleans from powerful hurricanes. Congress, the White House, the corps, and state and local agencies with authority over levees never fully addressed the complex challenge of shielding the city from storm surges. In the decades leading up to the fateful summer of 2005, virtually no one pushed strongly for bigger, better levees. Instead, agencies sought to cut costs, each looking out for its own short-term interests. "It seems no federal, state, or local entity watched over the integrity of the whole system, which might have mitigated to some degree the effects of the hurricane," the House Select Committee that investigated the response to the storm reported in March 2006. "When Hurricane Katrina came, some of the levees breached—as many had predicted they would—and most of New Orleans flooded to create untold misery."[13]

The geography of New Orleans exposes it to the vagaries of nature. The city sits in a low-lying river delta mostly below sea level in a dangerous hurricane zone. It is surrounded by water, nestled in a curve of the Mississippi River, with Lake Pontchartrain immediately to the north and Lake Borgne and Chandeleur Sound to the east. All of these bodies of water are, to varying degrees, open to the Gulf of Mexico. Day-to-day, that's not a problem: tidal flows vary by an average of less than a foot and a half.[14] But when a hurricane strikes, chaos ensues. The whirlwinds, forward motion and low pressure of a hurricane generate enormous storm surge waves. When one comes ashore near New Orleans, it rises up and can flow inland for miles. Even weak storm surges can inundate hundreds of square miles.[15]

From the early 18th through the mid-20th centuries, authorities focused protective measures on river flooding, which occurred more frequently and also was more predictable than storm surges. As a result, the river levees guarding New Orleans—massive structures averaging 25 feet high—are much safer than the city's hurricane levees.[16] They have not been seriously breached for more than a century. (Levees south of the city were dynamited in 1927 in an attempt to take pressure off New Orleans during the "Great Mississippi Flood"—an action later demonstrated unnecessary.)[17]

But in the mid-20th century, spurred by coastal development and a run of severe hurricanes, the federal government began to focus on hurricane flood control. By the early 1960s, the Corps of Engineers had devised a basic design for the levee system. Those plans sat on the shelf until Hurricane Betsy struck New Orleans on September 9, 1965, flooding the Lower Ninth Ward, parts of sparsely inhabited eastern New Orleans and some areas to the west. Within weeks, Congress sprang into action, ordering the corps to go ahead and build the new levee system. At a projected cost of $85 million and time frame of 13 years to complete, it was one of the most ambitious sea flood protection plans in the world.[18] (As of August 2005, the system was still not complete. The corps estimated it would be finished in 2015—50 years after its initial authorization—at a total cost of $738 million.)[19]

At the time, the understanding of storm surges was crude. To figure out how to design hurricane levees, engineers adapted the approach employed with river levees, which were designed to repel a "Standard Project Flood" of a set height and frequency—typically, a high-water peak that came along an average of once every 100 years. For storm surges, the corps worked with the U.S. Weather Bureau (later renamed the National Weather Service) to devise a "Standard Project Hurricane." Along the shore of Lake Pontchartrain, the flood from the SPH, as it was known, would rise to 11.5 feet above sea level. The corps calculated that such a flood would come along once every 200 to 300 years, depending on where you were in New Orleans.[20]

Though innovative for its day, this approach was seriously flawed. Based on data compiled in 1959, the SPH was later demonstrated to be a meteorological impossibility. Based in part on features from storms that had hit New Orleans in 1915 and 1947, its combination of atmospheric pressure and wind speeds could never occur in the same storm.

Worse, the SPH obscured the levee system's biggest weakness: It could not intercept a surge wave from a major storm. That became clear in the 1980s and 1990s, when the corps translated the SPH standards to the more easily understood categories of the Saffir-Simpson Hurricane Scale and ran computer models showing probable flood patterns. The results were alarming: The New Orleans levees could repress floods only from relatively weak storms—most in Categories 1 and 2, and only some in Category 3, with wind speeds up to 130 mph. Worse, the levee system going up around the city was turning it into a set of shallow bowls that would trap water that got inside. If a powerful storm hit and the levees were breached or overflowed, the water would remain inside. In short, it would turn the city into a lake.[21]

Emergency managers called it "the Big One." Yet even as its outlines grew clear, no politicians rushed to demand more protection, and with the risk of catastrophe still considered low, the public seemed indifferent.

"Nobody wanted to raise the level of protection," said Al Naomi, the levee system's project manager. Naomi oversaw a preliminary analysis, completed in 2000, on raising the levees to protect against a Category 5 hurricane. But funding for further study was rejected. (After Katrina, the study was quickly revived.) "There was no popular push to do anything about the levee system," he said. "Not a single person expressed an interest. Nobody wrote letters to their congressmen. People were happy and confident. They were overconfident."[22]

Yet, using the corps's own flood frequency estimates, the risk of disaster was consequential if estimated over decades, or the 40-year lifespan of the project. For example, in any 20-year period the risk of such a catastrophe was close to 10 percent—a virtual roll of the dice. And in fact, the actual risk was considerably higher; the corps just didn't bother to calculate it. The agency repeatedly opted to stick with the design parameters originally outlined in 1965, rejecting newer and more accurate information that pointed to rising risks.

In 1979, for example, the National Oceanic and Atmospheric Administration updated the SPH using new, more detailed flood data that showed more severe risks of storm surge flooding.[23] Scientists at the Louisiana State University Hurricane Center concluded that it would have resulted in a stronger SPH: Instead of the rough Category 3 equivalent, the new SPH would be a Category 4 storm with higher flood levels, especially along the city's exposed eastern flank.[24] But the corps did not incorporate the new data into its Lake Pontchartrain and Vicinity Hurricane Protection Project to protect most of New Orleans. Instead, it used it in the designs for newer projects in suburban areas across the Mississippi River.[25]

The corps also failed to account for the changes of a dynamic environment. The Mississippi River Delta, on which New Orleans sits, is gradually sinking into the Gulf of Mexico, some areas by as much as 5 feet in the past 50 years. As the ground sinks, the height of levees and floodwalls built on top of it falls, reducing the protection they provide. In 1983, the National Geodetic Survey issued updated—lower—measurements for ground elevations in south Louisiana. Two years later, Frederic M. Chatry, chief of engineering of the corps' New Orleans division, made a fateful decision to incorporate the data only into new structures.

"Modification of projects which have been completed will not be considered," Chatry wrote in an August 7, 1985, letter to his superiors. "The level of precision in the current data, and the practical difficulty and cost of changing such projects combine to mandate this course of action at least for the foreseeable future."[26]

There was a somewhat twisted rationale for sticking with the old data and ignoring the new. Using the new data would have required engineers to build higher levees, and that would have been a bureaucratic nightmare. It would have meant redesigning projects, getting more funding from a skeptical Congress and local agencies, and potentially delaying the completion of the system by years. Project manager Naomi described this as a kind of Catch-22: No matter which choice was made, it would produce unsolvable problems. When standards change, he said, "You have a large number of projects completed at one elevation. That means everything that's built must be rebuilt. Then in another 10 years, there's a new SPH. You never, ever finish, and you never have a consistent level of protection."

For the corps, whose bread-and-butter had always been big, complex navigation projects with rich rewards for contractors, building hundreds of miles of walls was a tedious, second-tier priority, said Ivor van Heerden, deputy director of the LSU Hurricane Center and leader of Team Louisiana. "Their whole focus is on navigation and keeping the port authorities happy, and in turn, the port authorities' lobbyists," he said. "When you come to the levees, there are no lobbyists, only the local agencies, and they want things done cheap."[27]

Agencies such as the local levee boards paid 30 percent of the cost of federal projects, and they pressured the corps' New Orleans district to complete the levee system as quickly and as inexpensively as possible—even at the price of safety.

In 1982, the General Accounting Office reported that state and local officials in Louisiana were complaining that the corps wasn't building levees fast enough. To speed things up, the GAO said, the officials wanted to cut corners even if that meant raising the risk: "Orleans Levee District officials believed that the corps' standards may be too high for what is really needed for adequate protection and for what is affordable by local sponsors. . . . They recommended the corps lower its design standards to provide more realistic hurricane protection to withstand a hurricane whose intensity might occur once every 100 years rather than building a project to withstand a once in 200- to 300-year occurrence. This, they believe would make the project more affordable, provide adequate protection, and speed project completion."[28]

The two decades of cutting corners came at a high cost. Along the Orleans drainage canal, for example, floodwalls were never completed adjacent to the pumping station, leaving gaps on both sides that allowed water to spill out during Katrina's aftermath. To the east, miles of levees protecting the Lower Ninth Ward and St. Bernard Parish were virtually washed away after being overtopped because they were constructed out of inexpensive, sandy fill taken from the adjacent canal.[29]

---

Cost—as well as environmental concerns—also forced changes in the corps' overall flood control strategy, again escalating safety risks. The corps originally set out to protect New Orleans with an approach called the "barrier plan." A 1962 corps report outlined it this way: "a barrier at the east end of the lake to exclude hurricane tides, coupled with construction or enlargement of protective works fronting developed or potentially developable areas."[30] From an engineering standpoint, building levees and gates across the entry points to Lake Pontchartrain would keep rising water far away from most of the city. The areas on the city's east side—the Lower Ninth Ward and the largely undeveloped area of eastern New Orleans—would still be directly exposed to water flowing in from the gulf; those areas would be protected by higher levees.

But the gate project soon stalled. Local environmental groups led by Save Our Wetlands won a federal injunction against the corps for failing to do a detailed environmental impact study, as required by law, on the effects of building gates and levees in sensitive marshes. The corps maintained that it would take too long to do such a study, and the whole project languished. By 1984, calculations showed building gates would be more expensive than simply raising levees around inhabited areas.[31] So the corps dropped the barrier plan and went with the "high level plan" instead.[32]

The fallback had one serious downside: It would allow a storm surge to flow into Lake Pontchartrain. As a result, when a storm hit, floodwaters would come much closer to the city's most populous areas. Thus, the consequences of failure at any point along dozens of miles of levees—even a single floodwall—became catastrophic. Then a virtually identical debate played out again—this time, over the 17th Street, London Avenue and Orleans Avenue drainage canals, which carried rainwater from the city's network of pumping stations to Lake Pontchartrain. When a storm surge in the lake entered the canals, water would rise a stone's throw from backyards and city streets. That certainly wasn't a new problem. As far back as 1871, city surveyor W. H. Bell had warned of it, reporting to his superiors that a severe storm could raise water levels in the canals high enough to flood the city.[33]

The corps proposed building floodgates at the mouths of each canal. Once again, it ran afoul of local agencies that feared the gates would cost more than building higher levees along the canals and that they might make it more difficult to pump rainwater out of the city during a storm. One by one, the gate proposals were abandoned. In 1992, it was settled, and with it, the fate of New Orleans: More walls would have to be built, and there were even more possible places for something to go wrong.[34]

Sure enough, it did. The soil beneath New Orleans is notoriously soft and squishy. A 1981 stability analysis of proposed upgrades to the 17th Street Canal levees noted that the designs didn't meet safety standards in several spots because of "the existence of very soft clays with minimal cohesion."[35]

Here, another controversial corps engineering standard came into play. Virtually all structures—buildings, bridges, elevators—are built to withstand more force than they will routinely encounter. This "factor of safety" cushion is supposed to provide for the unexpected—unforeseen forces, design or construction problems, or other things that could weaken the structure—so that it doesn't collapse at the first sign of trouble. The higher the factor of safety, the more secure the structure. The corps' factor of safety for levees and floodwalls is 1.3, meaning a given structure would have 30 percent more strength than it actually needed.[36] But engineers from the Independent Levee Investigation Team say that standard is hopelessly outdated—that it originated when levees protected mainly rural areas that flooded often, not big urban areas with hundreds of thousands of residents.[37]

Joseph Wartman, a geotechnical engineer at Drexel University and a member of the independent team, called 1.3 "way too low."

"That is widely recognized among practitioners, academics—everybody except some at the corps," he said. "The one we usually work with for a system like this is a factor of safety of 1.5 or 2. We lean toward 2 when you have a lot of uncertainty in the ground parameters, which is certainly the case here."[38]

The corps doesn't accept that conclusion. "I don't think—at least the evidence that I've read, there may be evidence out there that I haven't seen—that they used the wrong safety factor," said Maj. Gen. Don Riley, the corps' director of civil works.[39]

In any case, the 1.3 factor of safety was a margin of error too narrow to withstand the designers' mistakes. Not anticipating the gaps suggested by the 1985 sheet pile test was one of those errors. Another occurred at 17th Street, where designers used data from soil borings taken along the top of the levee. The weight of a levee packs and consolidates the soil underneath it; thus the surveyors missed a weak area a few feet away, where the bulge of the levee met the ground. There, about 15 to 20 feet below sea level, sat a layer of organic peat and, underneath it, soft clay. It was that layer that ultimately gave way, sliding out from underneath the floodwall's sheet pile foundation and opening up the breach that flooded most of Lakeview.[40]

Carlton Newman, a 47-year-old chef, was one of the few who lived near the 17th Street breach to return to the neighborhood in the year following the storm. "If we can send a spacecraft to Mars," he said, "we ought to be able to build a 14-foot wall on a glorified drainage ditch."[41]

An August 28, 2005, satellite photograph offers a bird's-eye view of Hurricane Katrina as the storm bears down on the Gulf Coast a day before making landfall.

*Photo courtesy of the National Oceanic and Atmospheric Administration*

The breach of the 17th Street Canal levee, seen here in an aerial photo taken nearly a week after Katrina struck, left the surrounding neighborhoods submerged.

*Photo courtesy of the U.S. Army Corps of Engineers*

Fast-rising floodwaters in New Orleans' Treme area on August 29, the day of Katrina's landfall, force a man and woman to flee their car, which has begun to float. The man places a baby in a carrier atop the roof out of harm's way as they abandon the vehicle.

*Photo by Rick Wilking / Reuters*

Petty Officer 2nd Class Eric Sciubba, an aviation survival technician with the Coast Guard, aids in the August 31 rescue of two elderly people from their flood-devastated east New Orleans neighborhood.

*Photo courtesy of the U.S. Coast Guard*

New Orleans residents still stranded on a rooftop on September 1, three days after Katrina hit, try to signal rescuers.

*Photo by David J. Phillip / Reuters*

As the National Guard patrols in the background, 81-year-old Louis Jones, left, and Catherine McZeal, 62, head down Poydras Street to the Superdome on September 1. The pair said they decided to see each other through the crisis because their relatives were not allowed into the area to help them evacuate.

*Photo © Michael Ainsworth* / Dallas Morning News / *Corbis*

Flood victims wade to the Superdome to be evacuated from the city by bus on September 2, four days after the storm.

*Photo by Bill Haber / AP Images*

A man in New Orleans' Treme section crosses the flooded street on the day that Katrina hit, while stranded residents watch from a porch.

*Photo by Rick Wilking / Reuters*

Caretaker Terry Jones, right, and others try to cool down her charge, overheated and exhausted Dorothy Duvic, 74, at the Ernest N. Morial Convention Center on September 1.

*Photo by Eric Gay / AP Images*

A woman sobs as she and others await help on September 1 at the Ernest N. Morial Convention Center.

*Photo by Eric Gay / AP Images*

With the help of search and rescue workers, medics and fellow National Guardsmen from his San Diego unit, Spc. Manuel Ramos, center, lifts Edgar Hollingsworth, 74, onto an ambulance gurney outside his home. The emaciated, severely dehydrated Hollingsworth was found inside his house two weeks after the hurricane hit.

*Photo by Bruce Chambers* / Orange County Register

Hurricane survivors waiting for medical attention lie along Interstate 10 near Causeway Boulevard outside New Orleans on August 31, two days after Katrina. They were among the thousands of people who gathered at the interchange locals call the "Cloverleaf" to await evacuation. Many had come from hospitals and nursing homes; others were residents dropped off there by rescuers who had plucked them from rooftops or floodwaters.

*Photo © John O'Boyle* / Star Ledger / *Corbis*

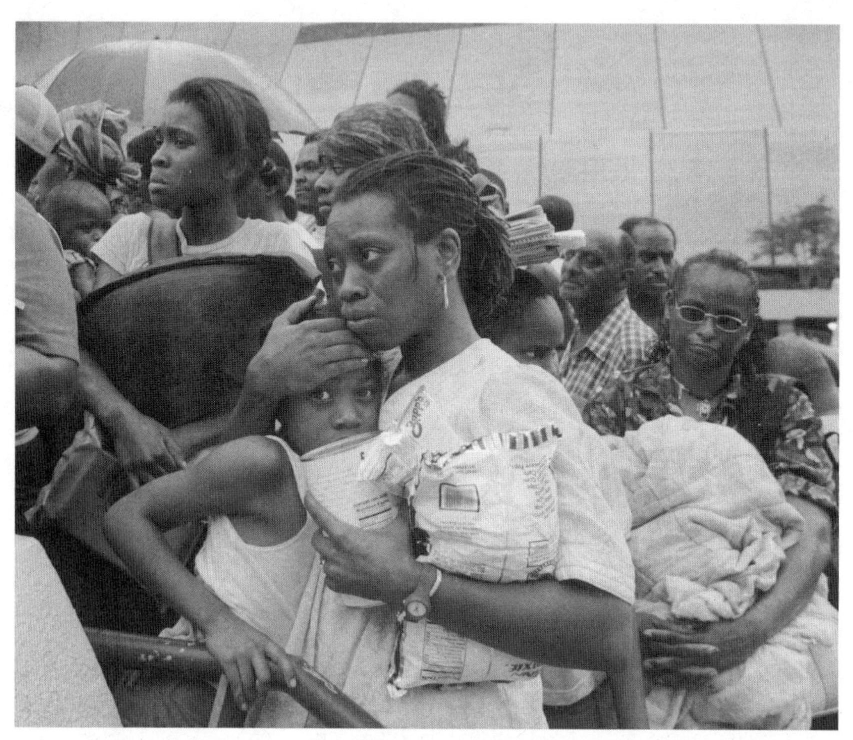

Cavel Fisher Clay, 33, and her daughters, Alexis Fisher, 14, back left, and Dejon Fisher, 8, center, wait with an agitated crowd of evacuees at the Superdome for buses bound for Houston three days after the storm.

*Photo © Michael Ainsworth* / Dallas Morning News / *Corbis*

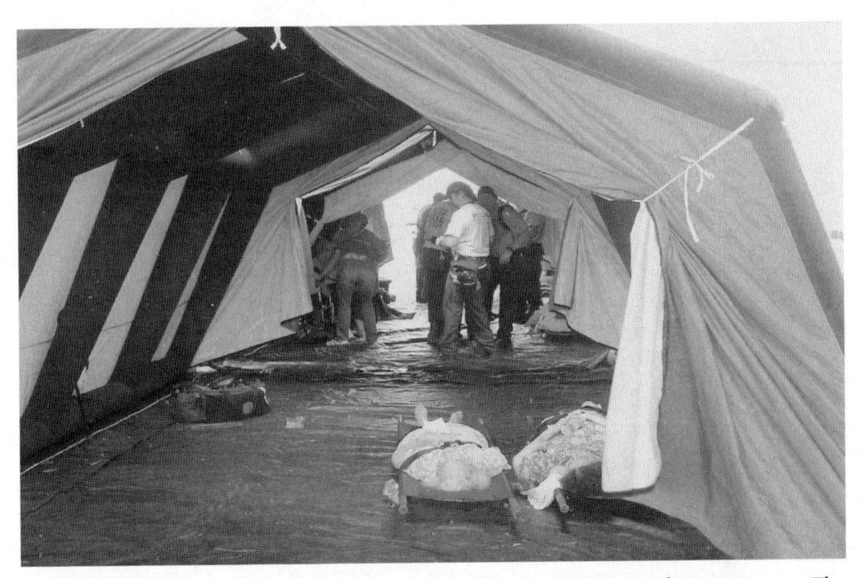

Elderly patients on stretchers await evacuation at the I-10 "Cloverleaf" staging area. The tents were set up by the Austin-Travis County (Texas) Emergency Medical Services Katrina Task Force to treat evacuees who were brought there by helicopter and ground transport.

*Photo courtesy of the Austin-Travis County EMS Department*

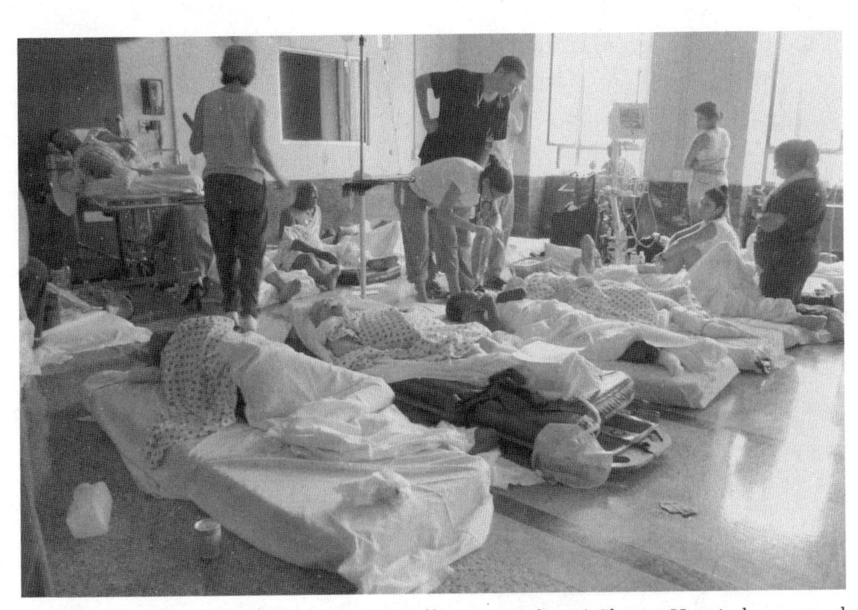

The day after Katrina's landfall, medical staff at New Orleans' Charity Hospital assess and treat incoming patients from area nursing homes and other hospitals. The patients were put on mattresses on the floor of the hospital's second-story auditorium.

*Photo courtesy of Ben deBoisblanc*

FOUR

# Emergency Preparedness

Jim Morris

*"Everything that happened in Katrina was preventable, and everything that happened was predictable."*

—George Haddow, former FEMA deputy chief of staff

Approach the perimeter of the Joint Field Office, the base of operations for federal and state disaster relief officials housed in a former department store in Baton Rouge, La., and you will encounter a conspicuous layer of security: somber men in matching tan shirts, large pistols on their hips. They are employed by Blackwater USA, a contractor perhaps best known for its paramilitary work in Iraq.

At first, the show of force at the JFO, as it's called, seems unnecessary. Then it begins to make sense. This, after all, is the nerve center for FEMA, the agency whose response to Hurricane Katrina came to symbolize government ineptitude. Hostility toward FEMA still smolders.

In the weeks immediately after the storm, "we had a number of bomb scares" at the JFO, explained Art Jones, Louisiana's disaster recovery director.[1] Hence the Blackwater guards.

Jones and others are quick to point out that the post-Katrina fiasco in New Orleans should not be pinned on any one person or agency. "This was a catastrophic disaster," Jones said. "The blame can be equally shared."

The fact remains, however, that FEMA stumbled badly both before and after the storm. Moreover, many of its missteps had been foreseen. In May 2006, a Senate Committee on Homeland Security and Governmental Affairs report on the disaster stated: "These failures contributed to human suffering and the loss of life. The causes of many of these failures were known long prior to Katrina and had been brought repeatedly to the attention of both DHS [the Department of Homeland Security] and FEMA leadership. Despite warnings, leadership failed to make vital changes."[2]

The months after Katrina yielded countless congressional hearings and several major reports in addition to the Senate committee's offering. The White House weighed in, as did the DHS inspector general and the House Select Bipartisan Committee to Investigate the Preparation for and Response to Hurricane Katrina. The reports varied in tone and level of detail, but common themes emerged:

- Lessons weren't learned from the 2004 "Hurricane Pam" disaster drill, which was based on a scenario remarkably similar to what unfolded during Katrina—a Category 3 storm striking New Orleans and causing widespread flooding. Such a storm, officials concluded after the drill, could put the lives of 60,000 people at risk and trap many thousands more.[3]
- Despite unusually strong warnings from the National Hurricane Center as Katrina was strengthening in the Gulf of Mexico, local, state and FEMA officials all failed to grasp the severity of the storm's potential impact on low-lying New Orleans.[4]
- FEMA's once-solid relationships with state and local officials had deteriorated, leading to poor coordination of efforts before and after Katrina.[5]

But perhaps the most significant finding was that "too many leaders failed to lead," as the House Select Committee put it.

"Critical time was wasted on issues of no importance to disaster response, such as winning the blame game, waging a public relations battle, or debating the advantages of wardrobe choices," the panel found.[6]

The leadership void undermined the public's confidence in the federal government's ability to rise to the challenge when a large natural disaster strikes. It also spawned the deep resentment that lingers in New Orleans a year after the storm—especially in neighborhoods like the Lower Ninth Ward, where many homes were pulverized by floodwaters and residents have grown weary of what they perceive to be FEMA's apathy and incompetence.

Dorothy and Lawrence Davis live in the Lower Ninth. Their house was damaged, but not destroyed. They said they were told by a FEMA representative that they had too little income to qualify for a loan to help them make repairs.

"We just said, 'Forget it. We'll make it on our own,'" Dorothy Davis said.[7]

James Lemann finally got a FEMA trailer in May 2006, eight months after Katrina made landfall. Like many residents of the Lower Ninth, however, he had no electricity and was using a generator for power. His generator was stolen about 10 days after the trailer arrived.

Asked if he felt the federal government had let him down, Lemann said, "It's not a matter of them letting you down. They ain't done nothing."[8]

State Representative Charmaine Marchand, who represents the Lower Ninth, said some of her constituents have complained that FEMA officials have been rude and unhelpful.

"The hostility is coming across through the phone," Marchand said. "You get conflicting information from FEMA and it's because their representatives are not properly trained."[9]

FEMA, established in 1979 to mobilize federal resources and coordinate with state and local officials during disasters, heard similar criticisms in 1992 after Hurricane Andrew blew ashore in South Florida, causing an estimated $26.5 billion in damage.[10] A study by the National Academy of Public Administration concluded that in the waning months of the George H. W. Bush administration, the agency was "a patient in triage" and that the president and Congress "must decide whether to treat it or let it die. . . . FEMA has been ill-served by congressional and White House neglect, a fragmented statutory charter, irregular funding, and uneven quality of its political executives."[11]

The same indictment could apply today. After experiencing a renaissance under the leadership of James Lee Witt during the Clinton administration, FEMA again finds itself bereft of money, high-level talent and—not to be underestimated—pride. Although many skilled and compassionate people are still employed by the agency, it has become a laughingstock, personified by Michael D. Brown, its seemingly inept leader during Katrina.

"I feel so sorry for those I've left behind," said Karen Keefer, who joined the agency at its birth and retired as a senior program officer in January 2006. "They really want to help people, and they're not being allowed to do it right."[12]

Current FEMA chief R. David Paulison declined through a spokesman to be interviewed for this book. However, both Brown and his predecessor, Joseph Allbaugh, said it is unfair to lay most of the blame for the Katrina debacle on an agency that lost much of its punch after being forced to merge with the roughly 180,000-employee Department of Homeland Security in 2003.

"I had 500 vacancies out of 2,500 people when Katrina hit," said Brown, who now runs a disaster preparedness consulting firm.[13] The effect of this "brain drain" was compounded by the magnitude of the storm, Louisiana's weak emergency management structure and a DHS bureaucracy that stifles decision-making by the FEMA director, he said.

According to Allbaugh, the agency has lost vital flexibility since DHS absorbed it.

"One of the strengths of FEMA was having the elasticity to call upon assets and personnel that other federal agencies had," said Allbaugh, also a consultant whose clients include firms in the disaster relief field. "Over time that elasticity has been eroded. The director doesn't have the latitude to make on-site decisions; the decisions are being made by the bureaucracies in Washington."[14]

---

As federal agencies go, FEMA is relatively young. It was created by President Jimmy Carter at the urging of the National Governors Association. The group's members had grown frustrated with the nation's disaster-response system, which at the time encompassed more than 100 agencies.

FEMA's mission was clear: it would advise communities on building codes and flood plain management plans, help train emergency managers, help equip first responders and provide aid to disaster victims.

All of this was easier said than done, however. Until it was folded into DHS, "FEMA was a combatant in many bureaucratic 'turf' wars," Richard Sylves and William R. Cumming wrote in a 2004 article in the *Journal of Homeland Security and Emergency Management*. "FEMA officials were often pitted against officials in competing federal agencies over 'who was in charge.' There was a longstanding division within FEMA over whether the agency would give priority to natural disaster management, as state and local governments generally preferred, or instead management of national security, continuity of government, and civil defense against nuclear attack duties, as several of its directors and three presidents [Ronald Reagan, George H. W. Bush and George W. Bush] preferred."[15]

FEMA veteran Keefer described the tone set during the Reagan years: "Everything was like the Red scare: The communists were coming. The office was being bugged. Close all your blinds, because [Soviet] satellites could pinpoint all your papers. They didn't pay too much attention to natural disasters."

According to Jane Bullock, who left FEMA in 2002 after 22 years and now works as an emergency preparedness consultant, "what we're seeing in Katrina was exactly what happened in the '80s. This is a total repeat. FEMA did the same thing under Bush II [as it did under Reagan]: They took three-quarters of the agency and devoted it to terrorism. The parallels are unbelievable."[16]

The 1980s were largely free of natural disasters, masking the agency's many weaknesses. But then came Hurricane Hugo in September 1989. The Category 4 storm tore through the Caribbean and made landfall in South Carolina. Reports of the number of deaths directly related to Hugo range from 49 to 56 people,[17] and the storm caused about $7 billion in damage in the United States alone.[18] FEMA's plodding response moved then-senator Ernest Hollings, D-S.C., to call the agency's leaders "the sorriest bunch of bureaucratic jackasses I've ever worked with in my life."[19]

Keefer was dispatched to South Carolina to deal with Hugo's aftermath. "It was a real mess. The magnitude was just too much for us, and it became politically hot," she said. "They sent me to open a disaster relief application center in downtown Charleston. We had National Guardsmen with rifles standing in the entranceway, making sure that the people who came in didn't cause a problem."

Less than a month after Hugo, the Loma Prieta earthquake struck California, killing 62 people in the San Francisco Bay area.[20] Again FEMA was criticized for its sluggish response.

FEMA's problems dealing with those disasters prompted the agency to develop a Federal Response Plan, which was finished in April 1992.[21] It spelled out how the federal government was to provide personnel, equipment, supplies and other assistance to state and local agencies.[22]

The plan, while comprehensive, didn't solve FEMA's problems. In July, not quite three years after the Loma Prieta quake, a damning report by the House Appropriations Committee was made public. Taking a shot at FEMA head Wallace Stickney—a one-time gubernatorial aide to New Hampshire's John H. Sununu, who had gone on to become White House chief of staff—the panel's report described the agency as a "political dumping ground . . . a turkey farm, if you will, where large numbers of positions exist that can be conveniently and quietly filled by political appointment." Stickney's inattentive management style had helped cause morale at FEMA to plunge to "an all-time low," according to the report.[23]

A few weeks later, FEMA was tested again as Hurricane Andrew, a rare Category 5 storm, took 65 lives in Florida, Louisiana and the Bahamas—40 of them in Miami-Dade County, Fla.[24] Federal aid was slow to reach storm victims, prompting Kate Hale, the county's emergency management director, to ask during a news conference, "Where in the hell is the cavalry on this one? They keep saying we're going to get supplies. For God's sake, where are they?"[25]

In January 1993, FEMA's inspector general delivered a post-mortem report on the response to Andrew that scolded the agency for adopting a "wait and see" posture. "Even in disasters with advance warning and accurate forecasting of severe results, FEMA does not use systems to predict response requirements and begin Federal response activities," the inspector general wrote. "The Federal response is to wait until the disaster occurs, then decide what Federal action is appropriate."[26]

A month later, a bill to shift FEMA's responsibilities to the Department of Defense was introduced in the House of Representatives.[27] Three months after that, however, FEMA won a reprieve with the naming of a charismatic new director, James Lee Witt.

Unlike some of his predecessors, Witt brought extensive disaster management experience to the job. He had been head of the Arkansas Office of Emergency Services in the late 1980s under then-governor Bill Clinton. According to Bullock, who became his chief of staff, FEMA's new chief went to Capitol Hill and told seething lawmakers, "'Give me six months.'"

"Witt focused on customer service and reorienting assets," Bullock said. "FEMA was spending too much time doing nuclear-attack planning. Literally, FEMA had no idea what its real mission was."

Mike Austin, a 20-year employee who was FEMA's senior policy advisor for international affairs when he retired in April 2005, called Witt's eight-year tenure at the beleaguered agency "a period of stability."

Until the new director arrived, "you hesitated saying you were with FEMA. You were on the defensive," Austin said. But once the new chief had shaken the place out of its torpor, he said, "we were more than happy to say we were with FEMA. There was a sense of internal pride."[28]

Witt had Clinton's unqualified support, so much so that the president elevated the FEMA director's job to Cabinet-level status in 1996.

"The joke was, if we wanted the space shuttle to do a fly-by, we could do it," said Mark Merritt, who served as Witt's deputy chief of staff at FEMA and now works for his disaster relief consulting firm, James Lee Witt Associates. "James Lee could pick up the phone and call the secretary of defense. They were equals."[29]

Shortly after his arrival, Witt set about reorganizing FEMA, dismantling what he called the "good old boy network" that had hamstrung the agency in times of disaster and discouraged the free exchange of ideas. Within two years, he said, he had opened a state-of-the-art operations center, replacing "a room with fold-up tables and fold-up chairs and wires hanging out of the ceiling."[30]

Clinton "made it very clear he was very supportive of everything we were doing," said Witt, whose firm was hired by Louisiana Governor Kathleen Blanco shortly after Katrina hit. "He also made it clear to other Cabinet secretaries that if FEMA needed help, they needed to provide it."

As a result, FEMA's on-the-ground performance improved dramatically. After the 1994 Northridge earthquake, for example, Witt immediately flew to California with two Cabinet secretaries—Federico Peña of the Department of Transportation and Henry Cisneros of the Department of Housing and Urban Development—and other top officials in tow. Smashed bridges and freeway overpasses were rebuilt in record time, Witt said, because the Transportation Department "threw out the [regulations] and red tape and did bids very fast."

At a news conference two days after the quake, two Republican politicians complimented FEMA's efforts. Los Angeles Mayor Richard Riordan praised Clinton for "the timely coordinated federal response that has come through your leadership." And Governor Pete Wilson remarked on the "really superb coordination and cooperation" among federal, state and local agencies.[31]

Witt said his philosophy was to keep state and local officials—and members of Congress—in the loop.

"Many times, especially when we were tracking hurricanes off the coast, we would invite a member of Congress from [a potentially affected] state to the ops center to get a full briefing," he said. "We made sure we discussed with the states, 'OK, what kind of help are you going to need evacuating? What kind of help are you going to need immediately after the storm?' We would pre-position teams . . . in state operations centers before anything happened."

When Clinton left office in 2001, Witt left with him. George W. Bush named Joseph Allbaugh as his FEMA replacement. He had been the former Texas governor's chief of staff and the national campaign manager for Bush-Cheney 2000.

Not long after he took over at FEMA, Allbaugh made it clear that his approach would be different from that of his predecessor. "James Lee did a great job of shifting the agency's focus somewhat to local preparedness," he said, adding that there was another side of the equation: response and recovery. That became his emphasis, and he sought more money for first-responder training and equipment.

"You can throw all the money in the world at preparedness, but you still have to be ready to go the minute the balloon goes up," Allbaugh said. "I have no apologies for rearranging [FEMA's] priorities."

One program that was eliminated was Project Impact, a $20 million preparedness campaign deemed ineffective by the White House—although it was credited with helping limit damage during the Seattle area's Nisqually earthquake in February 2001.

Money grew tight at FEMA once the September 11 terrorist attacks forced the agency to turn its attention back to civil defense. Two weeks after the first anniversary of the attacks, Allbaugh told the Senate Homeland Security Committee that in the previous year FEMA had spent or obligated more than $5 billion for relief activities in New York, Virginia and Pennsylvania.[32]

Looking back, Allbaugh said that FEMA had been "buying good will" throughout the 1990s, and some state and local recipients of agency funds grew disenchanted when funding dried up.

On March 1, 2003, FEMA became part of the new Department of Homeland Security—a move recommended by the Hart-Rudman Commission that the president initially opposed, then embraced. Allbaugh quit that day.

"I saw the handwriting on the wall," he said. "When you have three or four other decision-makers between the FEMA director and the president, information becomes clouded. . . . There's a loss of urgency. . . . It's just a no-win situation all the way around."

Allbaugh's deputy director, Michael Brown, succeeded him as agency chief. He had hired Brown as FEMA's general counsel in 2001; Brown's prior experience included working as a commissioner with the International Arabian Horse Association in the 1990s.

At DHS, FEMA became "the red-headed stepchild," said George Haddow, who had been Witt's deputy chief of staff and now runs a consulting firm along with former FEMA official Bullock. DHS "stole [FEMA's] money. They stole their people."

Long-cultivated relationships with state and local offices were damaged, Haddow said, after preparedness and grant-making functions were taken from FEMA and placed elsewhere within the new department. Witnesses told the Senate Homeland Security Committee that FEMA's budget "was far short of what was needed to accomplish its mission, and that this contributed to FEMA's failure to be prepared for a catastrophe," according to the panel's May 2006 report. For the past few years, it added, the agency has had an average employee vacancy rate of 15 to 20 percent—about 375 to 500 vacant positions out of 2,500.[33]

"We had no capability for hurricanes in '04 and '05 because they cut us back so severely," said former FEMA official Keefer. FEMA, she said, went from being "the star agency in the government" to "totally inept."

The stage had been set for the debacle in New Orleans.

---

Hurricane Katrina hit Louisiana on Monday, August 29, 2005. In the days before the storm, Leo Bosner, a watch officer at FEMA headquarters, was responsible for preparing "national situation reports" that sought to characterize the storm looming in the Gulf of Mexico.

Bosner's August 27 report stated that "Katrina is now a Category [3] hurricane and some strengthening is forecast during the next 24 hours." It noted that, among the areas expected to bear the brunt of the storm, "New Orleans is of particular concern because much of that city lies below sea level. . . . Lake Pontchartrain is a very large lake that sits next to the city of New Orleans and if the hurricane winds blow from a certain direction, there are dire predictions of what may happen in the city."[34]

(In an e-mail to the director of Florida's emergency management division that afternoon, FEMA chief Brown averred, "this one has me really worried . . ."[35])

In his August 28 report, Bosner described Katrina as a "dangerous Category 4" storm, and warned that it could become a Category 5 before it made land. "A direct hit," he wrote, "could wind up submerging the city in several feet of water. Making matters worse, at least 100,000 people in the city lack the transportation to get out of town."[36]

Bosner—president of Local 4060 of the American Federation of Government Employees, which represents about 500 FEMA employees—said that he and others on duty that weekend were stunned that a major evacuation of New Orleans wasn't already under way.

"We're going, 'What is this?' We knew that an awful lot of people in that city don't have cars, and just telling people to leave isn't going to work," he said. "We knew there had to be some massive evacuation [using] buses, which FEMA's authorized to do. It was just like watching a train coming down the track and the train's going to hit a kid on the track and nobody's getting the kid out of the way. At one point, we all turned to one another and said, 'Oh, shit, they're not going to do anything!'"[37]

There were more miscues after the fact. On Wednesday, August 31, when buses were desperately needed for evacuations, about 200 of them were held up in La Place, La., according to Terry Ebbert, homeland security director for the city of New Orleans. The reason? "FEMA contractors were checking the tread depth on the tires, making sure they were in compliance" with safety standards, Ebbert said.[38]

Brown said he couldn't verify that account, but found it believable. "That's the kind of bureaucratic bullshit you get into when you're trying to mobilize all of these contractors," he said. "That kind of stuff happens in every disaster."

Still, Brown said, DHS interference made matters worse. Not long after Katrina made landfall and daily after that, Governor Blanco asked him to find 500 buses to help evacuate some of the thousands of people sheltered at the Superdome.

"Those buses never showed up," Brown said, adding that he never learned why.

Similarly, Brown said, his plan to use commercial airliners for evacuations fell by the wayside when the Transportation Security Administration—also part of DHS—declared that it would have to screen every passenger first.

"I was like, 'Screw it. Put marshals on the planes, put people on the planes and fly them out,'" Brown said. "It's the damned bureaucracy. It's so absurd it's just unbelievable."

The upshot of FEMA's blundering is now well known. The various investigative bodies that picked apart the response to Katrina leveled all manner of charges at the agency. Among other things, the House Select Committee found that there had been "[a] complete breakdown in communications that paralyzed command and control and made situational awareness murky at best."[39]

The Senate Homeland Security Committee found that "Brown did not direct the adequate pre-positioning of critical personnel and equipment, and willfully failed to communicate with [Homeland Security Secretary Michael] Chertoff, to whom he was supposed to report."[40] And the DHS Office of the Inspector General observed that some "[d]ifficulties experienced during the response directly correlate with weaknesses in FEMA's grant programs," which were being administered elsewhere within DHS.[41]

For many, these failures translated into unspeakable losses. The Lower

Ninth Ward and other neighborhoods remained in ruins more than a year after Katrina, with many of their residents displaced, distraught and disgusted with all levels of government. FEMA has been the primary target of their rage.

James Lemann, a resident of the Lower Ninth, was among the 15,000 to 20,000 people who were trapped in the squalid conditions at the Superdome after the storm. He still marvels at the bedlam. "There was a total lack of any kind of coordination," he said. "You can't put nobody in those high-powered jobs if they can't perform."

Haddow, the former FEMA deputy chief of staff, maintained that "five days of chaos could have been mitigated to one day" had the agency been properly prepared.

"Everything that happened in Katrina was preventable, and everything that happened was predictable," he said.

But state and local officials are not blameless in the tragedy. Just 13 months before Katrina, many of them participated in the "Hurricane Pam" exercise, which was to have provided the basis for a catastrophic hurricane response plan for south Louisiana.[42]

Although the Pam scenario didn't include a levee failure in New Orleans, a National Weather Service computer model had predicted that a Category 3 storm's tidal surge could overtop the levees and immerse the city in 14 to 17 feet of water.[43]

The 2004 drill exposed deficiencies in New Orleans' evacuation planning, prompting city officials in the summer of 2005 to try to line up buses that could be used to get people out of town. As they did, however, officials found that there would not be enough qualified drivers to man the buses, rendering the plan unworkable.[44]

And, just as in the drill, the evacuation process before Katrina made landfall was flawed. While a half-million vehicles carried a million people out of the city before the storm,[45] the 100,000 or so New Orleanians who lacked transportation were stuck.[46] Ebbert, the city's homeland security director, told the Senate Homeland Security Committee that officials held off opening the Superdome to the general population on Saturday, hoping that people would evacuate rather than seek shelter in the city.[47]

At 8 a.m. on August 28, the day before Katrina hit, the Superdome was opened as a shelter of last resort for people who were mentally or physically handicapped or who had minor medical problems. Four hours later, Mayor C. Ray Nagin opened it to everyone.[48] The arena that hosted New Orleans Saints professional football games became an overcrowded hellhole, as did the city's convention center. Both were short of basic supplies.

"The [Ernest N. Morial] Convention Center was a refuge of last resort that came out of necessity. . . . We did not have the resources to provide food. Water ran until the pump stopped . . . so we had to get FEMA to provide bottled water," Nagin testified before the Senate panel in February 2006.[49]

Many evacuees were stranded at the Superdome and the convention center for up to five days until buses were finally secured. City officials later learned that, while school buses had sat flooded and unusable, about 200 Regional Transit Authority buses had been staged in an area safe from floodwaters, with open routes into and out of downtown—information that RTA officials had not relayed to Nagin.[50]

The New Orleans Police Department also suffered problems that had been predicted in the Pam drill. A number of patrol cars were submerged, as were police headquarters and most district stations; radio towers were knocked down by the wind or lost backup power, crippling communications.[51]

Timothy Bayard, commander of the NOPD vice and narcotics section, would later tell the Senate Homeland Security panel, "We did not coordinate with any state, local or federal agencies. We were not prepared logistically . . . we relocated evacuees to two locations where there was no food, water or portable restrooms . . . We did not utilize buses that would have allowed us to transport mass quantities of evacuees expeditiously . . . We did not have a backup communication system."[52]

State officials, for their part, seemed to be gambling that a monstrous storm wouldn't strike. Mark C. Smith, a spokesman for Louisiana's Office of Homeland Security and Emergency Preparedness, said as much in an interview: "It's not a secret that this was going to happen at some point. . . . It's not if, but when. And we all hoped it wouldn't happen during our tenure."[53]

Lt. Col. William Doran, chief of the state office's operations division, said it would have been "folly" for the state to engage in catastrophe planning without the cooperation of the federal government and other states. "The bulk of the response is going to come from the federals and from outside help, because you're going to be rapidly overwhelmed," Doran said.[54]

But Art Jones, Louisiana's disaster recovery director, said that confusion and buck-passing within a bloated DHS kept state and local officials from doing their jobs more effectively during Katrina. In one instance, Jones said, he had to pass through "nine levels of bureaucracy" to get a response to a request. "To get a decision you got to go to one [DHS official], then another one. You do that nine times before you get to the White House," he said.

Still, in its post-Katrina investigation, the Senate Homeland Security Committee faulted the city and state of Louisiana for failing to invest enough in emergency preparedness. It found that the state homeland security office was grossly underfunded, with a staff of 44 before the storm—about 60 percent of the national average for such an office. "The inadequacy of [the agency's] resources was a chronic issue, known to Louisiana officials well before Katrina," the panel wrote in its report.

Low pay had led to high turnover and made it difficult to recruit qualified staff, the committee found. "Planning, in particular, suffered," the report concluded.[55]

---

Ultimately, however, FEMA is supposed to take the lead during disasters of Katrina's scope. By any number of measures, it failed, and few experts believe it is any more capable today.

At his swearing-in ceremony on June 8, 2006, new FEMA director R. David Paulison promised that the agency would rebound.

"These are tough times for you, I know," Paulison, a career firefighter, told the workers present. "You hold your head high. We are going to make America proud of this organization."[56]

But he faces a colossal challenge, given the agency's budget cuts and talent drain.

"FEMA was steadily bled to death by its new siblings and a parent organization that focused on terrorism," Representative Tom Davis,

R-Va., chairman of the House Select Committee, said at a May 2006 news conference.[57] He and others in Congress feel the only way to solve FEMA's problems is to remove the agency from DHS and make it independent again.

At a hearing the day Paulison was sworn in, however, DHS head Chertoff warned that this would not be a panacea.

"Many of those who argue most strenuously for separating FEMA from DHS paint a portrait of FEMA's history that does not comport with reality," Chertoff said. "Essentially, they long for a return to glory days that in fact never existed. . . . FEMA was simply not tested in the 1990s on a scale [in] any way comparable to Hurricane Katrina."[58]

True enough. But most seem to agree that FEMA was, by any number of measures, a more competent agency in the mid- to late 1990s that probably would have made a better showing in New Orleans.

Witt said, "I don't know what Mike Brown or FEMA did before [Katrina] hit. It was a catastrophic storm. When I was [at FEMA], I would be on the phone with governors, mayors of cities that would be impacted, and I would ask them hard questions: 'Are your nursing homes evacuated? Are your hospitals evacuated? Your special-needs people? You tell us what you need.' I don't know if that happened this time. It should have. If it didn't, that's a terrible mistake."[59]

Although the rebuilding effort in New Orleans has made modest headway in the year since the hurricane, FEMA continues to absorb criticism from disheartened flood victims.

In late May 2006, for example, the Lower Ninth Ward was still without electricity and laden with debris, and FEMA trailers were just beginning to arrive. People like Annie and Greland Washington, who had been renting a house that is now uninhabitable, had abandoned the idea of getting help from FEMA and were in the process of cleaning up and rebuilding on their own with the help of nonprofit relief organizations like Common Ground.

The Washingtons said they applied for a FEMA grant but were deterred by the myriad restrictions. "I heard other people with rentals got $10,000 [grants] like nothing," Greland said, taking a break from gutting the green house at the corner of Egania and North Derbigny streets.[60]

Common Ground legal coordinator Soleil Rodrigue said that displaced New Orleanians, frustrated with FEMA, continue to turn to her group for food, water, cleaning supplies and other essentials.

"More [are coming] all the time," Rodrigue said. "The need is far beyond the capacity to respond."[61]

In June 2006, a judge who held in part for FEMA in a class-action housing lawsuit nonetheless felt compelled to comment on the agency's treatment of storm victims. In his decision, U.S. District Judge Stanwood R. Duval Jr. of the Eastern District of Louisiana chided FEMA "for what could be considered cagey behavior with regards to [its] ever-changing requirements."[62]

Wrote Duval: "Rather than hiding behind bureaucratic double-talk, obscure regulations, outdated computer programs, and politically loaded platitudes such as 'people need to take care of themselves,' as *the face* of the federal government in the aftermath of Katrina, FEMA's goal *should* have been to foster an environment of openness and honesty with *all* Americans affected by the disaster."[63]

That same month, FEMA was excoriated by government auditors. The Government Accountability Office released a much-cited report documenting up to $1.4 billion in improper and possible fraudulent individual assistance payments made by the agency. FEMA-issued debit cards, the GAO found, were used for diamond jewelry, a visit to a "gentleman's club," a Caribbean vacation and "adult erotica products," among other things.[64]

And at a June 14, 2006, hearing, Representative Bennie Thompson, D-Miss., a member of the House Homeland Security Committee, said the chicanery uncovered in FEMA's individual assistance program "pales in comparison to FEMA's contracting follies," and is a symptom of a "disease . . . an agency that issues sole-source contracts to politically connected companies instead of using competition to get the best deals and the best product."

According to media reports, FEMA's use of multiple layers of subcontractors in the Katrina cleanup, and the size of the contracts themselves, have greatly increased the odds of waste, if not outright fraud.

However, the federal relief effort has not been without bright spots.

In a June 1, 2006, news release, FEMA touted the cleanup and rebuilding work done on the Gulf Coast by the college-age members of the National Civilian Community Corps, part of the AmeriCorps public service program begun by President Clinton in 1994. To that point, the release said, the NCCC had provided an estimated $9 million in services and helped more than 1.1 million hurricane victims, handing out "1,714 tons of food and 2,790 tons of clothing, serving one million meals, and refurbishing 1,529 homes."[65]

But the release neglected to say that the White House had proposed killing the NCCC in March 2006.[66]

In a letter to several media outlets in June, three former directors of the program came to its defense, noting that its young workers had logged more than 500,000 hours on the Gulf Coast. "The National Civilian Community Corps should not be kicked to the curb," they wrote.[67]

FIVE

# Social Services

Katy Reckdahl

*"The Red Cross said, 'To heck with y'all, we're going to come in after the mess.'"*

—Louisiana State Representative Jeff Arnold, aide to former mayor Marc Morial

On Tuesday, August 30, 2005, the day after Katrina hit New Orleans, Brian Greene drove from Baton Rouge to his food bank's warehouse, located outside the city. As he was leaving the warehouse, his truck's radio was tuned to the area's only working station. "A public official was begging anybody, 'Give us water and food,'" Greene says. "I had four trucks of everything he needed, but I had no way of reaching him, no way of knowing where he was."[1]

Greene, then head of the Second Harvest Food Bank of Greater New Orleans and Acadiana, couldn't call anyone because phones weren't working. His trucks couldn't distribute food anywhere because all of his drop-off points within the city had been destroyed. His usual charity partners—the Salvation Army and the American Red Cross—hadn't yet arrived in New Orleans. "We were so disconnected," he says.

Greene's tale is typical of New Orleans' social service agencies. After the hurricane, they felt the urgency to help and knew their assistance was desperately needed. But they weren't able to respond.

On Wednesday morning, 37-year-old Patricia Ann Hills and 15 members of her extended family left their apartment in the Lafitte public housing development and began walking through chest-deep floodwaters.[2] Her husband Jeffrey hoisted their daughter Jeffreyell, 6, and their son Jeffrey Jr., 3, onto his shoulders as the family waded to the convention center. They'd spend two days there in 90-degree temperatures and squalid conditions before walking away on Friday morning, after their two youngest children started shaking from dehydration.

"There was no food and water—nothing," Patricia Ann Hills says. "My 15-year-old daughter Le-Ann said, 'Mama, we're going to die, huh?' I thought, 'People are suffering in here. People are dying. Where are the groups that help in times of crisis?'"

The city of New Orleans' hurricane plan specifically designates three social service agencies to do just that: the Red Cross, Salvation Army and Second Harvest.[3] According to the city's plan, after a disaster, the Red Cross and Salvation Army were to take the lead in establishing "feeding sites" across the city. Second Harvest was to "provide leadership" in acquiring and distributing food and water.[4]

That didn't exactly happen immediately after Katrina.

The need for food and water was most pressing in the days following the storm, when as many as 100,000 people remained in New Orleans,[5] nearly half of them in the Louisiana Superdome and the convention center.[6]

During that first week, local Salvation Army employees were trapped by floodwaters at the Center of Hope, the agency's homeless shelter in uptown New Orleans. They had stayed behind with 300 clients and neighbors who were unable to evacuate and had crowded into the center before the storm.[7]

The Southeast Louisiana Red Cross chapter was farther away and also unable to respond. Rather than set up facilities in the city in anticipation of the storm, it had heeded a decade-old policy of packing up and evacuating whenever a serious hurricane headed toward New Orleans. Workers and volunteers set up a temporary chapter headquarters in Covington, across Lake Pontchartrain from the city. Stranded there, they scrambled to care for 5,000 sheltered evacuees and communicated sporadically with the outside world using an older-model satellite phone.[8]

To fill the vacuum left by the missing agencies, a patchwork of neighbors worked together. Malik Rahim, a longtime community activist, started a collective called Common Ground that cleared blocked city streets and started to feed residents by bicycle brigade since gasoline was in short supply.[9] Impromptu kitchens sprang up at churches.

Of the city-designated agencies, Second Harvest may have rebounded best. The day after the storm, the food bank began sending supplies to New Orleans. It also worked with elected local officials and state legislators to create viable drop-off points in the city for Second Harvest trucks.[10] And staffers set up shop in a former Wal-Mart outside of Baton Rouge, where they processed food and water that came in on 18-wheelers from the U.S. Department of Agriculture and through the America's Second Harvest nationwide food bank network.[11] Still, says Greene, they couldn't do enough. "If you don't have a lot of guilt coming out of this, you didn't have a lot of responsibility," he says.

---

Pamela Smith, 50, a New Orleans resident, had evacuated and weathered the storm with her elderly mother in Bay St. Louis, Miss. When Smith returned to New Orleans in October 2005, she was surprised at the difference. "In Mississippi, the American Red Cross was very visible—everyone knew where they were located. Not here," says Smith.[12] She'd ask neighbors where the Red Cross was, and they wouldn't know. Around her neighborhood, she says, the national Salvation Army had a noticeable presence. Not the Red Cross.

It's been Red Cross policy since 1995 not to operate any hurricane shelters south of Interstate 12, which runs through southeast Louisiana, above the north shore of Lake Pontchartrain.[13] (New Orleans borders the lake's south shore.) The agency considers anything below I-12, including the entire city of New Orleans, a risk area. And it won't leave anyone—neither staff nor volunteers—in a risk area.[14]

Eric Cager, a well-known local activist who spent a night at the convention center before evacuating on foot, defends the Red Cross. "You can't expect the Red Cross to be your entire plan," he says.[15] The city, he says, knew that when a major hurricane threatens, the Red Cross evacuates staff members to the other side of Lake Pontchartrain. From that location, they couldn't realistically return soon after a major storm.[16]

A decade ago, when the Red Cross announced its policy, Louisiana state legislator Jeff Arnold was an aide to the mayor of New Orleans, Marc Morial. Morial appealed to Red Cross officials to reconsider, says

Arnold. "If you take a look at the flood map from 100 years ago, you know where the dry areas will be," he says. "Why couldn't the Red Cross locate a facility in one of those areas?"[17]

But the Red Cross had set its policy, and it was absolute. Since 1995, the Southeast Louisiana chapter hasn't left a single staffer behind, not in the Superdome[18] and not even in the city's Emergency Operations Center.[19] After Hurricane Katrina, that absence created a problem, because, according to Red Cross protocol, the local chapter is the designated liaison for municipal officials. (The agency's national headquarters deals with state and federal officials.)[20]

"The Red Cross does not support anyone pre-storm or during a storm here," says Terry Ebbert, homeland security director for the city of New Orleans.[21]

Kay Wilkins, chief executive officer of the Southeast Louisiana chapter of the Red Cross, says that she spoke with both the mayor and the governor after a pre-storm news conference and reminded them of the Red Cross' evacuation policy. "The plan was, as soon as it was safe, we would reenter the city of New Orleans," she says.[22] Her chapter did not return to downtown New Orleans until the end of September, one month after Katrina.[23]

Wilkins was not unfamiliar with disaster, at least on a small scale: Prior to Katrina, the Southeast Louisiana chapter was the fourth-busiest of the 800-plus American Red Cross chapters in dealing with disasters. On average, it responded to a fire within its 12-parish area every eight hours.[24]

But after Hurricane Katrina, even Red Cross shelters that were set up outside the city were threatened. Blocked roads left Wilkins and her staff stranded and overwhelmed by the needs of the shelters filled with evacuees from New Orleans. Red Cross chapters are instructed to run shelters for three to five days on their own, without any help from the Washington, D.C., headquarters.[25] As the storm approached, that seemed feasible, Wilkins says. But afterward, vendors couldn't get through to make deliveries, so the chapter began rationing food at its shelters. "I never imagined that people in Baton Rouge couldn't get to us," says Wilkins. "Because that's where all the resources were. They were 50 miles away."

Wilkins herself went an entire day without any communications. On

Tuesday morning, the day after Katrina struck, a local emergency worker gave her a marginally functioning satellite phone. That afternoon, she got through to Joe Becker, senior vice president of preparedness and response at the national Red Cross in Washington. Communications were spotty through that entire first week, says Wilkins. As a result, her chapter wasn't a big part of the recovery effort that happened—or didn't happen—in New Orleans.

---

Accounts differ as to why the national Red Cross didn't make its way into New Orleans in the week after the storm. On Friday, September 2, Marty Evans, the American Red Cross' then-president and CEO, appeared on CNN's *Larry King Live*. "Larry," she said, "we were asked, directed by the National Guard and the city and the state emergency management not to go into New Orleans because it was not safe."[26]

The social services agency asserts that it made repeated requests to enter the city. "On Thursday, September 1, and again on September 2 and 3, the Red Cross made offers to the state to provide food and water to those still in the city of New Orleans," says the Red Cross' Becker.[27]

Mark C. Smith, spokesman for the Governor's Office of Homeland Security and Emergency Preparedness, says he was unaware of those requests. More than anything else, he says, the state was focused on getting people out of New Orleans.[28] His office was also reluctant to send in aid organizations, because of media reports of "perceived civil disobedience" in the city.[29]

Ebbert, of the New Orleans homeland security office, says the city never barred the Red Cross or any other agencies. But he places the blame for the dire conditions in the convention center squarely on FEMA, not the Red Cross. "Those initial lifts of water and food, that's a government issue," he says. "We can't depend on a volunteer organization for initial response."

Vic Howell, head of the Red Cross' Baton Rouge chapter, was the agency's official liaison within the state's Emergency Operations Center. He remembers getting a call from Red Cross headquarters in Washington, D.C., on Friday, asking if the organization could get clearance to enter New Orleans to provide food and water. He brought the request to Jay Mayeaux, deputy director of Louisiana's homeland security office, who told Howell that the state couldn't divert resources to help them get in and couldn't provide security for their volunteers within the city.[30] Mayeaux was also concerned that a Red Cross presence could deter evacuation, says Howell. "He thought that if our red and white trucks started showing up with assistance, people might stick it out and not try to leave."

But a report issued by the Democratic staff of the U.S. House Committee on Homeland Security suggests that the charity's absence in New Orleans is part of a pattern. The report cites allegations that the Red Cross "is frequently late in responding to large-scale disasters—often arriving on the scene days after other relief organizations have arrived." It says the situation is "not only distressing but also highlights a critical gap in the Department of Homeland Security's plans for responding to large disasters."[31]

To support that argument, the report lists several examples in which other charities were able to respond more quickly than the Red Cross. One was in March 1990, when waters breached the levee in Elba, Ala., and flood victims waited six days for Red Cross relief workers to arrive.[32] The report describes the situation: "While the Red Cross waited to confirm the severity of the disaster, the Salvation Army was already on the ground working with local churches at the site to distribute food, pillows, and blankets to the flood victims."

Representative Bennie Thompson of Mississippi, the ranking Democrat on the committee, sees clear parallels with the situation in New Orleans. He's traveled there and talked with residents, he says, and knows that parts of the city were accessible, even right after the storm. "There were areas where things could have been set up," he says."[33]

A month after Katrina hit the Gulf Coast, the *Wall Street Journal* wrote about similar refrains in East Biloxi, Miss., and other affected areas: "Here in some of the poorest parts of Mississippi and much of the Gulf Coast, the Salvation Army is drawing praise for its swift arrival in the most distressed areas. . . . To some people here, the Red Cross, under growing criticism for letting bureaucratic hurdles slow down aid in the disaster area, suffers by comparison."[34]

Musician and lifelong New Orleans resident Kenny Claiborne remembers seeing the national Salvation Army roll into his downtown neighborhood, the Marigny. It was the first agency to arrive there with food. "We'd been hearing about the Red Cross and how it was going to pull in billions of dollars," he says, "and then here comes the Salvation Army—we were laughing about how it was paid for by that little red bucket."[35]

Before Katrina hit on Monday, the Salvation Army had moved its supplies outside of the hurricane's cone of probability, to Port Sulphur, La. (about an hour southeast of New Orleans), Jackson, Miss., and Tallahassee, Fla.[36] The day after the storm, the Salvation Army set up kitchens in a warehouse in La Place, 25 miles west of New Orleans. By the following day, it was sending several 5,000-meal canteens twice a day into less-damaged parts of the city.[37]

In Claiborne's neighborhood, the Salvation Army started driving down the streets on Wednesday, two days after the storm, announcing mealtimes through a public address system, he says. On Friday, the organization was setting up food canteens on Canal Street in front of the Sheraton Hotel, just a few blocks from the convention center.[38]

Entry into the metropolitan area wasn't a problem, says Salvation Army spokesman Mark Jones. "We were not hindered in any way getting into the city of New Orleans."[39]

In an Urban Institute report, *After Katrina: Public Expectation and Charities' Response*, philanthropic experts noted that agencies like the Salvation Army don't focus just on disasters, but provide a wide range of community-based social services. As a result, they are "constantly in touch with a wide variety of local service providers and funders."[40] After Katrina, said the experts, the Red Cross was hindered by "its relative lack of integration with local networks of social welfare agencies and public and private funders."[41]

U.S. Government Accountability Office teams visiting the Gulf Coast in October 2005 also observed that the Red Cross was not present in all areas. "The American Red Cross did not provide relief in certain areas because of safety policies; and thus, other charities, such as the Salvation Army and smaller charities, often helped to meet the needs of those areas," Cynthia Fagnoni, the GAO's managing director of education, workforce and income security issues, testified before a House subcommittee in December.[42] Fagnoni said that some of the Red Cross' absences could be explained by its policy against establishing shelters in areas that might flood. Still, needs went unmet, she said. "[V]ictims remained in areas where the American Red Cross would not establish shelters."[43]

Wilkins, head of the Southeast Louisiana Red Cross, says that she's grateful that one of the Red Cross' community partners could be in New Orleans. "That's why the social service network is so incredibly important," she says. In other parishes, Wilkins notes, the Red Cross was present and the Salvation Army was not.

Joe Leonard Jr., executive director of the national Black Leadership Forum, saw these differences firsthand as he drove through Alabama, Mississippi and Louisiana in the wake of Katrina. "I kept wondering why, everywhere I went, people loved the Salvation Army, but I didn't see the Red Cross at all," says Leonard. "I found out that the Salvation Army is trained to make decisions on the ground; they give them the independent autonomy to make a decision."[44]

Congressman Thompson believes race was a factor in the Red Cross' spotty response. In Mississippi, the Red Cross often set up in the white, middle- or upper-class parts of town, he says, not in black neighborhoods. The charity's absence in the majority-black New Orleans may be related to that, he says. "When you have people not accustomed to working with certain people, certain biases and fears set in. Even without a hurricane, the same people could travel through that area and still have the same fear."

The Red Cross' chief diversity officer, Rick Pogue, denies that the agency's efforts were in any way driven by race. "There were many areas of the Gulf Coast that we simply could not reach," he says.[45]

But in a June 2006 review, the Red Cross examined its performance during Katrina and acknowledged a number of weaknesses. The review found that the agency "had not developed strong and enduring local partner relationships" in some hurricane-prone areas and had missed opportunities to reach underserved areas and respond to diverse groups, such as minority communities. The report also concluded that many Red Cross chapters "were quickly overwhelmed by the outpouring of volunteer offers" and that many volunteers were turned away or placed on waiting lists.[46]

The charity turned its back on the people of New Orleans, says legislator Arnold. "It's a slap," he says. "The Red Cross said, 'To heck with y'all, we're going to come in after the mess.'"

---

The American Red Cross is a volunteer-based nonprofit organization, but it has a unique relationship with the United States government. In 1900, Congress granted it a charter to "mitigat[e] the sufferings caused by pestilence, famine, fire, floods, and other great national calamities."[47] The organization's structure reflects this relationship: at least eight of 50 board members are appointed by the White House and the president of the United States is considered the organization's honorary chairman.[48]

Historically, the relationship was even more cozy. Calvin Coolidge, who was president during the devastating Mississippi River flood of 1927, was also president of the Red Cross at the same time. In response to the flood, Coolidge named future president Herbert Hoover as chairman of the relief effort and issued a proclamation asking the public to donate $5 million to the Red Cross—"the agency designated by Government charter to provide relief in disaster."[49]

That designation continues today. The current National Response Plan mandates the Red Cross to provide disaster victims with shelter, food, emergency first aid, immediate financial assistance and disaster welfare information.[50] Under the plan, it is the only charity given primary agency responsibilities.[51]

It may also be the best-known charity in the country. As of January 2006, the Red Cross had raised $2 billion for Katrina relief, more than six times as much as the second-highest amount, $295 million, raised by the Salvation Army.[52] That's often the pattern after a disaster, according to Daniel Borochoff, president of the American Institute of Philanthropy, who testified before a congressional subcommittee in December 2005. "The Red Cross is the ultimate brand for charities, the Coca-Cola of charities," Borochoff said.[53]

After the September 11 terrorist attacks, about half of all donations went to the Red Cross' Liberty Disaster Relief Fund.[54] A few months later, the organization was harshly criticized when it decided to keep a significant amount of that money in reserve. It later backtracked and said that it would distribute all September 11 donations to victims' families.[55]

It was not the first time that the Red Cross had been challenged about withholding funds. As *The Washington Post* reported in November 2001, "[i]n at least a half-dozen disasters over the past decade, local officials have had to pressure the American Red Cross to give victims the donations the public intended for them." The article cited the case of San Francisco's mayor, who complained after a 1989 earthquake that the Red Cross had raised about $55 million in donations for victims, but had put much of it into a general disaster relief fund instead.[56] After the Red River flooded its banks in 1997, Minnesota's attorney general called the Red Cross into public hearings, demanding that it disclose how it spends donations for victims.[57]

Senator Chuck Grassley, R-Iowa, chairman of the Senate Finance Committee, has been investigating Red Cross operations since September 11. One of his particular interests is the Red Cross' inability to keep presidents, which he calls "the turmoil at the top of this [premier] charitable organization."[58] In 2001, soon after the September 11 controversy, the Red Cross president, Bernadine Healy, resigned.[59]

Shortly afterward, Red Cross board member Bill George e-mailed the board chairman, David McLaughlin, saying that the board's current direction was unclear. "I do not think the board can continue kidding itself that it wants a strong leader and then not giving that person the authority to lead," George wrote.[60]

In December 2005, after the nation's next big disaster—Hurricane Katrina—Healy's successor, Marty Evans, stepped down,[61] also under a cloud of criticism, this time about the organization's response to the hurricane.

The Red Cross acknowledges that it faced problems with the September 11 and Katrina relief operations. But, says spokesman Bob Howard, the issues were completely different. "With regard to 9/11, the issue was how fast we were spending the money and that the then president of the Red Cross wanted to use part of the funds for preparedness. . . . With Katrina, the issue was our ability to respond."[62]

During the first five months after the storm, nearly 75 percent of the Red Cross' projected $2 billion in donations was budgeted for emergency financial assistance; the agency gave an average of about $1,000 a family to 1.4 million families.[63] Food and shelter made up only 10 percent of the Red Cross' outlays during that same period. Its kitchens served up to 995,000 meals a day to hurricane victims across the Gulf Coast, and its shelters tallied 3.4 million overnight stays.[64]

But the financial assistance process was often rocky. The Southeast Louisiana chapter had to give vouchers to evacuees because it wasn't connected to the system that activates debit cards. And the need was so enormous that the Red Cross kept running out of plastic cards to activate.[65]

For a while, evacuees who were staying in hotels or with family members couldn't get assistance. "If you were not in a Red Cross shelter, you couldn't access any of their services," says Stephen Bradberry, the head organizer for the New Orleans chapter of ACORN (Association of Community Organizations for Reform Now), which advocates on behalf of moderate- and low-income people. Anyone outside those shelters was told to dial a toll-free number, which was always busy, says Bradberry.[66] The Red Cross' Wilkins says that it initially began issuing financial assistance in Red Cross–designated shelters because "we wanted to take care of our shelter population first." It started providing financial assistance outside those shelters a few weeks later.

At first, lack of electricity and computer access made it nearly impossible to keep track of the funds and allowed fraudulent claims to slip through. As of mid-December 2006, the Red Cross said it was investigating more than 12,000 cases of potential fraud by clients who got financial assistance and it had turned over more than 2,100 of the cases to law enforcement officials.[67] The agency also opened separate investigations into alleged waste and theft by volunteers in New Orleans[68] and by employees at a call center in Bakersfield, Calif., that fielded calls from evacuees.[69] By the end of 2006, a federal Hurricane Katrina Fraud Task Force investigating charity and insurance fraud and identity theft had brought criminal charges against 525 people across the country.[70]

New Orleans coffee shop owner Bob Patience didn't witness any criminal activities. But he says he's highly frustrated by the Red Cross' response within the city. "The Red Cross came too late, stayed too late and were ineffective when they were here," he says.[71]

Patience reopened his downtown shop, Marigny Perks, a month and a half after the storm, and was scrambling hard to recover financially. The Red Cross didn't seem to appreciate that, he says. Even as late as

December 2005, weeks after the coffee shop was fully up and running, customer Pat Glorioso recalls sitting at a table there and seeing a Red Cross volunteer offer food from a canteen he'd parked right in front of the shop.

"It's a coffee shop—they sell soup and sandwiches there," she says. "I thought it was a wasted effort."[72]

SIX

# Health Care

Jenni Bergal

*"We started to see the dehumanizing of the human victims in this. Every rescue group was just moving them to the next step, almost like cattle."*

—Gordon Bergh, former assistant director of the Austin-Travis County (Texas) Emergency Medical Services Department

Dr. Scott Delacroix was startled when he heard the woman's voice on the radio broadcast screaming for help at Charity Hospital in New Orleans.

He had worked with her there and couldn't believe that she and hundreds of patients and staffers were stranded at the hospital, which had lost power and was surrounded by floodwaters two days after Hurricane Katrina struck.

Delacroix, 26, a second-year urology resident at Louisiana State University School of Medicine, pulled on green scrubs and a white medical coat, grabbed some supplies, made sure his .38-caliber revolver was in the car and set out for Charity.

He never made it.

While he talked his way through police barricades and roadblocks, 15 feet of water blocked the only route to the hospital. There was no way his Toyota 4Runner with 190,000 miles on it—or any other vehicle—could get through.[1]

So Delacroix decided instead to head for a crowd that he saw gathered at the Interstate 10 and Causeway Boulevard interchange, a spot locals call the "Cloverleaf."

"It's a sight I'll never forget," says Delacroix, a New Orleans native with thinning blond hair and blue eyes who smokes Camel Lights. "There were probably 2,500 people there. There were medical supplies strewn around. There were patients everywhere. There were patients in wheelchairs, patients laying on broken gurneys, on cardboard boxes. Old people, 80, 90 years old, laying in the middle of the highway, laying there with their charts on top of them. Patients in hospital gowns, along with people who looked like they were in a rugby match, all covered in mud, wet, filthy."

"There were some nurses, a psychologist. They said, 'Are you a doc?' I just started triage."

The young doctor had stumbled upon a hastily thrown-together staging area where rescuers were dropping off thousands of people, many of them patients from hospitals and nursing homes, along with men, women and children who had been stranded on rooftops or found in floodwaters.

The Cloverleaf was a dry spot where helicopters could land and leave people, with the expectation that they would be evacuated. But many hadn't been. There was no food, and medical supplies were scarce. Evacuees were dehydrated and exhausted.

"Some were in pretty bad shape," Delacroix recalls. "I came across a guy who had open heart surgery two days before. A lot of old people didn't know where they were. An old couple was there holding hands. They laid there with [soiled] diapers on . . . for two days. They couldn't get out of there."

Delacroix discovered that he and a man from the Louisiana Department of Health and Hospitals in Baton Rouge were the only medical doctors there. Then the other doctor left, and aside from the nurses, the psychologist and some ambulance workers, "it was just me." Delacroix, who took charge of the medical care and triage efforts, later was joined by an emergency room doctor-in-training, a neurologist and another psychologist.

The chaos at the Cloverleaf embodied everything that went wrong with the health care system in New Orleans in the days following Katrina.

While thousands of medical professionals and volunteers scrambled to help the sick and infirm—some heroically risking their own lives—the poor planning, lack of communications, delays and shortages of supplies exposed a system that was itself in need of rescue.

Many hospitals lost emergency power because they kept generators and emergency power sources below ground, where they were flooded.[2] Before the storm, hospital officials had decided not to evacuate, leaving patients and staffers trapped in stifling heat with limited food and water. At the Superdome and convention center, where tens of thousands of residents, some with chronic illnesses, had evacuated, it took days to get medical supplies.[3]

Louis Armstrong New Orleans International Airport, which had been turned into the central evacuation hub and triage center, was the scene of confusion and death.

And the nation's disaster medical system, which was supposed to help evacuate patients and provide medical care, lacked supplies and was disorganized and mismanaged.

"Medical care and evacuations suffered from a lack of advance preparations, inadequate communications, and difficulties coordinating efforts," concluded a report by the House Select Committee investigating the Katrina response.[4]

---

The biggest problem at the Cloverleaf was that no one seemed to be in charge of the evacuation, says Delacroix.

The military's helicopter operation ran smoothly, but Delacroix saw no communication with the state and FEMA. "The military were the only ones who had their act together," he says. "The feds dropped these people off and thought the state or somebody would pick them up. . . . It was overwhelming."

Delacroix wrote a critical account of his experience that was published three weeks later on a Web site for physicians.[5] Dr. David Shatz, who headed a FEMA medical search and rescue team in New Orleans disputed Delacroix's account. In a letter to the Web site, he praised the agency's response and noted that a "huge FEMA operation" was treating patients less than a mile away.[6]

Shatz, a trauma surgeon at the Ryder Trauma Center at the University of Miami, said in an interview that he and his team of two paramedics triaged 1,700 people on the other side of the interstate bridge near the Cloverleaf, but that they may have left by the time Delacroix arrived.

"We were there before he [Delacroix] got there, but we moved on" to other areas of the city, Shatz said. "The whole FEMA system moved on. When he got there, those people had been sitting out there for two days."

Shatz said that overall, the FEMA medical teams did a "fairly decent job," but that they were overwhelmed, considering the magnitude of the disaster and the many communications and patient transportation problems they encountered.[7]

A White House investigation ended up confirming the confusion on the interstate, finding that "large numbers of people gathered or were deposited by search and rescue teams—who were conducting boat and helicopter rescue operations with neither a coordinated plan nor a unified command structure—atop raised surfaces, such as the I-10 cloverleaf downtown. People brought to the raised surfaces as they transitioned to safety had little shelter from the sun and were in ninety-eight degree heat."[8]

Gordon "Gordo" Bergh, then assistant director of the Austin-Travis County Emergency Medical Services Department, brought 40 medics, five ambulances, two medical casualty trailers, a communications center and other equipment to the Cloverleaf from Texas three days after the storm.

"It was absolute chaos. It looked like a bomb had gone off, given the debris field. They had no command system in operation," Bergh says of the scene when he arrived with the task force from his department and a half-dozen other Texas-based first responder agencies. "The only person who was managing things was Scott. He was absolutely a zombie."[9]

Bergh, who is now retired from the EMS department, recalls the disorganization among local, state and federal agencies handling the evacuation. As for the patients and evacuees, Bergh says some were bordering on anarchy, as more and more desperate people kept getting dropped off on the sweltering highway.

"We started to see the dehumanizing of the human victims in this," Bergh says. "Every rescue group was just moving them to the next step, almost like cattle. It was very poignant for me. People lost sight of the fact that they were there to help. . . . They weren't treating people as people."

Bergh was asked by a special assistant to President Bush to compose a letter describing his experience. In it, he wrote: "We did our best to make a difference but our single most challenging obstacle was not the scale of the disaster, certainly overwhelming in and of itself, nor was it threat to our safety be it sniper or environmental but rather a local/state/federal bureaucracy that seemed incapable of cohesive leadership. The end result was a vortex of indecision and inability to use common sense, creative problem solving, and work as a unified cross-functional team."[10]

Delacroix says he was somewhat relieved when his patients on the Cloverleaf got loaded onto helicopters and taken to the airport. At least they'll be getting great care, he thought.

What he later found out when he visited the airport, however, was that conditions weren't much different, except that there was air conditioning.

Upstairs in the ticketing area, he saw long lines of patients standing, waiting to be seen. Downstairs in the baggage claim area, he saw people lying everywhere.

"It was a cattle herd. There were dead people laying there. There were people laying on the ground," Delacroix says. "I thought, 'Damn, that sucks. We just evacuated all those people and they're just sitting down there.'"

The airport, normally filled with tourists flocking to the city for its funky music and Creole cooking, had turned into a major triage center for critically ill patients and others evacuated from hospitals and nursing homes. An estimated 3,000 to 8,000 patients were deposited there, and the House investigation later described the scene as "chaotic due to lack of planning, preparedness, and resources."[11]

Many of the patients scattered around the main terminal and ticketing area were elderly and infirm, but a number had been injured during the hurricane. The first floor was set up as a morgue. A commander of a U.S. Department of Agriculture Forest Service team called it "surreal."[12]

Thousands of others packed into the airport weren't sick; they simply had been deposited there by rescuers and were waiting to go somewhere, anywhere.

Mary Moises Duckert was among them. The slim, well-mannered, dark-haired native New Orleanian who wears pearls and flowered skirts chose to ride out Katrina with family members in her large two-story

Spanish-style house. A few days after the storm, she reluctantly evacuated in a swamp boat whose captain offered to help.

Duckert, her 17-year-old daughter Sarah, two other relatives and a frail, elderly man with health problems who lived nearby were taken to a ramp on the highway, then picked up by helicopter and deposited at the airport.

"It was horrible," says Duckert, 55, a paralegal for the Louisiana Supreme Court. "People were urinating on themselves. There was no order there. People were just laying in corners, along walls, dazed. It didn't matter who you were. You didn't know where you were going. Some were in stretchers and in wheelchairs. People were so helpless. Some had come out of nursing homes. They couldn't take care of themselves. Some were in diapers. There were not enough people to take care of them."[13]

Duckert bursts into tears when recalling the scene. "It does get to you. There were so many elderly people . . . Even the National Guard, the federal air marshals, everyone was stunned."

Duckert had brought a bag of chips and some peanut butter crackers from home. She started handing them out to grateful people around her, some of whom hadn't eaten in days. After more than 10 hours of waiting, Duckert's group was among those herded onto an Air Force cargo plane. Like so many other Katrina victims, they were not told where they were going. They ended up in San Antonio, dazed and disappointed at the government's incompetence and insensitivity.

"The medical teams weren't prepared. It was so overwhelming. . . . Everyone was traumatized," she says.

While the sheer number of patients at the airport had placed tremendous pressure on exhausted medical workers, Katrina exposed a much greater problem: a national medical disaster network that was badly flawed.

---

When state and local health care officials are overwhelmed after a disaster, they turn to the National Disaster Medical System, called NDMS, for help.

The program, created in 1984, is a partnership of federal, state and local governments and health care providers that, until recently, was housed in the Department of Homeland Security and administered by FEMA. It has dozens of regional teams composed of 35 volunteer doctors, nurses and emergency medical workers who are deployed immediately after a disaster to provide medical services and prepare seriously injured patients for evacuation.[14]

In preparation for Katrina, FEMA started activating the system four days before the hurricane made landfall. Initially it sent only nine of its 45 medical teams to the Gulf Coast region. Ultimately, nine of the teams wound up at the airport. Five went to the Superdome, but usually only one team was there on any given day.[15]

NDMS didn't work the way it was designed. Some teams weren't fully staffed. Some encountered delays in mobilizing and getting to their destination and found that staging areas were hundreds of miles away from where they were needed. Many had trouble with communications once they got to their destination. Supplies were lacking, and so was leadership.[16]

"Despite the treatment and evacuation of thousands, the medical operation at the New Orleans Airport was chaotic due to lack of planning, preparedness, and resources," the House investigation later found.[17]

An Oregon-based NDMS team sent to the airport reported that it lacked basic supplies to treat the sick. The team wrote that operations were "extremely disorganized" and that management decisions being made were "not based on the best interests of the patients."

"There didn't appear to be a clear plan for dealing with the approximately 25,000 evacuees who arrived at the airport," the Oregon team concluded. "There was insufficient food, water and sanitation. . . . The situation was very similar to those found in the Third World."[18]

At the Superdome, the federal medical teams found there were inadequate supplies, little security and no communications system.[19]

This wasn't the first time the federal disaster medical program had been the subject of criticism.

In 2002, when NDMS was under the jurisdiction of the U.S. Department of Health and Human Services, an internal report found major flaws, including "poor system readiness" and a lack of standards and guidelines.[20]

After NDMS was moved to Homeland Security in 2003, the program's budget was frozen at $34 million, of which $20 million was siphoned off for "unidentified services." It also lost two-thirds of its staff, according to a report by the Democratic staff of the House Committee on Government Reform.[21]

Ever since the transfer, HHS and Homeland Security have been battling over which agency should be running the medical program. During Katrina, the two agencies were often at odds.

"FEMA deployed NDMS teams without HHS' oversight or knowledge. FEMA administrative delays in issuing mission assignments exacerbated the lack of coordination . . . and created additional inefficiencies," a White House investigation found. It added that HHS "felt compelled to take emergency response actions without mission assignments, bypassing FEMA."[22]

Because of the inadequacy of NDMS' response during Katrina, the agency was moved back under the jurisdiction of HHS in January 2007.

Even before Katrina, NDMS had come under fire. It was chided in reports after the 2004 hurricane season, when some of its teams were found to have inadequate staff and supplies, including essential drugs, such as antibiotics and pain medications.[23]

The most explosive criticism of NDMS came in January 2005, in a report by a senior official prepared at the request of Tom Ridge, who was then the homeland security secretary.

Dr. Jeffrey Lowell, then senior medical advisor to Ridge, reported that NDMS lacked medical leadership and that its teams were responding to incidents for which they were not fully prepared. His report described the system as underfunded and understaffed and relying too much on volunteers who were not always available or qualified. It recommended a "radical transformation."[24]

A FEMA spokeswoman said NDMS' former chief, Jack Beall, did not want to comment about any past or current deficiencies for this book. "I don't think this is something we want to participate in," said public affairs officer Deborah Wing.[25]

Lowell, chief of the pediatric transplant program at Washington University School of Medicine in St. Louis, says that in his report he tried to present an unbiased analysis of NDMS' weaknesses and not get caught up in political "red-blue issues."

"Why is this important? Because in almost every event, the final common pathway is that people are going to get hurt or die," he says. "Our country needs and deserves more than what we've put together so far. It's not enough."[26]

---

New Orleans hospitals thought they were prepared for any disaster. For years, they had gone through plenty of hurricane planning sessions and emergency drills.

State law mandated that they be responsible for setting up their own emergency evacuation plans.[27] The Joint Commission on Accreditation of Healthcare Organizations required them to meet established disaster preparedness standards.[28] They were supposed to be equipped to communicate with outside agencies and each other during a disaster.

Ever since Hurricane Georges narrowly missed New Orleans in 1998, hospitals also were well aware of how vulnerable they were to losing electrical power from flooding because their emergency generators and power switches were below ground.[29] That's what happened in Houston in 2001, when Tropical Storm Allison caught that city off guard, causing nine area hospitals to close because of flooding. Many lost primary and backup power, leaving them without lights, air conditioning or ventilation after their basements were submerged. Since then, many have moved their equipment to higher ground.[30]

While the hospitals in New Orleans knew about such potential problems, most either didn't have the money or the will to make such changes. The Louisiana State University hospitals in New Orleans, for example, had been unable to get funding from the state legislature to move their generators above flood level.[31]

In 2004, local, state and federal medical emergency officials participated in the "Hurricane Pam" mock disaster exercise. Under the scenario, hospitals lost power and communications and became inoperable. They needed to evacuate patients, staff, family members and "refugees" who had come for shelter.[32]

Despite all the preparations, neither the hospitals nor the government was prepared for what would happen during Katrina. As the Louisiana Recovery Authority put it, the state's health care system may have had disaster plans, but it didn't have "an emergency preparedness culture."[33]

Hospitals around New Orleans were barely able to communicate with the outside world, as cell phones and satellite phones proved useless. Helicopters and ambulances were tied up trying to save residents stuck on rooftops and in floodwaters. Rumors of snipers hampered rescue efforts.[34] Hospitals were forced to fend for themselves and try to evacuate thousands of patients, family members and staffers.

The same was true with nursing homes, many of which were not evacuated. At St. Rita's Nursing Home in nearby St. Bernard Parish, 34 elderly residents drowned in the flood. The Louisiana attorney general's office charged the home's owners with negligent homicide.[35] They deny the charges.

Bethany Nursing Home in New Orleans, which also was surrounded by floodwaters, needed two high-water vehicles to evacuate residents. But Joseph A. Donchess, executive director of the Louisiana Nursing Home Association, says his group was told that the National Guard diverted the vehicles and that FEMA commandeered two nearby buses that were supposed to assist in the evacuation. Half a dozen patients died while awaiting evacuation. The surviving patients at Bethany didn't get out until three days later.[36]

In all, about 215 people died in New Orleans–area hospitals and nursing homes as a result of failed evacuations.[37] Only three acute care hospitals in the region, all of them outside the city, maintained operations during and after the storm.[38] Six of the nine hospitals in New Orleans remained shut down a year after Katrina because of severe damage from the storm.[39]

One is the massive Charity Hospital in the city's downtown, where Scott Delacroix was headed before he ended up at the Cloverleaf. The 21-story, art deco–style public hospital, built in 1939 and run by the Louisiana State University system, served mostly the poor and the uninsured. It was the only major trauma center within 300 miles.

City residents thought of it as a sanctuary in times of crisis. The perception was that "Big Charity," founded in 1736, and one of the oldest continuously operating hospitals in the country,[40] would stay open under any circumstances.

For years, disaster mode at Charity had meant that nonessential employees would be sent home and that the "Code Grey" medical team would arrive and camp out for a few days before being relieved by a new team. "In the past, we brought lots of chips and dip, and then went home," recalls Dr. Ben deBoisblanc, who headed the hospital's medical intensive care unit.[41]

This time, it was different.

After Katrina struck on Monday, the streets around Charity flooded. The power went out, and portable generators ran the ventilators, lights and fans in the intensive care unit. On Tuesday, the backup generators failed. "We were inside this huge, dark hospital. It was pitch-black and the walls were perspiring. The aromas reminded me of being in a medieval castle," deBoisblanc says. "We were told, 'FEMA is coming. Prepare.' FEMA didn't come Tuesday or Wednesday or Thursday."

In the ICU, about 50 patients were critically ill, and the medical staff had to ventilate them by hand with bags because there was no power. "We had no infusion pumps, no dialysis, no X-rays," deBoisblanc says. "We were taking care of these people without any technology, just basic skills."

Dr. Peter DeBlieux, Charity's director of emergency medicine, says patients and staff were becoming agitated. "It was 90-degree heat and 95 percent humidity. Tempers were short."[42]

By Wednesday, when staffers realized no one was coming to rescue them, they began feeling abandoned. Some went to the roof and held up signs, reading, "Help Please" and "Fly Us Out Please."

"For the first time ever, Charity turned people away," says DeBlieux. "We were unable to care for the people we had. We had to tell people to go elsewhere. We had to turn away hundreds. They went to the Superdome, the convention center."

Meanwhile, conditions at the hospital were becoming unbearable. Patients who died were placed in the external stairwell, in body bags. The bathrooms became filled with liquid and solid waste. "The smell was horrid. People were urinating in the stairwells, defecating in cardboard boxes," deBoisblanc says. "We had hand cleaner. It's all we had. People put it in their hair as mousse, underarm deodorant, in their shoes. They traded the containers like cigarettes in a prison."

Staffers were elated when they got word that help was coming. They prepared for evacuation, but patients couldn't be moved out because the water level was too high for the National Guard's trucks. The next day, evacuation was put on hold because of rumors of violence and potential harm to rescuers.[43]

It took the efforts of the Coast Guard, a private ambulance service and Jewel L. Willis, a timber management consultant whose daughter was an ER supervising nurse at Charity, to finally get the rescue efforts moving.

At first, Willis, 70, made it to the hospital with his grandson by flat boat, hoping to persuade his daughter to leave. She wouldn't abandon her patients. So his wife got in touch with an old friend from the Louisiana Department of Wildlife and Fisheries, who sprang into action. Willis says wildlife officials marshaled their boats and headed for the hospital and began evacuating patients.[44]

"It's criminal," he says. "Those people were abandoned there, by the system, by the state officials. The patients in the hospital should have been a priority. They should have been moved. The National Guard had the equipment. They should have evacuated them four or five hours after the water flooded the hospital area. The university [hospital] system is responsible, too. A phone call should have initiated this whole thing just a few hours after they flooded, especially after they realized they had no power and no water."

Nine of the approximately 400 patients at Charity died, some "directly because of the prolonged evacuation process," the Senate Homeland Security Committee's investigation later concluded.[45]

Other hospitals throughout the area also had lost emergency power and were struggling to find ways to get patients and staffers out. Across the street from Charity, about 200 patients and more than 1,200 staffers were stranded in the heat and darkness at Tulane University Hospital and Clinic. Officials from Hospital Corporation of America, a national chain that is the hospital's majority owner, realized that help wasn't going to come from the government. So the corporate headquarters in Tennessee hired 20 helicopters to evacuate patients, family members and staff. It took four days to get everyone out.[46]

At the private Memorial Medical Center in Uptown New Orleans, 260 patients, 500 employees and hundreds of family members rode out the storm after the hospital's basement flooded and the power went out.[47]

Thirty-five patients at Memorial, which is owned by Tenet Healthcare Corporation, a Dallas-based chain, died before they could be evacuated. Twenty-four of them were in a separate long-term care facility on the seventh floor operated by LifeCare Hospitals Inc.[48]

In July 2006, the Louisiana attorney general's office arrested a doctor and two nurses on charges of second-degree murder, alleging that they administered lethal doses of drugs to four of those patients at the hospital in the aftermath of Katrina.[49] The defendants have denied the allegations, which have created a stir in the medical community. Many area physicians have come to their defense, saying the three were helping, not harming patients. In late January 2007, the Orleans Parish coroner said that after reviewing all the physical evidence, he could not conclude that the deaths were homicides and instead classified them as "undetermined."[50]

While Memorial was able to evacuate 18 babies from its neonatal intensive care unit aboard Coast Guard helicopters on the day after the storm, its chief executive officer said other patients had to wait longer to be rescued. "Conditions at the hospital deteriorated rapidly," Renee Goux testified before a House subcommittee.[51] "The hospital's air-conditioning system broke down, causing temperatures to reach higher than 105 degrees . . . There was no plumbing; the toilets were overflowing. The smell of sewage was nauseating and it was unbearably hot. We started breaking windows to give our patients some ventilation. Communications were unreliable, although we were able to maintain sporadic contact with Tenet headquarters by cell phone and a satellite phone delivered by helicopter. Communication with emergency officials was nearly nonexistent."

Nyla Houston had just given birth by Caesarian section two days before Katrina. The baby, Landon, was having respiratory problems, so he was in the neonatal intensive care unit, says Houston, an elementary school teacher.[52]

Houston, 32, says she was terrified when staffers told her that they were airlifting her baby out but that she couldn't go with him. "I begged them to tell me, 'Where are you taking him? Why can't I go with him?' I was so upset, me and the other parents. At that time, it was total chaos."

Houston later was able to find out where Landon was taken and notify her parents so they could be there.

On Wednesday morning, hospital staffers told Houston and her husband, Tyrone, who was staying with her, to go to the roof of the high-rise hospital, where helicopters would be picking them up. "I was up there on the roof the entire day. It was very hot up there. I had just had surgery," she says. Then the Houstons were told to walk downstairs, where they got in a long line to take a swamp boat.

"I was dripping with sweat. I had run out of pain medication. I started crying," she recalls. "The nurse said, 'Don't worry, they'll take you to the Superdome.' I was hysterical. I said, 'I'll die if I go there!'"

A sympathetic police officer got the Houstons into a boat that dropped them off at higher ground, where they had to get out and walk. "The water came up to my knees. I had to walk about six blocks. I was in a lot of pain. I was crying and shaking," she says.

They came to a spot where many other Memorial patients had been evacuated. A bus picked them up and dropped them off at the convention center. "You could see thousands of people waiting. By the time we got there, it was dark. There were no lights outside. There was fighting, there was gunshots. It was a horrible scene."

Houston and her husband first sat outside, then, fearing for their safety, went indoors. There, they found trash everywhere and people lying on the floor. There was no food or water. "The toilets had backed up. There was human waste all over the floor. I just didn't use the bathroom. I hoped that I wouldn't wet myself," she says, starting to choke up.

The couple settled into a corner for the night. Houston says she was suffering from heavy vaginal bleeding but didn't see any medical staffers around. "There was no EMS, no ambulance, no medical supplies, nothing."

The next morning, they went outside, where Houston says she spotted an elderly patient from Memorial who had been awaiting evacuation on the roof the day before. The woman was in her wheelchair, next to a garbage can, with a blanket thrown over her. She was dead. "She still had the paper slippers on from the hospital," Houston says.

The Houstons knew they had to get away from the convention center and try to make it back to their home in Harvey, La., across the Mississippi River. It didn't matter that she was in terrible pain from her surgery or that they had no transportation.

Her husband found a catering delivery cart, placed her on it and rolled her toward the Crescent City Connection Bridge. On the other side was the West Bank where they lived.

But the road was uphill and he was exhausted, so they walked across. "It took me three and a half hours to get to the other side," Houston says. She was wearing blue flip-flops and her black Capri maternity pants.

The couple finally made it home, then to her grandfather's house in Baton Rouge. They decided that they couldn't live in New Orleans anymore. They had lost their confidence in the government and the health care system.

"I don't feel they're prepared if something like this would happen again," says Nyla Houston, who now lives with her husband and children in suburban Atlanta. "They ended up dumping me somewhere with no medical care. They traumatized me by separating me from my child. It didn't seem like they knew what they were doing at all."

---

Scott Delacroix chuckles as he picks at a po'boy sandwich and sips a Bud draft at a neighborhood restaurant after work on a hot summer night nearly nine months after the hurricane. He talks excitedly about how he lost his shoes and was covered with mud up to his knees and how he grew a full beard while he was at the Cloverleaf.

Suddenly, he turns somber. He says he's disgusted with the politicians and government officials, each blaming one another for the health care failures during and after Katrina. "It was a power struggle to get in front of TV cameras. It was political wrangling. I'm a Republican and I voted for Bush twice," he says. "There was no health care for these people."

"Planning doesn't do shit," he adds, shaking his head. "The resources weren't there. It's going to take a city in the East or West Coast, not the armpit of Louisiana, for people to realize the first responder system sucks. The lesson is: Don't be sick or old or disabled if a national disaster comes."

Cement stairs and a foundation are all that remain of one Lower Ninth Ward home in May 2006, nine months after Hurricane Katrina. The house had been located fewer than 400 feet from the breached Industrial Canal floodwall.

*Photo by John Perry*

President George W. Bush, center, surveys the rebuilding of a levee with New Orleans Mayor C. Ray Nagin and Louisiana Governor Kathleen Blanco on March 8, 2006.

*Photo © Brooks Kraft / Corbis*

A car rests beneath the remains of a house in the Lower Ninth Ward in May 2006, nine months after the storm. The scene was a typical one in the neighborhood, where Katrina washed many homes off their foundations and overturned vehicles.

*Photo by Jenni Bergal*

Kevin Lair stands amid the wreckage of his Lakeview neighborhood home in June 2006. The house, located near the breach in the 17th Street Canal floodwall, was destroyed by the force of the flooding. The house was torn down in August. Lair, who runs a real estate brokerage, says he never thought the levee would fail.

*Photo by John McQuaid*

A tree is the only thing left unscathed on one Lower Ninth Ward block where many homes were annihilated by Katrina's flooding. Half of the "Lower Nine" still lacked electricity or potable water nine months after the storm.

*Photo by Jenni Bergal*

Volunteers with Samaritan's Purse, a nondenominational Christian relief organization, gut a flood-ravaged home near the London Avenue Canal breach in April 2006. The workers wear Tyvek suits and respirators to protect them from mold and chemicals in the debris.

*Photo by Sara Shipley Hiles*

A hurricane-damaged luxury home near the 17th Street Canal's floodwall breach in the Lakeview neighborhood remains uninhabited in May 2006. Its owners put up a banner calling attention to the amount their insurance company paid on the claim—$10,113.34.

*Photo by Jenni Bergal*

Dr. Peter DeBlieux, director of emergency medicine at Charity Hospital, stands in a tent set up in the lobby of a former downtown department store that was being used as a temporary clinic in the spring of 2006. Charity and five other major New Orleans hospitals were still closed a year after Katrina. DeBlieux describes the city's health care system as "incredibly broken."

*Photo by Jenni Bergal*

The walls of this Lakeview house were ripped away during Katrina. In May 2006 an American flag, trophies, children's toys and other items are still on view for passersby.

*Photo by Frank Koughan*

Pearl Ellis and her son, Solomon, stand outside what is left of her Lower Ninth Ward home in May 2006. She evacuated before Katrina made landfall, and returned to find everything destroyed except for some Mardi Gras beads and knickknacks, at left. Her son, who didn't leave before the flooding, spent three days in the attic without food or water before he was able to chop a hole in the roof to escape to a neighbor's house. Ellis has no desire to come back. "It doesn't feel like home anymore," she says.

*Photo by Jenni Bergal*

One of the few things in Cora Charles' Lower Ninth Ward home that survived Katrina was the flag that had draped her brother's coffin, although portions were eaten away by the sludge that remained after the floodwaters receded. In May 2006, it hangs next to a small pre-Katrina photo of her home of 45 years. Charles did not have flood insurance. "I wasn't required to have it," she says.

*Photo by Frank Koughan*

The Treme Brass Band leads mourners on a march through the Lower Ninth Ward on Memorial Day 2006. The band played a dirge as the names of Katrina's dead were read aloud.

*Photo by Frank Koughan*

Lower Ninth Ward prayer service participants sing "We Shall Overcome" at the site of the Industrial Canal breach on Memorial Day 2006. They lay their hands on the repaired floodwall, praying that it will hold.

*Photo by Frank Koughan*

A chain-link fence across the street from St. Mary of the Angels School in the Upper Ninth Ward serves as the repository for the work gloves of Common Ground volunteers in April 2006. The school housed volunteers from the organization, which was formed shortly after Katrina to help feed residents, clean city streets and repair homes.

*Photo by Katy Reckdahl*

SEVEN

# Politics

## Curtis Wilkie

*"[I]t's very hard to escape the impression that nothing works in New Orleans. If there is another city where one encounters as much inefficiency and incompetence, I don't know it."*

—James Gill, *Times-Picayune* political columnist

Frustrated that New Orleans police were overwhelmed and that the federal response was sorely lacking in the three days following Katrina, Louisiana Governor Kathleen Blanco finally called on the National Guard to do the job the city couldn't—stop the looting and violence, and rescue thousands of stranded victims of the storm.

Using dramatic language directed at those she called the "hoodlums" terrorizing the city, the governor announced:

"Three hundred of the Arkansas National Guard have landed in New Orleans. These troops are fresh back from Iraq, well-trained, experienced, battle-tested and under my orders to restore order in the streets.

"They have M-16s and they are locked and loaded. . . . These troops know how to shoot and kill and they are more than willing to do if necessary—and I expect they will."[1]

The next day, thousands of additional National Guard troops arrived to help control the lawlessness and speed the evacuation of the city.

It was a startling predicament for an American city to be in, but in some ways, less so for the Crescent City. This was not the first time that New Orleans had suffered a crisis of leadership, nor the only instance of outside intervention in its affairs.

Chaos has a long history there.

---

From the time of its settlement nearly 300 years ago by French explorers, New Orleans existed as an island unto itself, virtually surrounded both by water and unsympathetic neighbors.

Though it grew to become the largest city in Louisiana, it failed to earn a proportionate amount of love outside its borders. Instead, it garnered a reputation as a hotbed of vice and corruption, where godlessness seemed to flourish in a carnival atmosphere in a region dominated by pious Protestants. Inside the state, the arrogance of its aristocracy and the profligate ways of its politicians engendered resentment.

Louisiana leaders often feuded with the city. Depression-era strongman Huey P. "Kingfish" Long went so far as to send in the National Guard over a political dispute. Six decades later, during his successful campaign for governor, Mike Foster called New Orleans "the jungle," suggesting that criminal elements controlled the streets.[2] Over the years, reports of featherbedding, exorbitant contracts and outlandish salary arrangements have led to repeated moves by the state to restrict the power of city institutions.

During the 20th century, New Orleans devolved from a wealthy, cosmopolitan port to a poorer place. Public school desegregation orders in the 1960s triggered flight by many whites to the suburbs. Two decades later, the oil and gas industry tanked, and petroleum giants who once walked Poydras Street disappeared.

At the beginning of this century, New Orleans was clinging to Entergy, its only Fortune 500 company at the time. By 2004, a survey by the U.S. Census Bureau reflected a New Orleans population that was two-thirds black, and in which 14 percent of families and 23 percent of individuals lived in poverty.[3]

The demographics of New Orleans changed dramatically from the time before the Civil War when it was the largest city in the South and one of the five largest in the country. Still, there was little to distinguish its modern-day approach to politics from the shady practices of the 19th century. According to *The Almanac of American Politics,* "This was one of the most corrupt American cities during Reconstruction and the Gilded Age, when its votes were regularly bid for and bought."[4]

Tales of shakedowns and payoffs have extended into the 21st century. It was simply the New Orleans way.

For all of its joie de vivre and funky charm, New Orleans has long suffered a litany of woes: its chronically troubled police department and legal system, its capricious politicians and widespread crime.

These were all apparent before Katrina hit. But it is also important to understand New Orleans' unique history to appreciate some of the forces that came into play in the aftermath of the storm. The city is, after all, located in a peculiar state where counties are called "parishes" and elections take place on Saturdays, where a legal system based on the Napoleonic Code rather than British common law prevails, and elected officials quite often seem to consider themselves above the law.

Like many urban settings in America, New Orleans has suffered from racial tensions. But the sociology of the city was built on a foundation far more complex than the black-white animosities found elsewhere in the South. With its strong French influence, original enmities pitted those of French ancestry against the English-speaking "Anglo-Americans." The city's main thoroughfare, Canal Street, divided two 19th-century communities—the French Quarter and the "American" neighborhood on the Uptown side of the boulevard. New Orleanians still refer to median strips, such as the wide swath in the middle of Canal Street, as "neutral ground."

The city was heavily Catholic, and church doctrine helped establish pockets of tolerance and inclusion. Even as it harbored slave markets, a class of black Creoles known as the *gens de couleur libres*—or free people of color—made up more than a quarter of the population, some living almost as comfortably as the white barony.

New Orleans surrendered without a shot in the Civil War, and compared to Alabama and Mississippi, Louisiana put up relatively minor resistance during the civil rights movement a century later. Abiding by a federal court ruling in 1944 that white-only Democratic primaries were illegal, Louisiana began to enfranchise blacks 21 years ahead of the Voting Rights Act of 1965.

While other southern cities became noted as racist strongholds, New Orleans was better known for Mardi Gras and its colorful collection of public officials that has presided over graft and inefficiency for decades.

Some say that notorious claim to fame has slowed the flow of federal dollars for post-Katrina reconstruction.

"There's a perverse pride here that we have crooked government," said James Gill, a political columnist for *The Times-Picayune.*

Louisiana is not necessarily more corrupt than, say, Rhode Island or New Jersey, he said. "But it's very hard to escape the impression that nothing works in New Orleans. If there is another city where one encounters as much inefficiency and incompetence, I don't know it."[5]

Since the city's inception, New Orleanians who adore their hometown have overlooked the defects in its government. A popular slogan that captures New Orleans' spirit deems it "The City That Care Forgot."

City officials never seem to learn from experience. When the Superdome was used as a shelter for more than 14,000 people during Hurricane Georges in 1998, furniture was stolen and property damaged.[6] Without adequate planning in the interim for security precautions, the scene at the sports facility was far more nightmarish after Katrina.

Federal officials who control the purse strings on aid to the hurricane-ravaged state have insisted that more care be taken with plans for rebuilding New Orleans. Mayor C. Ray Nagin and Governor Blanco were among the local officials who met about seven weeks after Katrina with U.S. House Transportation and Infrastructure Committee members and sought to assure them that there would be strict accountability concerning federal funds. The governor, participating via teleconference from Baton Rouge, pledged that in addition to the state's auditors reviewing Katrina aid spending, Louisiana would hire two "Big Four" accounting firms—one to check every hurricane-related expenditure and another to evaluate those findings.[7]

---

Fittingly, some of the earliest political disputes in New Orleans involved pirates. Jean Lafitte bedeviled Governor William Claiborne and the U.S. government in New Orleans shortly after the Louisiana Purchase in 1803, preying on merchant ships.

As Louisiana political commentator John Maginnis wrote in his book *The Last Hayride,* "What galled Claiborne even more than the elusiveness of Lafitte was his broad-based popular support in the city of New Orleans. [Claiborne] couldn't catch him in the swamp and he couldn't touch him in the city. There the strikingly handsome, impeccably dressed, glib and flirtatious Lafitte was a favorite among the French leaders and ladies."[8]

The pirate king who helped Andrew Jackson defend the city against the British in the War of 1812 proved the archetype, and the public's fascination with rogues—not just in New Orleans, but across Louisiana—has spanned two centuries. None was more colorful than Huey Long, who served as a governor and U.S. senator. A product of north Louisiana, Long built a political machine, blatantly venal, yet so formidable that his "Share Our Wealth" movement threatened the first re-election of a fellow Democrat, President Franklin D. Roosevelt. When Long was assassinated in 1935 in the state capitol he constructed and controlled, U.S. Treasury agents acting on Roosevelt's instructions were preparing an income tax evasion case against him.[9]

A few years later, Richard Leche, one of his gubernatorial successors, would declare, "When I took the oath of office, I didn't take a vow of poverty." The Democrat would later go to jail, convicted of bribery.[10]

In his landmark 1949 study *Southern Politics in State and Nation,* V. O. Key Jr. wrote of corruption in Louisiana: "Few would contest the proposition that among its professional politicians of the past two decades Louisiana has had more men who have been in jail, or who should have been, than any other American state."[11]

The trend seemed to spread in the latter part of the 20th century. Four-term governor Edwin W. Edwards, another Democrat, is now in repose in a federal penitentiary, convicted in 2000 of racketeering and extortion in connection with the award of a casino license. He heads a modern-day list of jailbird officials: an attorney general, an elections commissioner, an agriculture commissioner, three insurance commissioners, a congressman, a federal judge, a state Senate president, and various legislators, judges and other municipal and parish officers.[12]

Even as New Orleanians voted for mayor on May 20, 2006, FBI agents were raiding the Capitol Hill office of U.S. Representative William Jefferson, a Democrat whose district includes most of the city. The FBI, which had confiscated $90,000 in cash from the freezer at Jefferson's home during an August 2005 search, alleged that the eight-term congressman had taken hundreds of thousands of dollars in bribes from a firm seeking contracts in Africa.[13] Jefferson has denied any wrongdoing. Despite the controversy, he won a runoff election for his congressional seat in December 2006.

---

Although New Orleans might share a penchant for corruption with other locales in Louisiana, the city's poor relationship with its neighbors is unique. The schism between the city and the rest of the state was best illustrated by a crisis in 1927, when the rising Mississippi River threatened to spill into New Orleans after the "Great Flood" inundated tens of thousands of acres upriver in the Mississippi Delta.

As torrential rains drenched the city and the river churned near the tops of the levees, the wealthy powerbrokers of Uptown New Orleans prevailed upon the governor to blow up a levee below the city to relieve the pressure. Engineers used 39 tons of dynamite to blast open a crevasse 12 miles south of New Orleans.[14]

The city was spared, but the rural parishes of St. Bernard and Plaquemines—which would be deluged 78 years later by Katrina—were flooded. The hunters, trappers and watermen in the outlying parishes who suffered the loss of home and livelihood never forgave the New Orleans elite, and the incident became a part of local mythology. When levees were breached during Katrina, washing away poorer neighborhoods, rumors were rampant that white plutocrats were responsible for the deed.[15]

After the 1927 flood, Huey Long was elected governor with the strong backing of rural parishes—and despite the strong opposition of New Orleans' Regular Democratic Organization (known as the "Old Regulars")—and he set out to exploit the divide.

Once the election of his allies put the legislature firmly in his grasp, Long broke the back of the administration of New Orleans Mayor T. Semmes Walmsley (in office 1929–1936). He challenged the oil and gas interests, the bankers and the landed gentry of New Orleans, imposing heavy taxes on businesses and corporations to pay for roads, bridges and textbooks for his followers outside Orleans Parish. Long prevented the city from receiving appropriations to repair its roads, seized its liquor licensing authority and put the New Orleans police force under the thumb of a state commission.[16] He boasted that he had "taken over every board and commission in New Orleans except the Community Chest and the Red Cross."[17]

By 1934, Long had been elected to the U.S. Senate, but held near-dictatorial powers in Louisiana. In his boldest thrust, he had National Guard troops deployed to New Orleans in the wake of a disputed mayoral election. With bayonets drawn, the troops battered down the doors of the voter registration office and set up machine gun positions in an annex to City Hall.[18]

Though Walmsley ultimately won re-election, Long's legislative grip on city government induced the mayor to agree to resign in the middle of his term and the Old Regulars to support Long in the legislature. Despite the senator's assassination in September 1935, Walmsley still resigned in 1936 and was replaced as mayor by a Long stalwart, Robert S. Maestri.

Some years after Huey Long's murder, his brother Earl Long—a maverick governor widely known as "Uncle Earl," who served terms in office from 1939–1940, 1948–1952 and 1956–1960—renewed the conflict with City Hall.

In the period following World War II, the Mafia had gained a foothold in the city. Gambling and prostitution prospered openly, burnishing New Orleans' reputation as a wicked city. Though he depended upon the support of rural Louisiana voters who distrusted New Orleans, Uncle Earl actually was also allied with mob interests, who helped finance his campaigns.[19]

Long's chief adversary was deLesseps "Chep" Morrison, a freshly minted mayor (1946–1961), who exuded progressive energy. Morrison was able to flush some of the Mafia influence out of the city. He was helped by the highly publicized hearings in 1950 and 1951 held by a Senate committee investigating organized crime, chaired by Tennessee Democrat Estes Kefauver.

Long languished on the sidelines as reformer Robert Floyd Kennon occupied the governor's mansion for a term that began in 1952. But Uncle Earl was restored to office in the next election. As governor, he was determined to cripple the city again and nearly succeeded, introducing a number of legislative measures that both diminished the mayor's powers as well as New Orleans' revenue from taxes.[20]

As James Gill wrote in his book on the socio-political structure of the city, *Lords of Misrule: Mardi Gras and the Politics of Race in New Orleans,* "Louisiana voters, though they may suffer periodic fits of reformist zeal, have a limited tolerance for good government."[21]

---

New Orleans' dichotomy concerning race came to the fore in the early 1960s when the city's public schools faced desegregation orders.

In an attempt to circumvent *Brown v. Board of Education,* the Orleans Parish School Board (the parish is made up entirely of the city) abdicated its responsibility. The state legislature passed radical measures to abolish the school board, forbid any student transfers between black and white schools and close any school under a desegregation order.

To add to the drama, Leander Perez, the Democratic political boss of downriver Plaquemines Parish, materialized in New Orleans to exhort white crowds: "Don't wait for your daughters to be raped by these Congolese," he shouted. "Don't wait until the burr-heads are forced into your schools. Do something about it now!"[22]

National television captured ugly scenes of white mothers spitting and hissing at black students. There were bloody clashes in the streets between black and white citizens in a city once thought of as the most tolerant place in the Deep South. The episode opened racial wounds that persist to this day.

It also led to a significant change in the makeup of the city. Many upwardly mobile whites left, moving to suburbs in neighboring Jefferson Parish. Working-class whites also departed, relocating to nearby St. Bernard Parish. The percentage of blacks in the city population grew.

By 1978, the changing demographics had created a shift in the city's political balance as well, leading to the election of Democrat Ernest "Dutch" Morial, a Creole, as New Orleans' first black mayor. Today, the city's convention center—the site of both shelter and suffering after Katrina—bears his name.

Morial's victory would become the cornerstone of a black hegemony at City Hall, with him and each of his successors winning two four-year terms, the maximum allowed by law. In 1986, Sidney Barthelemy, another Creole, followed Morial in office. In 1994, Marc Morial—the former mayor's son—won the first of his own two terms. His successor in 2002 was Nagin, to whom the catastrophic concerns of Katrina would later fall.

Relations between the city and Jefferson Parish have smoldered for years. In an early instance of racial profiling, Jefferson Parish Sheriff Harry Lee, a Chinese American, announced in December 1986 that his deputies would begin stopping any black motorist driving "a rinky-dink car in a predominantly white neighborhood."[23] He rescinded the order the next day after public outcry over his remarks.

About two months later, the suburban parish's council approved blocking a road at the line between Jefferson and Orleans parishes. Two barricades were erected at the request of suburban residents, who asserted that criminals driving into their neighborhood from the city were responsible for burglaries and assaults.

Mayor Barthelemy had New Orleans city workers tear down the pair of barriers with bulldozers before the week was out, saying that the Jefferson Parish Council had no right to block a state highway and that he believed the move was racially motivated.[24] Jefferson Parish officials, who denied any racial motivation, initially said the barriers would be rebuilt. A day later, however, the plan was scuttled. Robert Evans Jr., the suburban parish's president, said the state Department of Transportation and Development had sent a letter stating that the roadblock was illegal.[25]

Evans and Barthelemy met the next day and together made the official announcement that the barriers would not go back up. Their public fence-mending included talk of a joint effort to fight crime.[26]

But almost 20 years later, Jefferson Parish officials would wall off the parish again after Hurricane Katrina.

With looting and violence erupting in New Orleans after the storm, the neighboring parish wanted no part of the city's problems. Evacuees trying to cross the Mississippi by bridge into the suburban town of Gretna were turned back by police.

"People with guns are going to be guarding our borders," declared

Aaron Broussard, the parish president. "They're not getting into the east or the west of us. . . . We're locking this parish down."[27]

---

The chaos of Mardi Gras lasts about two weeks, but city government in New Orleans is a study in dysfunction 52 weeks a year.

In order to create public jobs to feed the political patronage system, bureaucracies are piled on top of each other in a soulless City Hall edifice on Perdido Street.

"Every leader has people stuffed into public jobs," said Kristina Ford, who served as executive director of the City Planning Commission for eight years. "Lots of people essentially addressed envelopes."[28]

Others held plush positions and reaped unseemly rewards.

Under Marc Morial's administration, his uncle, Glenn Haydel, took advantage of a sweet consulting contract with the Regional Transit Authority. Charged with systematically swindling the authority for years, Haydel pleaded guilty in 2006 to stealing $550,000.[29] In 2004, Haydel's wife, Lillian Smith Haydel, an insurance executive, admitted giving kickbacks to an Orleans Parish school official in order to get lucrative contracts.[30]

While some of the exploitation is criminal, old-time patronage is a bigger drain on the city treasury.

New Orleans has no less than seven elected tax assessors, each in command of a district fiefdom and an office staff that sits in judgment on real estate values and property taxes. Attempts to streamline the cumbersome apparatus have failed over the years. After Katrina, however, voters approved a constitutional amendment in November 2006 that will consolidate the offices and call for the election of a single assessor—but not until 2010.[31]

Disgruntled legislators also wrestled over proposals to merge other duplicative offices in the predominantly Democratic city. In New Orleans, there are two elected sheriffs, two clerks of court and three court systems. Urging passage of bills that would cut the city's bureaucracy, Representative Emile "Peppi" Bruneau, a Republican, declared, "It's time to join the rest of the United States of America."[32]

Another of the legislature's targets for consolidation was southeastern Louisiana's levee boards.[33] The Orleans Levee District is one of 19 levee boards functioning in parishes that border the Mississippi River or have significant bayou territory. It has 200 employees and is governed by a board of a dozen political appointees who hand out lots of contracts.

"It's a big deal to get on the board. There's a lot of money collected, a lot of patronage. New Orleans was the worst," said Maginnis, a longtime Louisiana political commentator.[34]

Peirce F. Lewis wrote in his 2003 book, *New Orleans: The Making of an Urban Landscape,* "Louisiana's levee boards are considerable creatures, mandated by the state constitution not merely to keep their districts dry, but given all kinds of money and power to accomplish that purpose."[35]

The levee board handed out contracts from "a huge pork barrel," according to former planning commission executive director Ford, who served briefly as a mayoral appointee to the board in the 1990s.

"They left responsibility for safety to the [U.S. Army] Corps of Engineers. The levee board made sure the grass was mowed," said Ford, adding that she felt some of the board bylaws allowed "scurrilous" activities.

"It was possible to have meetings in Puerto Rico or somewhere offshore, to have junkets," she said.

Rather than concerning themselves with the maintenance of the levees, Ford recalled that the board spent an inordinate amount of time overseeing the arrival of a gambling boat, Bally's *Belle of New Orleans* on Lake Pontchartrain.[36]

---

In New Orleans government, bloat is especially visible in law enforcement, making the system another of the state's consolidation targets.[37]

In addition to the 1,500-member New Orleans Police Department, Orleans Parish funds a civil sheriff as well as a criminal sheriff—and both have separate teams of deputies. All sorts of uniformed officers roam the city, including members of two other police forces that patrol the levees and the docks.

In the 1990s, it seemed as if the city police department, plagued by a history of corruption, could sink no lower. Officers repeatedly drew charges of brutality, extortion and wanton beatings. And four members of the force were charged in connection with murders within a 12-month span that began in April 1994.[38]

In an effort to overhaul the department, Mayor Marc Morial brought in a new police superintendent, Richard Pennington, a 26-year veteran of the Washington, D.C., force.

"Pennington was the right person at the right time," said Mary Howell, a civil rights attorney who handles many police misconduct cases. "He was not part of any clique or faction," she said, but after his arrival in late 1994, he didn't have a long time to cleanse the department before the political interference began again.[39]

Pennington ran for mayor in 2002, and ended up in a runoff with Nagin, a Cox Communications executive. But voters had tired of growing allegations of corruption and cronyism in the Morial administration, and Pennington was perceived as a Morial loyalist. He was defeated by Nagin, who had no experience in government but was seen by many as a fresh alternative.

Nagin encountered problems with the police department long before Katrina struck, including allegations that reform-minded commanders had been transferred, allowing rogue officers to drift back to brutish customs.

The situation worsened with the erratic police performance in the days after the storm. Though many officers performed heroically during Katrina, the department sustained new blows to its esteem after dozens deserted their posts. Some were investigated by the Louisiana attorney general's office in connection with the theft of vehicles from a downtown dealership.[40] In the glare of heavy media coverage, television cameras captured policemen in the French Quarter pummeling a man in his 60s.

"Nagin is out of his element. His handling of the police department has been a complete disaster," Howell said. "It's really sad."[41]

Less than a month after Katrina, beleaguered Superintendent P. Edwin "Eddie" Compass III, Nagin's choice to succeed Pennington, resigned. The mayor replaced him with Warren Riley, a much-admired professional policeman who had worked his way up the NOPD ranks, as had Compass.

The city's judicial system likewise had been a mess long before the main courthouse took on several feet of water. With layers of elected judges in charge of dozens of divisions, it presented a labyrinth that invited petty corruption. A backlog of cases caused overcrowded jails and cluttered court dockets, among other problems.

After Katrina, those problems became even more staggering. For nine months there were no criminal jury trials. The shortage of public defenders in Orleans Parish—fewer than 30 to handle the cases of thousands of indigent defendants—led Arthur Hunter, an Orleans Parish Criminal District Court judge, to temporarily suspend all cases involving public defenders in February 2006, warning that the prisoners' constitutional right to a speedy trial was being violated.[42]

In late July 2006, with hundreds of defendants still jailed without representation and a backlog of some 6,000 cases, the judge issued an order stating that if something was not done about the problem by August 29—the one-year anniversary of Katrina—he would begin releasing inmates on a case-by-case basis, though charges against them would not be dropped.[43]

Nine months after Katrina, Orleans Parish District Attorney Eddie Jordan was asked in a televised interview on CNN if he thought the justice system was broken. "I think it is certainly crippled," he replied.

Jordan said that he thought the problem was serious, but called the possible release of thousands of prisoners awaiting trial "outrageous."

"The wholesale release of inmates on the streets of New Orleans would create enormous havoc," he said.

"I think that the solution is to provide these poor inmates with the lawyers they deserve," Jordan said. He added that he thought developing creative alternatives to fund the public defender system crisis would be better than "to simply open the jail doors."[44]

---

Hurricane Katrina caused the postponement of the city's February 2006 elections until late April, and many wondered how the mass evacuation of the city would come into play at the polls.

In the weeks after Katrina, Nagin endured heavy criticism for a lack of planning, the late order to evacuate, inadequate shelters and woeful rescue missions. In his re-election campaign, he made an all-out bid to gain the votes of black constituents. In January, he even spoke of a vision of New Orleans remaining a "chocolate city"—a comment for which he apologized days later.

Given an opportunity to install a fresh slate of city officials, New Orleanians elected a few new faces, but returned many incumbents to office—most significantly Nagin. The mayor won a hard-fought May runoff election against Lieutenant Governor Mitch Landrieu, garnering about 52 percent of the vote and edging his opponent by slightly more than 5,300 ballots.[45] As political analyst Silas Lee noted, the vast majority of Nagin's votes came from black supporters who managed to return to the city to cast ballots or voted absentee. "New Orleans politics has always been racially divided," Lee said, "and it's exacerbated even more."[46]

It was a sharp contrast to Nagin's runoff election effort four years earlier, when he ran as a pro-business candidate and won about 59 percent of the overall vote—and more than 80 percent of the white vote.[47]

In both the pre- and post-Katrina political landscape, Nagin has found himself in a spot familiar to many of his predecessors in office: at odds with the governor.

In the 2003 governor's race, Nagin endorsed Republican Bobby Jindal over fellow Democrat Blanco. Relations between them had been strained ever since, and the generally united front that he and Blanco presented in the wake of the hurricane gave way to sniping and finger-pointing. The pair met after Nagin's victory in May and agreed that, as a *Times-Picayune* article put it, "they need to repair their dysfunctional relationship and work together to rebuild the city."[48]

However, Nagin's GOP political alliances could no doubt keep them at odds. Jim Carvin, who managed his campaign, said that the mayor got no direct help from the national Republican Party in his re-election bid. But the consultant also noted that Nagin "could throw his weight around" politically in the upcoming governor's race.[49] Jindal, now a congressman, has announced that he will challenge Blanco in the 2007 governor's race. If the mayor were to back him again, it could play into GOP plans to take the governor's office, and next, Democrat Mary Landrieu's U.S. Senate seat in 2008—a prospect that could add yet another contentious chapter to the long history of political wrangling between New Orleans mayors and Louisiana officials.

Despite Katrina's impact, Gill, who has covered New Orleans politics for years, said he is unconvinced that government will be transformed for the better.

"I do not see any evidence that there will be less graft or incompetence," he said. "It would be nice to think otherwise, but there's absolutely no evidence of that."[50]

EIGHT

# Housing and Insurance

Frank Koughan

*"People want to rebuild. People want to do the right thing, but they don't have the resources to do it. Really, it's not clear to any of us who's in charge."*

—LaToya Cantrell, Broadmoor Improvement Association president

Nearly nine months after President George W. Bush strode before the cameras and pledged the first $60 billion of "one of the largest reconstruction efforts the world has ever seen,"[1] Fred Yoder sits in one of his Lakeview neighborhood's three functioning restaurants, tallying his share of the federal largesse.

"I got a $2,300 check from FEMA through this whole thing . . . I got zero else," he says, digging into his plate of red beans and rice. Yoder, a well-off construction executive, doesn't particularly care if the government gave him money. "What bothers me is the *illusion* that they did," he says.[2] Money, he says—even for those who don't have any—is hardly the biggest obstacle to rebuilding his city. For him and countless others trying to reconstruct their homes and their lives after Katrina, a bigger obstacle is the lack of official leadership.

"On any given day I'll have at least 400 e-mails from people asking me for help. They just want direction," he says. Though actively involved in the effort to plan and rebuild his beloved neighborhood—one of the lowest lying and hardest hit in the city—Yoder is just a volunteer. "'Where do we stand on the levees, Freddy? Is it safe to move back? Should I rebuild? *Tell me what to do.*' There needs to be somebody other than me that can give people these directions."

The leadership void is especially surprising given that the hurricane's tumultuous aftermath was not. Thirteen months before Katrina hit, the Homeland Security Department's "Hurricane Pam" exercise predicted that 135,853 New Orleans households would suffer long-term displacement in the event of a Katrina-like catastrophe.[3] The estimate was not too far off: In its wake, Katrina left about 105,000 properties badly damaged.[4]

New Orleanians didn't need a federal study to tell them they were at risk. Everyone knew the city was built on vulnerable ground. In fact, in 1722, just four years after New Orleans was founded by the French, a hurricane wiped it out. The settlement was considered too militarily and commercially important to abandon, and the French rebuilt on the highest available site. "They weren't looking to build a great civilization," says University of New Orleans history professor Arnold Hirsch.[5] But as the population grew over the years, so did the boundaries, spreading out over lower and lower terrain.

Why build a large city in such an obviously dangerous place? "I don't think anyone was out taking depth measurements," says Hirsch. "They knew they were on low ground; they knew when it rained it was difficult to drain and it was like living in a swamp. But the value of having the port business there apparently outweighed all the negative considerations. It just grew."

And once a city has been built—wherever it's been built—that's it. There's no turning back.

"It's your home, for goodness' sake. It's where you were raised. It's where your children are," says Yoder, 57. "It's not a simple matter of saying just, 'Why don't you move?' It's not me. It's my four children. And their spouses. And all of their families . . . All of those people aren't going to move. They have too many things that hold them here."

Some residents did weigh the negative considerations and take precautions. Many of the city's older homes are raised well above the ground. Flooding was common, but it seldom was threatening to life or property. As time went on, a system of levees and other flood-control measures emboldened residents to build in areas that previously were uninhabitable. Still, gazing up to watch ships passing by along the waterways was a constant reminder of the city's precarious location.

But the levees held. Year after year, storm after storm, they did their job. Only during Hurricane Betsy did the levees fail, and even that was 40 years before Katrina came along.

Months after Katrina, frustration is evident everywhere. In the endless wait for trailers, which FEMA won't deliver until the water and electricity are restored. In the piles of debris still lining the sidewalks. In the relief checks getting lost in the mail. In the pace of the demolitions—only about 200 houses by early August of 2006.[6] In the rules, regulations and applications from the government, and in the endless wait for one of the city's overworked, high-priced contractors. "It's almost like we're prisoners," says resident John Lockwood, who hasn't seen his builder in two weeks.[7] And as more people begin to rebuild, the delays will only get worse.

Tensions are high because the stakes are so high: Without people, there is no city, and without housing there are no people. Until housing is fixed, nothing else will be. "Housing is the foundation of our recovery," Governor Kathleen Blanco said during a news conference in May 2006, "and that's what's making our recovery so slow."[8]

The governor was announcing the long-awaited approval by the U.S. Department of Housing and Urban Development of her rebuilding commission's blueprint for the state.[9] But hers is just one of the city's recovery efforts. Mayor C. Ray Nagin appointed his own commission, Bring New Orleans Back. It developed a highly restrictive plan, which the mayor largely rejected. (Under part of the plan that went into effect, the commission established 13 district planning committees responsible for creating plans for their own areas of New Orleans.)[10]

The city council, meanwhile, already had hired its own experts to coordinate grassroots rebuilding plans in 49 neighborhoods. The mayor's plan and the city council's efforts were eventually combined and will be submitted to the Louisiana Recovery Authority, the agency that promises to put the biggest cash grants into the hands of homeowners. The grants—more than $10 billion[11]—needed congressional approval, which didn't happen until mid-June 2006.[12] The Recovery Authority was hoping to write checks by August.[13]

Caught in the middle are the hundreds of thousands of displaced New Orleanians. Some are scattered in distant hotel rooms, others camped in trailers on their own front lawns, their temporary shelter provided by FEMA. Without clear leadership or direction, people are forced to make decisions: stay or go, rebuild or not. But many, unsure of their resources or what's to become of their neighborhood, remain stuck in a state of uncertainty.

"It's unacceptable," says LaToya Cantrell, president of the Broadmoor Improvement Association, which is working to devise a rebuilding plan for her shattered neighborhood. "People want to rebuild. People want to do the right thing, but they don't have the resources to do it.

"Really, it's not clear to any of us who's in charge."[14]

---

Nothing has paralyzed the reconstruction process more than uncertainty over how high off the ground houses would be required to be built.

In areas susceptible to flooding, it's FEMA's job to determine the base flood elevation—a level above which all new houses must be constructed if their owners wish to receive federal money. In New Orleans, the base flood elevations in many areas of the city were below sea level and hadn't been adjusted in more than 20 years.[15] After Katrina, most people assumed there would be big changes in the elevation rules, so they were reluctant to start building before the new rules were released.

Months and months passed, to the dismay of officials and homeowners alike. Finally, in mid-April 2006, FEMA announced that the base flood elevations would remain unchanged.[16]

The delay was "kind of frustrating," says Lakeview's Yoder. On the other hand, people were relieved that they wouldn't have to put their homes on stilts—especially Yoder, who, in his determination to inspire his hesitant neighbors, had already finished rebuilding.

FEMA concluded that the failure was not with the elevations, but with the levees, which are now undergoing reconstruction. The levees, however, won't be finished for at least four hurricane seasons.[17] So until then, FEMA declared that the bottom floor of any houses being built must be at least 3 feet higher than the surrounding ground—even if that ground is already at a safe height.[18] Many local officials questioned FEMA's logic.

"It just didn't make any sense," says Walter Leger, chairman of the Louisiana Recovery Authority's housing task force.[19] Leger questioned FEMA representatives. Could he comply with the rule by digging a 3-

foot trench around his house? Yes, he was told. Once the levees are certified, will the 3-foot rule expire? Yes. So some homeowners might have to spend tens of thousands of dollars to comply with a rule that is going to expire soon? That's true, said FEMA.[20]

Leger is a lawyer by trade. "I'm thinking, 'My God, I wish this was a trial—I'm winning!' But I'm not winning; I'm losing." The mayor's office estimates that 50,000 homes could be affected by the 3-foot rule, though many were damaged beyond salvation anyway. People whose houses sit on ground-level concrete slabs will be the most harshly affected. Concrete slabs are impossible to raise, and those houses would have to be torn down.[21]

FEMA mitigation specialist Butch Kinerney agrees that the 3-foot rule sounds "crazy" but says that until the levees surrounding the city are certified by the U.S. Army Corps of Engineers to withstand a 100-year flood, the agency has no choice but to impose the rule.[22]

Nathan Kuhle, 31, wasn't required to raise his Broadmoor house, but as he inspects the freshly poured, 9-foot concrete pillars now holding it up, he feels the peace of mind is worth the $45,000 price tag.[23]

Two-thirds of the cost of lifting his house was covered by FEMA as part of his flood insurance. "Our thought is that if it floods again, one, we'll be dry, and two, it may increase the resale value," he says.

Across the street, David Winkler-Schmit, 42, is keeping his home right where it is. "It's a conscious decision," he says. "My house is 86 years old and has never taken a drop of water" before Katrina. As for his friend across the street, Winkler-Schmit just laughs. "If this area floods again like that," he says, holding his palm 6 feet in the air, "nobody's going to be here anyhow!"[24]

It's an argument Kuhle concedes. "It's a bit of a gamble," he says.

Living in New Orleans has always been a gamble. After Betsy inflicted unprecedented damage in 1965, the federal government sought to improve the odds—not just for residents, but also for the taxpayers who finance disaster relief. One result was the creation of the National Flood Insurance Program.

For the people of New Orleans, the program has brought much-needed relief. Taxpayers have not fared quite as well. As the cost of Katrina started to climb into the billions, politicians and pundits laid the blame on the victims: Because they hadn't bothered to buy flood insurance, the argument went, the taxpayers have to bail them out.[25]

In fact, two-thirds of the homeowners in New Orleans had flood insurance, making it one of the most heavily insured areas in the nation.[26] And that, ironically, is why the taxpayers have to bail them out.

The reason the federal government insures flood-prone homes is that the private insurance industry generally won't.[27] To increase participation, the government subsidizes many of the premiums,[28] even in risky states like Louisiana. What's more, while FEMA, which runs the program, requires that the homes it insures be above the base flood elevation, houses built before the regulations went into effect are exempt.[29] In New Orleans, which had many older properties, the majority of flood-insured homes sat below the required flood elevations.[30]

So for the federal flood insurance program, Katrina was the dreaded perfect storm: a catastrophic flood in a densely populated, heavily subsidized, highly subscribed area where the vast majority of insured properties were below elevation. The program—which over the years had remained basically solvent[31]—capsized into the red. It paid Louisiana policyholders more than $13 billion in fiscal 2005,[32] an amount greater than all the payouts from all the floods since the program began in 1968.[33]

Payouts for all states affected by Katrina and 2005 hurricanes Rita and Wilma could exceed $23 billion. The Government Accountability Office put it succinctly: The federal flood insurance program "cannot absorb the total costs of paying these claims,"[34] which means the American people will.

Yet, despite the expense to taxpayers, the people of New Orleans have not benefited as much as they could have. Most homeowners who had insurance were underinsured. And many of the people living in the most heavily flooded areas had no flood insurance at all—because they were told they didn't need it.[35]

---

Cora Charles has lived in her house in the Lower Ninth Ward for 45 years. The living room walls are newly Sheetrocked, but grass stains are still visible atop the windowpanes, marking how high the water climbed. "Oh, it was horrible," says the 70-year-old, recalling her first glimpse of the waterlogged room.[36] The house was a total loss. She had no flood insurance.

"I wasn't required to have it," she says. "They said, 'You're not in the flood zone.'"

It's true. According to FEMA's flood maps, much of the area where Charles lives, filled with flood-devastated houses as far as the eye can see, an area that flooded again during Rita and decades ago during Betsy—the very flood that led to the creation of the National Flood Insurance Program—is in fact not considered a flood zone.[37]

"The administration of the program has been awful," says J. Robert Hunter, director of insurance for the Consumer Federation of America, who ran the program in the 1970s.[38] The flood maps, he says, were supposed to be updated every three years to keep up with development. But in 2001, the GAO reported that 63 percent of the nation's flood maps were more than a decade old.[39] In New Orleans, the flood elevations haven't changed since 1984.[40] And, of course, when setting the flood zones' boundaries, FEMA assumed that the levees would hold.[41] Cora Charles' neighborhood, protected by levees, was considered safe.[42]

A few feet up Jerry Schulin's front doorway—which is itself a few steps up off the ground—is a pencil mark, indicating sea level. The house, just a block and a half from the 17th Street Canal, looks empty, until a dirty hand extends up from below what used to be the living room. "Welcome to the dungeon,"[43] Schulin says, grinning. Though he's 74, Schulin has cleaned and gutted the entire house by himself, and he is about to rebuild it. He relishes the task, but he also doesn't have much choice: He didn't have flood insurance.

Though his house sits near the bottom of a clearly mapped flood zone,[44] Schulin didn't have to have insurance because the program requires at-risk properties to be insured only for the value of their mortgages.[45] Like many older residents, Schulin paid his off years ago. Throughout the city, homes worth hundreds of thousands of dollars may have been insured for only the few thousands dollars left on their mortgages.[46] Even this requirement is not directly enforced by FEMA. Instead, the law requires lenders to ensure the policies are in place. Federal investigators have concluded that no one really knows whether the lenders do so.[47]

The federal flood insurance program is administered by private insurance companies, which are paid about a third of the premiums for writing the policies and adjusting the claims.[48] These are the same companies that provide homeowners with insurance for non-flood-related damage such as wind. And while residents have generally been pleased with the way their federal flood insurance claims have been handled, many believe the companies have been far less generous with wind claims, which come out of their own pockets.

---

Pat Hemard's house in the far eastern reaches of the city was more than just damaged by Katrina. "I can't even find it," he said in a phone call from Florida. "There's just a few pilings left. And I think a toilet."[49] He had already evacuated when the hurricane hit, but he believes its winds simply tore his house apart. His clothes were scattered in nearby treetops, high above the water line. He didn't have federal flood coverage, but his house and its contents were insured for about $180,000. His insurance company decided his claim quickly: It gave him nothing.

That's because the remains of Hemard's home were eventually covered in 14 feet of water. Hemard, however, believes his house had blown away at least half a day before the water came. While homeowner policies in hurricane country have traditionally covered only wind, not flood, people across New Orleans are discovering a bit of fine print called an "anti-concurrency clause," which absolves the insurer of liability if any of the damage was caused by flood.

"He said, 'Sorry, Charlie.' Those were his exact words," Hemard says of his adjuster.

"It's so frustrating," says Julie Quinn, a Republican who sits on the state Senate's Insurance Committee.[50] "They're using every trick and loophole in the book." Quinn's own roof was ripped open by the hurricane, allowing the rain to pour in. "The flood came in second, but they didn't want to pay for anything below the water line," she says.

Sworn into office just three weeks before Katrina hit, Quinn quickly discovered the power of the Louisiana insurance industry. "The first thing I learned was that the consumers have no lobbyists up there," she says. "And the insurance companies have several—several—hundred-thousand-dollar-a-year paid lobbyists." Three members of the state House

Insurance Committee list "insurance agent" as their day job.[51] Meanwhile, of the four previous state insurance commissioners, three have been sent to federal prison.[52] In such an environment, Quinn says, it's not surprising that many pro-consumer bills—including one aimed at the anti-concurrency clause—are killed.

Loretta Worters, vice president of communications for the Insurance Information Institute, sympathizes with the frustrated homeowners but notes that court rulings have upheld the anti-concurrency clause. "The bottom line is, the insurance companies paid what people were owed, what their contracts stated. The contract is very clear," she says.[53]

Louisiana homeowners have lodged so many complaints that Insurance Commissioner Jim Donelon has ordered a review into how two companies, Allstate and Travelers, have handled claims in the state.[54] The GAO has begun its own study of insurance payouts.[55] And Quinn says representatives of Fannie Mae, worried that they may someday own thousands of uninsured, foreclosed properties in Louisiana, were in Baton Rouge in late spring of 2006, urging legislators to get tough with insurers.[56]

While the bulk of the damage in New Orleans was indeed caused by flooding, the city was battered by hurricane winds before the levees were breached. Fighting to recoup their non-flood damages is among the many challenges New Orleans homeowners are facing.

Cora Charles' insurance company gave her $2,000 for damage to her roof. Jerry Schulin got $13,000 for his roof, but says the insurance company refused to pay for damages that occurred above the water line. He tried to argue, but says the company wouldn't return his calls. "I kinda gave up on it," he shrugs.

LaToya Cantrell's adjuster assessed the wind damage to her Broadmoor home at just $3,000. "We battled and battled and got it reassessed," she says. After three months of fighting, Cantrell got a check for $37,000. "They're protecting their money," she says. "The intent is just to lowball."

And, some say, to stall. Louisiana policyholders have only one year from the date of a disaster to litigate their claims. (Their neighbors in Mississippi get three years.)[57] Under normal circumstances, a year might be plenty of time, but in the year following Katrina, things have hardly been normal in New Orleans. With so many people still undecided about whether to rebuild, hauling the insurance company to court is an afterthought at best.

Insurers have opposed lawmakers' attempts to extend the one-year deadline. "I think the concern by some companies is that you set a precedent on this storm," says the insurance industry's Worters. As the deadline approached, Louisiana's insurance commissioner issued a directive ordering insurers to give hurricane victims an additional year. Nearly all of them complied.[58]

Should a generous insurance check arrive in the mail, however, the joy can be short-lived. Mortgage companies often demand that the checks be signed over to them, and in some cases the checks are even made out to them. The state Recovery Authority's Leger says the practice is widespread, even though a homeowner isn't obligated to comply.

After Hurricane Andrew in 1992, the insurance industry played a critical part in jump-starting the recovery.[59] Since Katrina, however, Senator Quinn worries that the insurance industry is slowing the city's recovery. "It's a business, I understand. But it's really hurting. It's not just hurting the people, it's hurting our state, because we can't get settled," she says. Unless something changes, she believes, "not as many people will be able to afford to rebuild. There'll be more blighted housing. The city will be smaller. And it's really going to happen anyway, because these cases have already taken too long."

Worters counters that as of July 2006, 90 percent of Katrina-related claims had been settled and that the industry has lived up to its obligations. "The insurance industry never has been responsible for flooding," she says, adding, "It's natural to strike out at the people you think should be covering you, but that's not the way it works."

---

An enormous number of New Orleanians have been spared the frustrations of rebuilding. For the renters who lived in 49 percent of the city's damaged housing, there's no wait for a FEMA trailer or debate about elevating the house. Their living quarters are gone, and they won't be fixed until their landlords—who are likely consumed with rebuilding their own homes first—get around to it. FEMA and HUD have been providing rental assistance vouchers, but the cost of the city's surviving rental units has skyrocketed,[60] making their return all but impossible. The Louisiana Recovery Authority plans to make rebuilding loans available to landlords, with the amount increasing as the landlord lowers the rent,[61] but even this money wasn't expected to start flowing before late summer 2006 at the earliest.[62]

One landlord that is not eligible is HUD. Since 2002, when the federal agency took control of the city's housing authority, it has been responsible for housing 14,000 of New Orleans' poorest families, thousands of whom were crowded into 10 major housing projects.[63] Katrina drove nearly all of them out; afterwards, some residents started returning.[64] HUD, critics say, is in no great hurry to bring them back.

In the months after Katrina, HUD secretary Alphonso Jackson singled out the city's River Garden housing development as the future of public housing in New Orleans,[65] an ominous sign for displaced residents afraid of being excluded from the city's future.

Now a charming collection of homes in traditional New Orleans styles, River Garden used to be the St. Thomas Housing Project, home to 800 low-income families. The project was turned over to a local developer who replaced it with a "mixed-income" village that few St. Thomas residents could afford. The complex contains only 122 affordable housing units. "I guess to some people that's considered a success," says housing advocate William Quigley of Loyola University New Orleans College of Law, who is suing HUD on behalf of displaced tenants. "But if you're really interested in low-income housing, it's a disgrace."[66]

While bringing people back is the city's top priority, public housing tenants have returned to find their homes surrounded by chain-link fences or shuttered with metal grating.[67] By early 2007, only 1,145 units had been reopened, according to Adonis Exposé, communications director for the Housing Authority of New Orleans.[68] At the same time, Quigley says, many projects seem to have weathered the storm better than private houses. "It's a public policy disaster from the perspective of the working poor," Quigley says, and a blow to the city's economy, since so many businesses depend on the low-wage workers who can afford only public housing.

In mid-June 2006, Secretary Jackson announced that four huge housing projects, totaling 5,000 units,[69] would be torn down and developed into River Garden–like developments. Jackson did not say how many, if any, low-income housing units would be included.[70]

After St. Thomas became River Garden, HUD resettled many of the displaced families in private apartments using vouchers. Many, according to Quigley, landed in New Orleans East. That area was virtually destroyed by Katrina. River Garden, on the edge of the historic Garden District, sustained only minor damage.

For all the obstacles New Orleans residents face in rebuilding their lives—bureaucratic, financial and physical—the biggest one, says Walter Leger, could be psychological: fear of what the current hurricane season might bring. "A lot of people are waiting on that," he says. "Every conscious thought we have in terms of planning business, housing and personal lives is centered around the possibility of hurricanes and floods now."

But Leger also sees signs of hope. For months after the flood, he says, people greeted one another with the question, "How'd you do?" The answer was always expressed in terms of the depth of the water or the scale of the loss.

"It's starting to evolve into, 'How are you doing?'" he says with a grin. "'How's your house going?' 'You coming back?' That's a huge change."

# Epilogue

*"If we don't get moving on restoration soon, the rest of the country will forget about the Louisiana coast. By then, our chances may have run out."*

—Tim Searchinger, attorney with Environmental Defense

More than a year and a half has passed, and New Orleans is not even close to recovering from Katrina. Those struggling to plot the city's future are plagued with questions about how to rebuild, what to do about its housing and its schools, whether its health care system will ever be the same, whether hundreds of thousands of residents will return.

And the most haunting question of all—whether the city will flood again—is unanswered because of doubts about whether the levees and floodgates will be able to hold off another storm like Katrina—or worse.

Low-lying New Orleans remains acutely vulnerable to storm surge flooding, while meteorologists say hurricane activity is on the upswing.[1] Without a major upgrade of the city's levee system, flooding from a big storm could once again inundate the city.

The U.S. Army Corps of Engineers has embarked on an ambitious program to quickly repair breaches, fortify dozens of miles of levees with rocks and concrete and build floodgates at the entrances to drainage canals where floodwalls collapsed and the rushing water devastated large swaths of the city.[2]

The corps is spending about $6 billion on the fixes. The first phase of emergency repairs was originally supposed to be done by June 1, 2006, but the work on floodgates and a temporary pumping station at the 17th Street Canal dragged all the way through the end of hurricane season.[3] Because of the slow pace of restoration work on the permanent pump stations, a corps analysis showed that if the gates were closed during a storm surge, trapped rainwater could cause significant flooding in some neighborhoods.[4]

As federal, state and independent investigators have detailed, even undamaged parts of the levee system may be defective. With the land sinking underneath them, many levees are not as high as they're supposed to be. At other points, new projects abut older ones at different heights.

"You have a place that may be at 18 feet, and another place that's at 15 feet, and all it means [is] if the water comes up over 15 feet it's going to go over that lower section," said Dan Hitchings, director of the corps' Task Force Hope, which is overseeing the improvements. The plan is to rebuild the system so that it meets all the original specifications—something the corps was never able to do before disaster struck. Once those repairs are complete, officials say, flooding from a storm comparable to Katrina would be substantially limited.

Still, many of the original problems that led to the inundation of New Orleans remain. The standards the corps is using for its repairs date to the 1960s and have never been upgraded, despite advances in the science of hurricanes and storm surges. The corps gauges these fortified levees will protect New Orleans only against weaker storms—Category 1, 2 and some Category 3 hurricanes. A direct hit by anything more powerful—a strong Category 3 or a Category 4 or 5 storm—would still swamp the city.[5]

Hassan Mashriqui, an assistant engineering professor who does computer modeling of storm surges for the Louisiana State University Hurricane Center, said he believes the corps' estimates are overly optimistic. Mashriqui's modeling shows the repaired levees are vulnerable to flooding even from some Category 2 storms.[6]

The city will get some additional protection on top of the repairs. Congress has authorized the corps to further enhance the levee system to protect against a 100-year storm surge—a flood height that is expected to occur once in 100 years. That will raise some levees higher.[7]

The corps also is studying the feasibility of providing protection against storm surges from Category 5 hurricanes. That would involve raising the levees significantly higher than the 100-year benchmark and building floodgates at channels into Lake Pontchartrain. A completed draft of the plan is due in late 2007.[8] Early estimates put the cost at $23.5 billion.[9] A new levee system also would have to be integrated with planned coastal restoration programs, which themselves have a price tag of around $14 billion.[10]

For those hoping for a hefty federal financial commitment, the initial signs have not been encouraging. A preliminary evaluation by the corps released in July 2006 sparked outrage from Louisiana political leaders because it contained no specific project recommendations. Earlier drafts did include a list of projects, but an assistant secretary of the Army removed them.[11] Environmental groups fear the report's emphasis on higher levees will ultimately result in a giant wall across Louisiana that would harm the environment.[12] The corps denies that its plans are that extensive.

The corps' ability to build such a complex system is also debatable. Outside investigators say that faulty corps management practices helped create the original, deeply flawed system. Yet corps investigations of the levee failures have focused on the science and engineering problems, not the decision-making and management practices that led to them.

"It disturbs me," said Ivor van Heerden, deputy director of the LSU Hurricane Center and the leader of the state's levee investigation. "The corps is not about to change the way it does business."[13]

Congress is taking steps to try to make the corps bureaucracy more accountable. In July 2006, the Senate passed an amendment that would require the corps to submit controversial projects or those costing more than $40 million to an independent peer review panel. The measure still had to be reconciled with a similar, less stringent version passed by the House.[14]

Meanwhile, FEMA, the agency charged with coordinating the response to disasters such as Katrina, has its own set of problems, including an exodus of experienced staffers and a diminished role within the federal government.

Some officials believe FEMA is no better prepared than it was during Katrina. In fact, one of its former directors, Michael Brown, said in an interview, "I honest to God believe it's worse."[15] The agency is buried in the massive Department of Homeland Security and the bureaucracy is more impenetrable than ever, according to Brown.

FEMA's woes have been thoroughly chronicled in congressional reports, academic studies and media exposés. But opinions differ on how to fix them.

Terry Ebbert, New Orleans' homeland security director, argues that the military should be immediately pressed into service whenever a catastrophe of Katrina's proportions looms. FEMA, ex-Marine Ebbert explained, is an "ever-changing, compliance-driven organization," meaning it is more concerned with following rules—even silly ones—than getting the job done. The military, in contrast, "gets the word 'mission,'" Ebbert said.[16] "You've got to have overwhelming force and instantaneously respond to show that you're in control," he added.

Others are loath to interfere with the Posse Comitatus Act of 1878, which restricts the role of the military in domestic matters. Among them are former FEMA directors Brown and Joseph Allbaugh, who say it would be a serious mistake to use the Department of Defense as a first responder.[17] Brown worries that troops could be put in a situation "where they might have to kill American citizens" if things got out of hand.

Most seem to believe that the answer lies not with the military, but with a stronger, more independent FEMA, along the lines of the agency James Lee Witt built in the 1990s. In late 2006, Congress ordered a remake of FEMA, giving it the sort of autonomy within the Homeland Security Department enjoyed by the Coast Guard and the Secret Service. The department announced its plans for the reorganization in early 2007, saying it would consolidate a variety of offices and programs in hopes of making FEMA stronger and less diffuse.[18]

Whether the new version of FEMA will perform more adeptly remains to be seen. But this much is clear: The congressional mandate is an acknowledgment that the agency, as configured during much of the Bush administration, doesn't work—and leaves the nation unprepared for another disaster.

In Louisiana, state and local officials say that they, too, recognize deficiencies in their emergency preparedness and are working to correct them.

The state's Office of Homeland Security and Emergency Preparedness, which had been part of the Military Department, now reports directly to the governor. Bills passed during the 2006 legislative session more than doubled the preparedness office's staff to about 100, including regional coordinators positioned to work closely with city and parish emergency officials. The office and the Louisiana State Police also have upgraded their communications systems.

In May 2006, more than 300 federal, state and local emergency management officials, as well as representatives of nonprofit groups and private businesses, met in downtown New Orleans for a hurricane simulation exercise to practice their response and identify gaps in emergency planning. The Louisiana Department of Transportation and Development has contracted for enough buses to evacuate 30,000 people and released a plan to shelter 65,000 people, including 2,000 with special medical needs.

Mayor C. Ray Nagin said that a Category 2 or stronger hurricane threatening New Orleans would most likely prompt an evacuation order 30 to 36 hours before predicted landfall and that there would be no shelter of last resort, such as the Superdome or the convention center.[19] The city's new evacuation plan provides for 12 pickup points around the city for those without transportation. The elderly and disabled would be taken to a transportation center, where they would board trains bound for shelters elsewhere in Louisiana.[20] To avoid a shortage of bus drivers, which occurred during Katrina, the mayor has declared them essential personnel who would be required to report for duty during an emergency.[21]

But not everyone is optimistic about the city's plans. Many residents say they have little confidence in the government's ability to protect them from another storm. "Getting people out of trailers and out of houses if they're not ambulatory or if they don't have transportation, none of us have the foggiest idea how they're going to get out," said neighborhood activist Phyllis Parun, who lives in the Bywater area. "None of us have any more faith in the city than we had before Katrina . . . or in the state."[22]

The state, for its part, has taken action to shore up New Orleans' defenses by pushing to consolidate the area's hodgepodge of levee boards into two districts—one governing the east bank of the Mississippi River and one responsible for the west bank. The measure, passed in a special legislative session, is designed to push out incompetent political appointees. Voters approved the state constitutional amendment in fall 2006. It also requires members of the levee boards to have professional backgrounds that enable them to work more closely with the Army Corps.[23]

The move, backed strongly by Governor Kathleen Blanco, was one of several interventions by the state to force changes in the city's cumbersome bureaucracy, including gradually merging the civil and criminal district courts into one system and paring down the multitude of judges and court clerks.

Facing the potential release of hundreds of criminal defendants who had been awaiting trial for months without representation, the mayor announced measures in August 2006 meant to get the city's stalled criminal justice system moving. Among them were getting prosecutorial help with the backlog of cases from neighboring Jefferson Parish, as well as pro bono legal aid from the Louisiana State Bar Association for inmates lacking public defenders.[24] But as recently as January 2007, the shortage of public defenders was still a problem in New Orleans' courts.

There still isn't any assurance that local leaders in New Orleans will improve on their poor performance during Katrina—any more than state or national leaders will.

Though yoked to each other by Katrina, the city of New Orleans and the state government have a history of alienation that seems fated to continue. The animosity between the city and the rest of the state was personified by the shaky working relationship between Nagin and Blanco in the period following Katrina, and it persists today.

Is there any reason to think that the two leaders will cooperate better in the event of another disaster?

"I wish I knew," said Lawrence Powell, a history professor at Tulane University and a longtime student of Louisiana politics.

Powell said both are "in the same foxhole," though "the governor calls the shots" because the state seized control over the city in so many areas. "But everything is so uncertain," he said.[25]

---

One thing that is certain is that New Orleans is a vastly different city today than it was before Katrina. Many of its neighborhoods remain in ruins and its population has plummeted by more than half. Many of its hospitals and nursing homes no longer exist.

Before the storm, there were nine acute care hospitals in New Orleans. A year later, there were three and only one was running at full capacity. Only 13 of 24 nursing homes in New Orleans were operating.[26] The nearest major trauma center was 300 miles away.[27]

"We can't handle another hurricane right now. We can't even handle a bus crash," said Dr. James Moises, an emergency room physician at Tulane University Hospital and Clinic who is president of the Louisiana chapter of the American College of Emergency Physicians.[28]

Many doctors left the area and never returned. A University of North Carolina study shows that more than 4,400 active physicians from the New Orleans metropolitan area were displaced by Katrina and that more than 35 percent of them were primary care doctors.[29]

"The system is incredibly broken right now. Another hurricane isn't going to improve it," said Dr. Peter DeBlieux, Charity Hospital's emergency medicine director. "This is going to continue to be a very bad place to receive your health care for a very long time, whether you're insured or not. It impacts everyone."[30]

The hospitals that are still open maintain that they are better prepared for a hurricane than they were when Katrina struck. Many are improving their communications systems, which were practically nonexistent during Katrina. Some staffers will be getting prepaid cell phones with area codes outside of the New Orleans region. Some will be given ham radios. Hospitals also are stocking up on supplies, acquiring boats and contracting with helicopter companies.[31]

The state's medical director says that some hospitals in high-risk flood areas will have to be evacuated if a major hurricane strikes.[32] The state will rely on the National Disaster Medical System, the federal program that is supposed to swoop in and assist local and state governments with medical services and patient evacuations.

But years of audits show that the federal program remains plagued by coordination and management problems. And the program is not designed to assist in evacuating nursing home residents.

A July 2006 GAO report also found that the federal program's evacuation procedures are limited because they do not include using ambulances or helicopters to move patients from health care facilities to airports. Making matters worse, there are no other federal programs that help with that type of transportation, the auditors wrote.[33]

Cynthia Bascetta, the GAO health care director who wrote the report, said it's unlikely that the federal medical evacuation process had been improved much since Katrina. "If it's a catastrophe and the state and locals can't handle it, I don't think there's clarity yet about what the federal role is in getting people out," she said.[34]

Michael Hopmeier, a consultant who evaluates disaster systems and has served as a special adviser to the U.S. surgeon general and the Homeland Security Department, said that during Katrina, the National Disaster Medical System had difficulty conducting mass evacuations and transporting patients out of the area. Those problems are systemic and aren't quickly fixed, he said. "We can be telling each other that we're fully prepared, but fundamentally we don't have a clue," he said. "Based on every test so far, we've pretty much failed."[35]

The Red Cross, for its part, acknowledges that its performance during Katrina was lacking.[36] It couldn't provide food or shelter in some crucial areas, including the city of New Orleans. In those places, hurricane victims turned to the Salvation Army and smaller, more nimble groups with local ties.

The Red Cross' own inexperienced volunteers and staffers were often unable to match up local charities and volunteers with people who needed help, according to John Davies, president and CEO of the Baton Rouge Area Foundation, which connects nonprofits, donors and community leaders.[37]

Those types of problems aren't new to the people who run the Red Cross.[38] The organization's former chief executive officer, Bernadine Healy, told the *Chronicle of Philanthropy* that the organization is "not preparing a skilled volunteer work force."[39]

Daniel Borochoff, president of the American Institute of Philanthropy, said he doesn't blame Red Cross leaders; he believes much of the power resides within the chapters, which compete with each other for resources and are resistant to change. Marty Evans was ousted as CEO a few months after Katrina, largely because she wanted to alter that power structure, Borochoff said.[40]

The Red Cross also has been criticized for setting up relief operations in white, upper-class communities more often than in low- or moderate-income black neighborhoods. In preparation for the next crisis, it's asking its local chapters to partner with churches and community groups, and it's trying to recruit more volunteers in minority communities, working nationally with largely black groups such as the African Methodist Episcopal Church and the NAACP.[41]

But many community leaders view those partnerships and diversity efforts as being largely for show, and they believe that the Red Cross' lack of local connections may doom its response in the future.

"It's all P.R.," said Joe Leonard Jr., executive director of the Black Leadership Forum, a national federation of civil rights organizations. Leonard, who saw the need for more black volunteers when he drove across the South in the week after the storm, said that unless the Red Cross puts together a more diverse volunteer base and does more outreach, minority communities will once again get short shrift in the next disaster. "We would see the same result. The exact same," Leonard said.[42]

Despite all the efforts to blunt the impact of another hurricane, in some ways New Orleans remains stubbornly stuck in the past.

After Galveston, Tex., was destroyed by a storm surge in 1900, the townspeople spent the next decade raising not only their houses, but the land underneath them as well, in some cases by as much as 13 feet.[43] After Grand Forks, N.Dak., flooded in 1997, city officials bought out 635 properties in the flood plain, demolished the houses and turned the land into green space.[44]

New Orleans, in contrast, is doing everything it can to rebuild the city exactly as it was. All options remain on the table, and the only limits on rebuilding are a homeowner's lack of resources. FEMA advisories left the city's base flood elevations unchanged and required damaged homes to be raised just 3 feet off the ground. Until they are adopted by local governments—which may take a year or more—they are merely guidelines, which any homeowner with a building permit may disregard without penalty.[45]

And they don't apply at all to homes that sustained less than 50 percent damage—a determination that has often proved to be difficult to make and easy to adjust.[46] New Orleanians have been flocking to the building permit office at City Hall, asking that their damage assessments be reduced to 49 percent. Knowing that doing so will make it easier for its citizens to rebuild, officials have been happy to comply.[47]

Armed with a 49 percent assessment, a homeowner is free to rebuild a house precisely where it was and to insure it through FEMA's National Flood Insurance Program.[48] Should the city flood again, the program could once again be bankrupted.

J. Robert Hunter, a New Orleans native who directed the program in the 1970s, said he would have disqualified the entire city based on its willingness to lower homeowners' damage assessments, though he understands the frustration with the 3-foot rule, which he calls "stupid."

"What the hell does it mean? I mean, either you can build back where you were because the levee's going to hold, or the levee's not going to hold and you should build up at least to where the water was up to on your house," Hunter said.

Which option would he choose? "If you look at the history of the corps' waterworks projects, dams and levees, they always fail sooner or later. It's just a matter of time. These things don't work."[49]

Katrina taught New Orleans that it needs multiple lines of defense, said Greg Miller, a project manager with the corps. "If you only have levees, and you have a problem like we saw in Katrina, you have a catastrophic failure," Miller said.[50]

The first lines of defense against hurricanes are barrier islands, coastal marshes and other natural features. The inner layers of protection are levees, floodgates and buildings elevated out of reach of floodwaters. Wetlands absorb the energy of storms that otherwise could barrel directly into buildings and levees. There is also evidence that they lower the height of hurricane storm surges.[51]

Corps senior project manager Al Naomi said that wetlands would be an essential part of any future hurricane protection plan for New Orleans. "I don't think anybody wants to build a Category 5 levee with no wetland in front of it. I think that would be a horrible mistake," he said.[52]

Environmental restoration is a relatively new mission for the corps, which is more accustomed to trying to tame nature by building large flood-control structures. In fact, the corps' structures are partly responsible for killing the wetlands in the first place. The same Mississippi River levees that keep New Orleans dry also shunt fresh water and land-building sediment into the Gulf of Mexico.

In recent years, the corps has tried to reverse some of the damage. At a place called Caernarvon, for example, 15 miles downstream from New Orleans, culverts siphon a share of the Mississippi into the marsh. The freshwater flow chases away salt water creeping in from the gulf. Since the project started in 1991 at a cost of $26 million, freshwater marsh plants have increased sevenfold and more than 400 acres of new land have formed.[53] That is a modest gain, given the 1.2 million acres of coastal land that have been lost in Louisiana since the 1930s.[54] But many scientists think it's a step in the right direction.

At best, however, "restoring" coastal Louisiana would not mean putting it back the way it was before European settlers arrived. And some question whether Congress will ante up $14 billion to restore Louisiana's coast.

As 2006 drew to a close, President Bush signed a bill that expanded oil drilling in the Gulf of Mexico and gave a share of the added revenue to Gulf Coast states. Louisiana could gain as much as $650 million a year starting in 2017. Although politicians have trumpeted the law as a lifeline for fragile wetlands, Louisiana also can use the funds for new levees. And some observers believe that passage of the revenue-sharing law may make it harder to persuade Congress to fund major restoration efforts.

"We are 18 months out from Katrina, but neither Congress nor the state has provided the kind of serious money needed to restore the wetlands," said Tim Searchinger, an attorney with the conservation group Environmental Defense. "We have a small window of opportunity here. If we don't get moving on restoration soon, the rest of the country will forget about the Louisiana coast. By then, our chances may have run out."[55]

Even if Louisiana succeeds in reinforcing its levees and coastline, scientists say two ongoing problems beyond its control will erode the gains. One is that the Mississippi River Delta will continue to sink. The other is that global warming will continue to raise sea levels and potentially make hurricanes even more intense.

"I would give the area probably a couple hundred years," said Roy Dokka, a geologist at Louisiana State University. He cautioned that another storm could be even worse than Katrina. "The fact is, this was a large hurricane, but it's not as big as it's going to get. . . . There's a Category 5 out there."[56]

# Notes

## ONE: THE STORM

1. Interviews with Pearl Ellis, May 3 and May 10, 2006.

2. New Orleans *Times-Picayune*, "Wise Suggestion" and "Single Commander Need, La. Guard Adj. Gen. Holds," September 21, 1965.

3. James H. Gillis, "Flood Barrier, M'Keithen Aim," New Orleans *Times-Picayune*, September 15, 1965.

4. National Hurricane Center, *Tropical Cyclone Report: Hurricane Katrina, 23–30 August 2005*, December 20, 2005 (updated August 10, 2006), p. 9, http://www.nhc.noaa.gov/pdf/TCR-AL122005_Katrina.pdf.

5. Louisiana Department of Health and Hospitals, "Hurricane Katrina: Reports of Missing and Deceased," August 2, 2006, http://www.dhh.louisiana.gov/offices/page.asp?ID=192&Detail=5248.

6. U.S. Department of Housing and Urban Development, Office of Policy Development and Research, *Current Housing Unit Damage Estimates*, February 12, 2006 (revised April 7, 2006), p. 23.

7. Louisiana Recovery Authority, "LRA Releases Estimates of Hurricane Impact," news release, January 12, 2006; *Tropical Cyclone Report*, p. 12; and Insurance Information Institute, "The Ten Most Costly World Insurance Losses, 1970–2005," http://www.iii.org/media/facts/statsbyissue/catastrophes.

8. National Oceanic and Atmospheric Administration, Atlantic Oceanographic and Meteorological Laboratory, Hurricane Research Division, *Atlantic Hurricane Database Re-Analysis Project*, "Hurricane Andrew's Upgrade," October 2002, http://www.aoml.noaa.gov/hrd/hurdat/andrew.html.

9. *Tropical Cyclone Report*, p. 8; National Weather Service Storm Prediction Center, "Converting Knots to MPH," http://www.spc.noaa.gov/misc/tables/kt2mph.htm.

10. U.S. Army Corps of Engineers, *U.S. Army Corps of Engineers Response to Hurricanes Katrina and Rita in Louisiana, Environmental Assessment*, EA #433, April 17, 2006, p. EA-4.

11. U.S. Senate Committee on Homeland Security and Governmental Affairs, *Hurricane Katrina: A Nation Still Unprepared*, May 2006, p. 1-1, http://hsgac.senate.gov/index.cfm?Fuseaction=Links.Katrina.

12. National Oceanic and Atmospheric Administration, "The Galveston Storm of 1900—The Deadliest Disaster in American History," http://www.noaa.gov/galveston1900.

13. U.S. House Select Bipartisan Committee to Investigate the Preparation for and Re-

sponse to Hurricane Katrina, *A Failure of Initiative: The Final Report of the Select Bipartisan Committee to Investigate the Preparation for and Response to Hurricane Katrina*, 109th Cong., 2nd sess., February 15, 2006, p. xi, http://katrina.house.gov/full_katrina_report.htm.

14. Comptroller General of the United States, *Report to the Congress: Cost, Schedule, and Performance Problems of the Lake Pontchartrain and Vicinity, Louisiana, Hurricane Protection Project*, PSAD-76-161, August 31, 1976, p. 1.

15. U.S. General Accounting Office, *Report to the Secretary of the Army: Improved Planning Needed by the Corps of Engineers to Resolve Environmental, Technical, and Financial Issues on the Lake Pontchartrain Hurricane Protection Project*, GAO/MASAD-82-39, August 17, 1982, p. 9.

16. Minutes of the Center for the Study of Public Health Impacts of Hurricanes Advisory Board meeting, September 15, 2003, pp. 21–23.

17. *A Failure of Initiative*, pp. 81–83.

18. *A Nation Still Unprepared*, p. 1-1.

## TWO: THE ENVIRONMENT

1. Louisiana Department of Natural Resources, Office of Coastal Restoration and Management, "Louisiana Coastal Facts," http://dnr.louisiana.gov/crm/coastalfacts.asp.

2. U.S. Geological Survey, National Wetlands Research Center, "Without Restoration, Coastal Land Loss to Continue," news release, May 21, 2003, http://www.nwrc.usgs.gov/releases/pr03_004.htm.

3. Ibid.

4. Interview with Greg Miller, April 24, 2006.

5. Interview with John Lopez, director, Lake Pontchartrain Basin Foundation, Coastal Sustainability Program, May 10, 2006.

6. Interview with Roy Dokka, May 5, 2006.

7. America's WETLAND, "America's Wetland in a Nutshell–FAQ's," http://www.americaswetlandresources.com/background_facts/basicfacts/FAQs.html.

8. Interview with Kerry St. Pé, April 27, 2006.

9. Interview with Paul Kemp, associate professor, research/special programs, Louisiana State University, School of the Coast and the Environment, May 11, 2006.

10. U.S. Senate Committee on Homeland Security and Governmental Affairs, *Hurricane Katrina: A Nation Still Unprepared*, May 2006, p. 9-1, http://hsgac.senate.gov/index.cfm?Fuseaction=Links.Katrina.

11. Ibid., p. 9-2.

12. Ibid.

13. Interview with Al Naomi, May 9, 2006.

14. Dokka interview, May 5, 2006.

15. Louisiana Coastal Wetlands Conservation and Restoration Task Force and the Wetlands Conservation and Restoration Authority, *Coast 2050: Toward a Sustainable Coastal Louisiana* (Baton Rouge: Louisiana Department of Natural Resources, 1998), p. 31.

16. Kemp interview, May 11, 2006.

17. Louisiana Coastal Wetlands Conservation and Restoration Task Force, *Coastal Wetlands Planning, Protection and Restoration Act: A Response to Louisiana's Land Loss,* April 10, 2006, p. 8.

18. Interview with Denise Reed, April 26, 2006.

19. Louisiana Coastal Area Study, "LCA Frequently Asked Questions," July 2004, http://www.lca.gov/nearterm/Q_A_6July%2004.pdf.

20. National Research Council, Committee on the Restoration and Protection of Coastal Louisiana, *Drawing Louisiana's New Map: Addressing Land Loss in Coastal Louisiana* (Washington, D.C.: National Academies Press, 2006), p. 11, http://books.nap.edu/catalog/11476.html.

21. Ibid., p. 3.

22. Matthew Brown, "Katrina May Mean MR-GO Has to Go; Channel Made Storm Surge Worse, Critics Say," New Orleans *Times-Picayune,* October 24, 2005.

23. Lopez interview, May 10, 2006.

24. Environmental Defense, Taxpayers for Common Sense, National Wildlife Federation and National Taxpayers Union, *Katrina's Costly Wake: How America's Most Destructive Hurricane Exposed a Dysfunctional, Politicized Flood-Control Process,* p. 2.

25. John McQuaid and Mark Schleifstein, "Washing Away," New Orleans *Times-Picayune,* June 23, 2002.

26. Ibid., June 24, 2002.

27. Joel K. Bourne Jr., "Gone with the Water," *National Geographic,* October 2004, http://magma.nationalgeographic.com/ngm/0410/feature5/?fs=www3.nationalgeographic.com.

28. Interview with Lynnell Rovaris, April 29, 2006.

29. Orleans Parish District Court, *John Johnson et al. v. Orleans Parish School Board et al.,* case no. 93-14333, ruling by Judge Nadine Ramsey, January 12, 2006.

30. Interview with Tom Harris; Darin Mann, spokesman, Louisiana Department of Environmental Quality; Don Williams, Environmental Unit deputy leader, U.S. Environmental Protection Agency; Jon Rauscher, Environmental Unit leader, EPA; and David Gray, spokesman, EPA, April 24, 2006.

31. Ibid.

32. Natural Resources Defense Council, "State, Federal Officials Paper over Toxic Contamination in New Orleans, Misleading Returning Residents about Health Risks, Groups Say," news release, February 23, 2006.

33. Interview with Don Williams, July 25, 2006.

34. Interview with Sam Coleman, July 21, 2006.

35. Interview with Wilma Subra, April 11, 2006.

36. *Johnson v. Orleans Parish School Board.*

37. Interview with Suzette Bagneris, July 18, 2006.

38. E-mail from Adonis Exposé, Housing Authority of New Orleans, to author, July 31, 2006.

39. Interview with Elodia Blanco, May 4, 2006.

40. U.S. Coast Guard, "Oil Pollution Containment and Recovery Continue," news release, September 15, 2005.

41. Erik D. Olson, senior attorney, Natural Resources Defense Council, *The Environmental Effects of Hurricane Katrina*, testimony before the U.S. Senate Committee on Environment and Public Works, October 6, 2005, p. 3.

42. U.S. Environmental Protection Agency, "Response to 2005 Hurricanes: Murphy Oil Spill," http://www.epa.gov/katrina/testresults/murphy.

43. Coast Guard news release, September 15, 2005.

44. Interview with Darin Mann and Chuck Brown, assistant secretary, Office of Environmental Services, Louisiana Department of Environmental Quality, May 10, 2006.

45. Ibid.

46. Louisiana Department of Environmental Quality, "Chef Menteur Landfill OK'd to Receive C&D Waste," news release, April 13, 2006.

47. City of New Orleans, Mayor's Office of Communications, "Mayor Does Not Renew Executive Order; Landfill Set to Close August 14th," news release, July 13, 2006.

48. Interview with Adam Babich, May 8, 2006.

49. Mann and Brown interview, May 10, 2006.

50. U.S. Geological Survey, "USGS Reports Latest Land-Water Changes for Southeastern Louisiana," fact sheet, February 16, 2006.

51. Interview with Michael Poirrier, April 27, 2006.

52. Michael Poirrier and Elizabeth Spalding, "Special Report: The Importance of *Rangia cuneata*, Clam Restoration to the Holistic Rehabilitation of the Lake Pontchartrain Estuary," University of New Orleans.

53. "America's WETLAND in a Nutshell—FAQ's."

54. Interview with Mark Davis, April 11, 2006.

## THREE: THE LEVEES

1. U.S. Army Corps of Engineers, Lower Mississippi Valley Division, *E-99 Sheet Pile Wall Field Load Test Report*, June 1988.

2. Ibid, p. 1.

3. Mete Oner, William P. Dawkins, Reed Mosher and Issam Hallal, "Soil-Structure Interaction Effects in Floodwalls," *Electronic Journal of Geotechnical Engineering*, 1997, http://www.ejge.com/1997/Ppr9707/Ppr9707.htm.

4. U.S. Army Corps of Engineers, "IPET Releases Results on London Avenue Canal Breaches," news release, May 2, 2006, http://www.hq.usace.army.mil/cepa/releases/ipetmay06.htm.

5. Interview with Mike Sharp, centrifuge modeler, U.S. Army Engineer Research and Development Center, Vicksburg, Miss., June 16, 2006.

6. Louisiana State University Hurricane Center, "Katrina Flooding Summary."

7. U.S. Army Corps of Engineers, Interagency Performance Evaluation Task Force, *Performance Evaluation of the New Orleans and Southeast Louisiana Hurricane Protection System*,

draft final report, June 1, 2006, vol. 1, p. I-7, http://www.asce.org/files/pdf/executivesum mary_v20i.pdf.

8. John Simerman, Dwight Ott and Ted Mellnik, "Majority of New Orleans Deaths Tied to Floodwalls' Collapse," Knight Ridder News Service, December 31, 2005.

9. *Performance Evaluation,* vol.1, p. I-3.

10. Interview with Bruce Davenport, June 13, 2006.

11. Interview with Kevin Lair, June 14, 2006.

12. Interview with Robert Bea, June 2, 2006.

13. U.S. House Select Bipartisan Committee to Investigate the Preparation for and Response to Hurricane Katrina, *A Failure of Initiative: The Final Report of the Select Bipartisan Committee to Investigate the Preparation for and Response to Hurricane Katrina,* 109th Cong., 2nd sess., February 15, 2006, p. 97, http://katrina.house.gov/full_katrina_report.htm.

14. National Oceanic and Atmospheric Administration, "Tides and Currents," http://tidesandcurrents.noaa.gov/tides06/tab2ec4.html.

15. National Hurricane Center, "Hurricane Preparedness: Storm Surge," http://www.nhc.noaa.gov/HAW2/english/storm_surge.shtml.

16. John McQuaid and Mark Schleifstein, "Evolving Danger" (in "Washing Away" series), New Orleans *Times-Picayune,* June 23, 2002, http://www.nola.com/washingaway/risk_1.html.

17. John M. Barry, *Rising Tide: The Great Mississippi Flood of 1927 and How It Changed America* (New York: Simon and Schuster, 1998), pp. 256–58.

18. U.S. Government Accountability Office, *History of the Lake Pontchartrain and Vicinity Hurricane Protection Project,* testimony of Anu Mittal, director, Natural Resources and Environment before the Senate Committee on Environment and Public Works, GAO-06-244T, November 9, 2005, p. 1, http://www.gao.gov/new.items/d06244t.pdf.

19. Ibid.

20. "Evolving Danger."

21. Ibid.

22. Interview with Al Naomi, June 13, 2006.

23. National Oceanic and Atmospheric Administration, *Meteorological Criteria for Standard Project Hurricane and Probable Maximum Hurricane Windfields, Gulf and East Coasts of the United States,* NOAA Technical Report NWS 23, September 1979.

24. Interview with Ivor van Heerden, deputy director of the Louisiana State University Hurricane Center, June 22, 2006.

25. *Performance Evaluation,* vol.1, p. I-12.

26. Letter from Frederic M. Chatry to Lower Mississippi Valley Division, U.S. Army Corps of Engineers, August 7, 1985.

27. Van Heerden interview, June 22, 2006.

28. U.S. General Accounting Office, *Improved Planning Needed by the Corps of Engineers to Resolve Environmental, Technical, and Financial Issues on the Lake Pontchartrain Hurricane Protection Project,* GAO/MASAD-82-39, August 17, 1982, p. 9.

29. Independent Levee Investigation Team, *Investigation of the Performance of the New*

*Orleans Flood Protection Systems in Hurricane Katrina on August 29, 2005*, final report, July 31, 2006, chap. 6, http://www.ce.berkeley.edu/~new_orleans/.

30. U.S. Army Corps of Engineers, *Lake Pontchartrain, Louisiana and Vicinity, Interim Survey Report*, November 21, 1962, p. 1.

31. U.S. Army Corps of Engineers, *Lake Pontchartrain, Louisiana, and Vicinity Hurricane Protection Project Reevaluation Study*, July 1984, syllabus p. 2 and errata sheet.

32. *History of the Lake Pontchartrain and Vicinity Hurricane Protection Project*, p. 3.

33. Jennifer Haydel, "The Wood Screw Pump: A Study of the Drainage Development of New Orleans," *Loyola Student Historical Journal*, 1995, http://www.loyno.edu/history/journal/1995-6/haydel.htm.

34. Bob Marshall, John McQuaid and Mark Schleifstein, "For Centuries, Canals Kept New Orleans Dry. Most People Never Dreamed They Would Become Mother Nature's Instrument of Destruction," New Orleans *Times-Picayune*, January 29, 2006.

35. Modjeski and Masters Consulting Engineers, "Slope Stability Analysis: 17th St. Canal Levees," January 28, 1981.

36. *Investigation of the Performance of the New Orleans Flood Protection Systems*, chap. 12, p.14.

37. Ibid.

38. Interview with Joseph Wartman, June 8, 2006.

39. Interview with Don Riley, June 27, 2006.

40. *Performance Evaluation*, vol. 5, pp. 7–38.

41. Interview with Carlton Newman, June 14, 2006.

## FOUR: EMERGENCY PREPAREDNESS

1. Interview with Art Jones, June 13, 2006.

2. U.S. Senate Committee on Homeland Security and Governmental Affairs, *Hurricane Katrina: A Nation Still Unprepared*, May 2006, p. 14-1, http://hsgac.senate.gov/index.cfm?Fuseaction=Links.Katrina.

3. Ibid., p. 4.

4. Ibid.

5. Ibid., p. 6.

6. U.S. House Select Bipartisan Committee to Investigate the Preparation for and Response to Hurricane Katrina, *A Failure of Initiative: The Final Report of the Select Bipartisan Committee to Investigate the Preparation for and Response to Hurricane Katrina*, 109th Cong., 2nd sess., February 15, 2006, p. 360, http://katrina.house.gov/full_katrina_report.htm.

7. Interview with Dorothy and Lawrence Davis, May 23, 2006.

8. Interview with James Lemann, May 23, 2006.

9. Interview with Charmaine Marchand, May 24, 2006.

10. Tropical Prediction Center, National Hurricane Center, *The Deadliest, Costliest, and Most Intense United States Tropical Cyclones from 1851 to 2004 (and Other Frequently Requested Hurricane Facts)*, NOAA Technical Memorandum NWS TPC-4, August 2005, p. 8.

11. *A Nation Still Unprepared*, p. 14-1.

12. Interview with Karen Keefer, May 13, 2006.

13. Interview with Michael Brown, July 7, 2006.

14. Interview with Joseph Allbaugh, June 29, 2006.

15. Richard Sylves and William R. Cumming, "FEMA's Path to Homeland Security: 1979–2003," *Journal of Homeland Security and Emergency Management* 1, no. 2, p. 1.

16. Interview with Jane Bullock, May 9, 2006.

17. National Hurricane Center, *The Deadliest Atlantic Tropical Cyclones, 1492–1996*, NOAA Technical Memorandum NWS NHC 47, May 28, 1995 (revised April 22, 1997).

18. NOAA Technical Memorandum NWS TPC-4, p. 8.

19. ABC News, *World News Tonight*, transcript, September 28, 1989.

20. California Governor's Office of Emergency Services, "California Governor's Office of Emergency Services: Origins and Development, a Chronology, 1917–1999," http://www.oes.ca.gov (see "About OES" page).

21. U.S. General Accounting Office, *Disaster Management: Recent Disasters Demonstrate the Need to Improve the Nation's Response Strategy*, testimony of J. Dexter Peach, assistant comptroller general, before the U.S. Senate Committee on Armed Services' Subcommittee on Nuclear Deterrence, Arms Control and Defense Intelligence, May 25, 1993, p. 3.

22. Federal Emergency Management Agency, *Federal Response Plan*, 9230.1-PL, April 1999 (supersedes FEMA 229, April 1992), p. 2.

23. Bill McAllister, "Appropriations Report Calls FEMA 'A Political Dumping Ground,'" *Washington Post*, July 31, 1992.

24. National Hurricane Center, *Preliminary Report: Hurricane Andrew, 16–28 August 1992* (February 7, 2005 addendum), table 3a, http://www.nhc.noaa.gov/1992andrew.html.

25. David Barstow, Bill Adair, Chris Lavin and Carol Marbin, "Federal Troops Fly Today to Disaster That Is Dade," *St. Petersburg Times*, August 28, 1992.

26. Federal Emergency Management Agency, Office of the Inspector General, *FEMA's Disaster Management Program: A Performance Audit after Hurricane Andrew*, January 1993, p. 18.

27. *A Bill to Transfer the Functions of the Director of the Federal Emergency Management Agency to the Secretary of Defense*, H.R. 867, February 4, 1993.

28. Interview with Mike Austin, May 4, 2006.

29. Interview with Mark Merritt, May 24, 2006.

30. Interview with James Lee Witt, May 9, 2006.

31. White House Press Office, "Remarks by the President and Federal, State and Local Officials in Discussion on Disaster Relief," transcript, January 19, 1994.

32. Joseph Allbaugh, director, Federal Emergency Management Agency, *Review of EPA and FEMA Responses to the September 11, 2001 Attacks*, testimony before the Senate Committee on Environment and Public Works, September 24, 2002.

33. *A Nation Still Unprepared*, p. 15.

34. Federal Emergency Management Agency, National Response Coordination Center,

"Katrina Becomes a Category Three Hurricane, Aims Towards Gulf Coast," *FEMA National Situation Report,* August 27, 2005.

35. E-mail from Michael Brown, director, Federal Emergency Management Agency, to Craig Fugate, director, Florida Division of Emergency Management, August 27, 2005.

36. Federal Emergency Management Agency, National Response Coordination Center, "Hurricane Katrina Upgraded to Category 4," *FEMA National Situation Report,* August 28, 2005.

37. Interview with Leo Bosner, May 2, 2006.

38. Interview with Terry Ebbert, May 25, 2006.

39. *A Failure of Initiative,* p. 359.

40. *A Nation Still Unprepared,* p. 7.

41. Department of Homeland Security, Office of Inspector General, Office of Inspections and Special Reviews, *A Performance Review of FEMA's Disaster Management Activities in Response to Hurricane Katrina,* March 31, 2006, p. 2.

42. Sean R. Fontenot, former chief, Planning and Preparedness divisions, Louisiana Office of Homeland Security and Emergency Preparedness, *Preparing for a Catastrophe: The Hurricane Pam Exercise,* testimony before the U.S. Senate Homeland Security and Governmental Affairs Committee, January 24, 2006; and *A Failure of Initiative,* pp. 81–83.

43. *A Nation Still Unprepared,* p. 8-3.

44. Ibid., p. 8-6.

45. James Varney, "Contraflow Evacuation a Hurricane Triumph; 1 Million Moved in 38-Hour Period," New Orleans *Times-Picayune,* May 28, 2006.

46. City of New Orleans, Office of Emergency Preparedness, *Comprehensive Emergency Management Plan 2004,* p. 55.

47. *A Nation Still Unprepared,* p. 11-12n80.

48. Ibid., pp. 5-6 and 5-7.

49. C. Ray Nagin, mayor, City of New Orleans, *Hurricane Katrina: Managing the Crisis and Evacuating New Orleans,* testimony before the Senate Committee on Homeland Security and Governmental Affairs, Federal News Service transcript, February 1, 2006.

50. *A Nation Still Unprepared,* p. 22-5.

51. *A Failure of Initiative,* p. 164.

52. Timothy Bayard, commander, New Orleans Police Department, Vice Crimes/Narcotics Section, *Hurricane Katrina: Urban Search and Rescue in a Catastrophe,* testimony before the Senate Committee on Homeland Security and Governmental Affairs, January 30, 2006.

53. Interview with Mark C. Smith, May 18, 2006.

54. Interview with William Doran, May 18, 2006.

55. *A Nation Still Unprepared,* p. 6-5.

56. Associated Press, "Paulison Sworn in as FEMA Chief," June 8, 2006.

57. U.S. Reps. Tom Davis, Don Young, James Oberstar, Bill Shuster, Zach Wamp and U.S. Del. Eleanor Holmes Norton, "Removing FEMA from the Department of Homeland Security," news conference transcript, CQ Transcriptions, May 9, 2006.

58. Michael Chertoff, secretary, U.S. Department of Homeland Security, *National Emergency Management Structure,* testimony before the Senate Committee on Homeland Security and Governmental Affairs, CQ Transcriptions, June 8, 2006.

59. Witt interview, May 9, 2006.

60. Interview with Annie and Greland Washington, May 23, 2006.

61. Interview with Soleil Rodrigue, May 23, 2006.

62. U.S. District Court, Eastern District of Louisiana, order, *Beatrice B. McWaters et al. v. Federal Emergency Management Agency et al.,* civil action no. 05-5488, sec. "K" (3), June 16, 2006, p. 29.

63. Ibid., p. 30.

64. Gregory D. Kutz, managing director, and John J. Ryan, assistant director, Forensic Audits and Investigations, U.S. Government Accountability Office, *Hurricanes Katrina and Rita Disaster Relief: Improper and Potentially Fraudulent Individual Assistance Payments Estimated to Be Between $600 Million and $1.4 Billion,* testimony before the House Committee on Homeland Security's Subcommittee on Investigations, GAO-06-844T, June 14, 2006, p. 22.

65. Federal Emergency Management Agency, "Americorps*NCCC Dedicated to Gulf Coast Recovery," news release, June 1, 2006.

66. *New York Times,* "Keeping the Faith with AmeriCorps," June 17, 2006.

67. Brig. Gen. Don Scott (retired), Lt. Gen. Andrew P. Chambers (retired), Col. Fred Peters (retired), "Congress Should Rethink Cutting Service Corps," letter to the editor, *Washington Times,* June 5, 2006.

## FIVE: SOCIAL SERVICES

1. Interview with Brian Greene, former executive director; Natalie Jayroe, president; and Jenny Rodgers, development and public relations director, Second Harvest Food Bank of Greater New Orleans and Acadiana, April 27, 2006.

2. Interview with Patricia Ann and Jeffrey Hills, May 5, 2006.

3. City of New Orleans, *Comprehensive Management Plan 2005,* Hurricane Annex, phase III, sec. IIB.

4. Ibid.

5. Interview with Terry Ebbert, May 12, 2006.

6. U.S. Senate Committee on Homeland Security and Governmental Affairs, *Hurricane Katrina: A Nation Still Unprepared,* May 2006, p. 401, http://hsgac.senate.gov/index.cfm?Fuseaction=Links.Katrina.

7. Interview with Maj. Richard Brittle, New Orleans area commander, Salvation Army, May 12, 2006.

8. Interview with Kay Wilkins, chief executive officer, southeast Louisiana chapter, American Red Cross, May 3, 2006.

9. Interview with Malik Rahim, April 29, 2006.

10. Greene, Jayroe and Rodgers interview, April 27, 2006.

11. Ibid.

12. Interview with Pamela Smith, May 1, 2006.

13. Wilkins interview, May 3, 2006.

14. Ibid.

15. Interview with Eric Cager, May 1, 2006.

16. Ibid.

17. Interview with Jeff Arnold, May 12, 2006.

18. Wilkins interview, May 3, 2006, and e-mail to author, July 31, 2006.

19. Wilkins interview, May 3, 2006.

20. Interview with Bob Howard, communication manager, American Red Cross, Hurricane Recovery Program, May 22, 2006.

21. Ebbert interview, May 12, 2006.

22. Wilkins interview, May 3, 2006.

23. E-mail to author from Wilkins, May 9, 2006.

24. Wilkins interview, May 3, 2006.

25. Ibid.

26. CNN, *Larry King Live*, transcript, September 2, 2005.

27. E-mail from Joe Becker to author, via Bob Howard, May 18, 2006.

28. E-mail from Mark Smith to author, May 24, 2006.

29. Ibid.; and interview with Mark Smith, May 24, 2006.

30. Interview with Vic Howell, May 25, 2006.

31. U.S. House Committee on Homeland Security Democratic Staff, *Trouble Exposed: Katrina, Rita, and the Red Cross: A Familiar History,* December 15, 2005, p. 3.

32. Ibid., p. 4.; and U.S. Army Corps of Engineers, Mobile District, "District Responds to Elba Flooding," vol. 20, no. 1, April–May 1998.

33. Interviews with U.S. Rep. Bennie Thompson, May 30 and August 2, 2006.

34. Chad Terhune, "Along Battered Gulf, Katrina Aid Stirs Unintended Rivalry. Salvation Army Wins Hearts, Red Cross Faces Critics; Two Different Missions," *Wall Street Journal,* September 29, 2005.

35. Interview with Kenny Claiborne, April 19, 2006.

36. Interview with Mark Jones, public relations director, Salvation Army, Alabama, Louisiana and Mississippi Division, May 1, 2006.

37. Ibid.

38. E-mail from Mark Jones to author, May 15, 2006.

39. Jones interview, May 1, 2006.

40. Steven Rathgeb Smith, "Rebuilding Social Welfare Services after Katrina," *After Katrina: Public Expectation and Charities' Response,* Urban Institute, May 2006, p. 7, http://www.urban.org/UploadedPDF/311331_after_katrina.pdf.

41. Ibid., p. 6.

42. Cynthia Fagnoni, managing director, Education, Workforce and Income Security Issues, U.S. Government Accountability Office, *Hurricanes Katrina and Rita: Provision of Char-*

*itable Assistance,* testimony before House Committee on Ways and Means' Subcommittee on Oversight, GAO-06-297T, December 13, 2005, p. 9, www.gao.gov/new.items/d06297t.pdf.

43. Ibid., p. 10.

44. Interview with Joe Leonard Jr., July 12, 2006.

45. E-mail from Rick Pogue to author, via Bob Howard, June 5, 2006.

46. American Red Cross, *From Challenge to Action: American Red Cross Actions to Improve and Enhance Its Disaster Response and Related Capabilities for the 2006 Hurricane Season and Beyond,* June 2006, pp. 8–9, http://www.redcross.org/static/file_cont5448_lango_2006.pdf.

47. U.S. Senate Bipartisan Task Force on Funding Disaster Relief, *Federal Disaster Assistance,* March 15, 1995, p. 47; and American Red Cross, "The Federal Charter of the American Red Cross," http://www.redcross.org/museum/history/charter.asp.

48. American Red Cross, "A Brief History of the American Red Cross," http://www.redcross.org/museum/history/brief.asp.

49. Congressional Research Service, *Disaster Response and Appointment of a Recovery Czar: The Executive Branch's Response to the Flood of 1927,* October 25, 2005, p. 5, http://www.fas.org/sgp/crs/misc/RL33126.pdf.

50. U.S. Department of Homeland Security, *The National Response Plan,* December 2004, http://www.dhs.gov/xlibrary/assets/NRP_FullText.pdf.

51. American Red Cross, "American Red Cross Key Part of National Response Plan," news release, January 6, 2005, www.redcross.org/pressrelease/0,1077,0_489_3922,00.html.

52. U.S. House Select Bipartisan Committee to Investigate the Preparation for and Response to Hurricane Katrina, *A Failure of Initiative: The Final Report of the Select Bipartisan Committee to Investigate the Preparation for and Response to Hurricane Katrina,* 109th Cong., 2nd sess., February 15, 2006, p. 343, http://katrina.house.gov/full_katrina_report.htm.

53. Daniel Borochoff, president, American Institute of Philanthropy, *Hearing to Review the Response by Charities to Hurricane Katrina,* testimony before House Committee on Ways and Means' Subcommittee on Oversight, December 13, 2005, http://waysandmeans.house.gov/hearings.asp?formmode=view&id=4573.

54. Congressional Research Service, *Homeland Security: 9/11 Victim Relief Funds,* updated March 27, 2003, pp. 9–10, http://www.law.umaryland.edu/marshall/crsreports/crsdocuments/RL31716.pdf.

55. Ibid., p. 12.

56. Mary Pat Flaherty and Gilbert M. Gaul, "Red Cross Has Pattern of Diverting Donations: Practice Was Used at Least 11 Years," *Washington Post,* November 19, 2001.

57. Office of Minnesota Attorney General, *After the Floods: Unmet Needs and Unspent Donations,* December 1998, pp. iii, 5.

58. U.S. Sen. Chuck Grassley, chairman, Senate Committee on Finance, "Grassley Questions Red Cross Board on Its Practices, Effectiveness," news release, December 29, 2005, http://www.senate.gov/~finance/press/Gpress/2005/prg122905.pdf.

59. American Red Cross, "President and CEO Dr. Bernadine Healy to Leave the American Red Cross," news release, October 26, 2001, www.redcross.org/press/other/ot_pr/011026healy.html.

60. E-mail from Bill George to David McLaughlin (via Ruth Sorrells) et al., October 29, 2001, Senate Committee on Finance file of American Red Cross communications, p. 221, http://finance.senate.gov/sitepages/breakdown/fileoneb.pdf.

61. American Red Cross, "Marsha J. Evans Steps Down as American Red Cross President and CEO," news release, December 13, 2005, http://www.redcross.org/pressrelease/0,1077,0_314_4983,00.html.

62. E-mail from Bob Howard to author, May 31, 2006.

63. American Red Cross, *Challenged by the Storms: The American Red Cross Response to Hurricanes Katrina, Rita and Wilma*," 2006, pp. 6–7, http://www.redcross.org/sponsors/drf/stewardship/HurrStewRep06.asp.

64. Ibid.

65. Wilkins interview, May 3, 2006.

66. Interview with Stephen Bradberry, May 3, 2006.

67. Interview and e-mail from Bob Howard to author, January 22, 2007.

68. American Red Cross response to letter from Sen. Chuck Grassley, March 29, 2006, pp. 8–12, http://www.senate.gov/~finance/press/Gpress/2005/prg040306a.pdf.

69. American Red Cross, "Red Cross Statement on Fraud at the Bakersfield Call Center," news release, December 27, 2005, http://www.redcross.org/pressrelease/0,1077,0_314_5029,00.html; and e-mail from Bob Howard to author, May 12, 2006.

70. Interview with Jaclyn Lesch, spokeswoman, U.S. Department of Justice, January 22, 2007.

71. Interview with Bob Patience, May 3, 2006.

72. Interview with Pat Glorioso, May 1, 2006.

## SIX: HEALTH CARE

1. Interviews with Scott Delacroix Jr., May 8 and June 27, 2006.

2. U.S. Senate Committee on Homeland Security and Governmental Affairs, *Hurricane Katrina: A Nation Still Unprepared*, p. 24-8, http://hsgac.senate.gov/_files/Katrina/FullReport.pdf.

3. U.S. House Select Bipartisan Committee to Investigate the Preparation for and Response to Hurricane Katrina, *A Failure of Initiative: Final Report of the Select Bipartisan Committee to Investigate the Preparation for and Response to Hurricane Katrina*, 109th Cong., 2nd sess., H. Rep. 109–377, p. 275, http://katrina.house.gov/full_katrina_report.htm.

4. Ibid., p. 4.

5. Scott Delacroix Jr., "In the Wake of Katrina: A Surgeon's First-Hand Report of the New Orleans Tragedy," *Medscape General Medicine*, September 19, 2005.

6. David Shatz, "Readers' Responses to 'In the Wake of Katrina: A Surgeon's First-Hand Report of the New Orleans Tragedy,'" *Medscape General Medicine*, December 8, 2005.

7. Interview with David Shatz, July 24, 2006.

8. The White House, *The Federal Response to Hurricane Katrina: Lessons Learned,* February 2006, p. 39, http://www.whitehouse.gov/reports/katrina-lessons-learned.pdf.

9. Interview with Gordon Bergh, May 22, 2006.

10. Letter from Gordon Bergh to Blake Gottesman, September 12, 2005.

11. *A Failure of Initiative,* p. 288.

12. *A Nation Still Unprepared,* p. 24-17.

13. Interviews with Mary Moises Duckert, May 10–12, 2006.

14. U.S. Department of Homeland Security, "National Disaster Medical System," http://www.oep-ndms.dhhs.gov/; and Congressional Research Service, *Hurricane Katrina: The Public Health and Medical Response,* September 21, 2005, pp. CRS-4, CRS-12 and CRS-13.

15. *A Failure of Initiative,* p. 271; Federal Emergency Management Agency, "NDMS Serves the Needs of Hurricane Victims," news release, December 8, 2005; and e-mails from Deborah Wing, FEMA public affairs officer, to author, June 27 and 29, 2006.

16. *A Nation Still Unprepared,* pp. 24-15, 24-16, 24-18 and 24-39; and *A Failure of Initiative,* p. 293.

17. *A Failure of Initiative,* p. 288.

18. OR-2 Disaster Medical Assistance Team, "Hurricane Katrina—After Action Report," September 25, 2005.

19. U.S. House Committee on Government Reform – Minority Staff, Special Investigations Division, *The Decline of the National Disaster Medical System,* December 2005, pp.16–17; and *A Nation Still Unprepared,* p. 24-9.

20. Rosemary Speers, Ted Jaditz, Monica Giovachino and Deborah L. Jonas, *Assessing NDMS Response Team Readiness: Focusing on DMATs, NMRTs, and the MST,* CNA Corporation, October 2002, pp. 2, 4.

21. *The Decline of the National Disaster Medical System,* p. 7.

22. *The Federal Response to Hurricane Katrina,* p. 47.

23. Monica Giovachino, Elizabeth Myrus, Dawn Nebelkpof and Eric Trabert, *Hurricanes Frances and Ivan: Improving the Delivery of HHS and ESF#8 Support,* CNA Corporation, February 2005, pp. 17, 19; and William L. Devir, *2004 Hurricane AAR's: NDMS Conference,* Federal Emergency Management Agency, May 3, 2005, pp. 22, 34.

24. Jeffrey A. Lowell, *Medical Readiness Responsibilities and Capabilities: A Strategy for Realigning and Strengthening the Federal Medical Response,* Department of Homeland Security, January 3, 2005, pp. 2, 6, 6-3 and 6-8.

25. Interview with Deborah Wing, July 17, 2006.

26. Interviews with Jeffrey Lowell, April 19 and April 25, 2006.

27. Louisiana Recovery Authority, *Emergency Preparedness and Disaster Planning White Paper,* March 2006, p. 9.

28. Dennis O'Leary, president, Joint Commission on Accreditation of Healthcare Organizations, *A Review of Federal Bioterrorism Preparedness Programs from a Public Health Perspective,* testimony before the House Committee on Energy and Commerce's Subcommittee on Oversight and Investigations, October 10, 2001, p. 50.

29. Bring New Orleans Back, Health and Social Services Committee, *Report and Recommendations to the Commission,* January 18, 2006, p. 13.

30. John R. Paradise, "Atmospheric Pressure," *NFPA Journal,* July/August 2002; and Risk Management Solutions, "Tropical Storm Allison, June 2001," pp. 9–10.

31. *A Nation Still Unprepared,* p. 24-34n2.

32. Federal Emergency Management Agency, *Combined Catastrophic Plan for Southeast Louisiana and the New Madrid Seismic Zone: Scope of Work FY 2004,* p. 6.

33. *Emergency Preparedness and Disaster Planning White Paper,* p. 6.

34. *A Nation Still Unprepared,* p. 24-5.

35. Louisiana Office of the Attorney General, "Nursing Home Owners Surrender to Medicaid Fraud Control Unit Investigators," news release, September 14, 2005.

36. Joseph A. Donchess, executive director, Louisiana Nursing Home Association, *Challenges in a Catastrophe: Evacuating New Orleans in Advance of Hurricane Katrina,* testimony before the Senate Homeland Security and Governmental Affairs Committee, January 31, 2006; and *The Federal Response to Hurricane Katrina,* p. 180.

37. *A Failure of Initiative,* p. 302; and PricewaterhouseCoopers, "Report on Louisiana Healthcare Delivery and Financing System," April 2006, p. 20.

38. *Report and Recommendations to the Commission,* p. 2.

39. U.S. Government Accountability Office, "Hurricane Katrina: Status of the Health Care System in New Orleans and Difficult Decisions Related to Efforts to Rebuild It Approximately 6 Months after Hurricane Katrina," GAO-06-576R, March 28, 2006, p. 5; and Louisiana Department of Health and Hospitals, "Health Standards: Hurricanes Katrina and Rita," http://www.dhh.louisiana.gov/offices/page.asp?id=112&detail=5185.

40. Donald R. Smithburg, executive vice president, Louisiana State University Health Care Services Division, *Hospital Disaster Preparedness,* testimony before the House Committee on Energy and Commerce's Subcommittee on Oversight and Investigations, January 26, 2006.

41. Interviews with Ben deBoisblanc, April 19 and May 8, 2006.

42. Interview with Peter DeBlieux, May 8, 2006.

43. *A Failure of Initiative,* p. 285.

44. Interview with Jewel Willis, May 30, 2006.

45. *A Nation Still Unprepared,* p. 24-10.

46. Hospital Corporation of America, "HCA Completes Air Lift Evacuation at Tulane University Hospital and Clinic; Assists Nearby Hospitals," news release, September 2, 2005.

47. Renee Goux, chief executive officer, Memorial Hospital Tenet Healthcare Corporation, *Hospital Disaster Preparedness,* testimony before the House Committee on Energy and Commerce's Subcommittee on Oversight and Investigations, January 26, 2006.

48. Interviews with Steven Campanini, senior director of media relations, Tenet Healthcare Corporation, August 29 and September 5, 2006; and Tenet Healthcare Corporation, "Tenet Responds to CNN Broadcast of March 8," news release, March 15, 2006, http://www

.tenethealth.com/TenetHealth/PressCenter/PressReleases/Tenet+Responds+to+CNN+Broadcast+of+March+8.htm.

49. Louisiana Office of the Attorney General, "Attorney General Charles C. Foti, Jr. Announces Arrests in Memorial Hospital Deaths Investigation," news release, July 18, 2006; and Orleans Criminal District Court, *State of Louisiana v. Anna M. Pou, Lori L. Budo and Cheri A. Landry,* affidavit.

50. Jeffrey Meitrodt, "N.O. Coroner Finds No Evidence of Homicide," New Orleans *Times-Picayune,* February 1, 2007.

51. Goux testimony, January 26, 2006.

52. Interviews with Nyla Houston, May 25 and June 28, 2006.

## SEVEN: POLITICS

1. Agence France-Presse, "Iraq-Tested Soldiers in New Orleans with Shoot to Kill Orders," September 2, 2005.

2. Bill Walsh, "Foster Remark Causes Furor," New Orleans *Times-Picayune,* November 10, 1995.

3. U.S. Census Bureau, *2004 American Community Survey,* "New Orleans City, Louisiana—Selected Economic Characteristics: 2004," http://factfinder.census.gov.

4. Michael Barone and Richard E. Cohen, *The Almanac of American Politics,* 2006 ed. (Washington, D.C: National Journal, 2005), p. 739.

5. Interview with James Gill, May 9, 2006.

6. Amanda Ripley et al., "Where the System Broke Down; the Mayor: Did C. Ray Nagin Do Everything He Could to Save His City?" *Time,* September 19, 2005, p. 36.

7. Bruce Alpert, "La. Will Be Honest, Panel Is Told," New Orleans *Times-Picayune,* October 19, 2005.

8. John Maginnis, *The Last Hayride* (Baton Rouge, La.: Gris Gris Press, 1984), p. 2.

9. Richard D. White Jr., *Kingfish: The Reign of Huey P. Long* (New York: Random House, 2006), p. 224.

10. Tyler Bridges, *Bad Bet on the Bayou* (New York: Farrar, Straus and Giroux, 2001), p. 4.

11. V. O. Key Jr., *Southern Politics in State and Nation* (New York: Knopf, 1949), p. 156.

12. *Bad Bet on the Bayou,* p. 4.

13. U.S. District Court, District of Columbia, search warrant (office of U.S. Rep. William Jefferson), case no. 06-231 M-01, issued May 18, 2006.

14. *Kingfish,* p. 26.

15. Douglas Brinkley, *The Great Deluge* (New York: Morrow, 2006), p. 45.

16. *Kingfish,* pp. 201–2.

17. Ibid., p. 202.

18. Garry Boulard, *Huey Long Invades New Orleans: The Siege of a City, 1934–36* (Gretna, La.: Pelican, 1998), pp. 116–19.

19. James Gill, *Lords of Misrule: Mardi Gras and the Politics of Race in New Orleans* (Jackson: University Press of Mississippi, 1997), p. 196.

20. Ibid, p. 197.

21. Ibid, p. 196.

22. Wayne Parent, *Inside the Carnival: Unmasking Louisiana Politics* (Baton Rouge: Louisiana State University Press, 2004), p. 109.

23. Bill Walsh, "Talking Back," New Orleans *Times-Picayune,* March 16, 1994.

24. Kevin McGill, "Bulldozed Barriers Will Be Rebuilt; Mayor Vows Court Fight," Associated Press, February 22, 1987.

25. Janet McConnaughey, "Suburb Backs Down on Vow to Rebuild 'Wall,'" Associated Press, February 23, 1987.

26. Associated Press, "Truce Called in Conflict over Street Barricades," February 24, 1987.

27. Lee Hancock and Michael Grabell, "'Desperate SOS' Amid Hunger, Thirst and Lawlessness, Frustrations Boil Over in New Orleans," *Dallas Morning News,* September 2, 2005.

28. Interview with Kristina Ford, May 17, 2006.

29. U.S. District Court, Eastern District of Louisiana (New Orleans), *United States v. Glenn A. Haydel,* criminal docket, case no. 2:05-cr-00242-SRD-DEK-1.

30. U.S. District Court, Eastern District of Louisiana (New Orleans), *United States v. Lillian T. Smith Haydel,* criminal docket, case no. 2:04-cr-00197-MVL-ALC-1.

31. Louisiana State Legislature, Act 863, 2006 Regular Session, http://legis.state.la.us.

32. Doug Simpson, "Katrina Disaster Seen as Chance to Fix New Orleans Bureaucracy," Associated Press, February 9, 2006.

33. Louisiana State Legislature, Act 43, 2006 First Extraordinary Session, http://legis.state.la.us.

34. Interview with John Maginnis, week of May 7, 2006.

35. Peirce F. Lewis, *New Orleans: The Making of an Urban Landscape,* 2nd ed. (Charlottesville: University of Virginia Press, 2003), p. 68.

36. Ford interview, May 17, 2006.

37. Louisiana State Legislature, Act 621, 2006 Regular Session, http://legis.state.la.us.

38. New Orleans *Times-Picayune,* "Cops Charged with Murder," chronology, March 5, 1995.

39. Interview with Mary Howell, week of May 7, 2006.

40. James Varney, "N.O. Cops Reported to Take Cadillacs from Dealership," New Orleans *Times-Picayune,* September 29, 2005.

41. Howell interview, week of May 7, 2006.

42. Mary Foster, "New Orleans Faces Public Defender Crisis," Associated Press, February 23, 2006.

43. Gwen Filosa, "Judge Vows to Free Untried Inmates," New Orleans *Times-Picayune,* July 29, 2006.

44. CNN, *CNN Live Today,* transcript, May 24, 2006.

45. Louisiana Secretary of State, "May 20, 2006 Orleans Run-Off Election: Mayor, City of New Orleans," http://www.gcr1.com/electionscentral.

46. Interview with Silas Lee, week of June 11, 2006.

47. Baodong Liu, "Whites as a Minority and the New Biracial Coalition in New Orleans and Memphis," *PS: Political Science and Politics* 39, no. 1 (January 2006), p. 71.

48. Laura Maggi and Frank Donze, "Nagin, Blanco Meet to Bury Hatchet," New Orleans *Times-Picayune*, May 24, 2006.

49. Interview with Jim Carvin, June 13, 2006.

50. Interview with James Gill, July 28, 2006.

## EIGHT: HOUSING AND INSURANCE

1. Speech by President George W. Bush, Jackson Square, New Orleans, September 15, 2005, http://www.whitehouse.gov/news/releases/2005/09/print/20050915-8.html.

2. Interview with Fred Yoder, May 30, 2006.

3. Innovative Emergency Management, *Southeast Louisiana Catastrophic Hurricane Functional Plan*, September 20, 2004, p. 83.

4. U.S. Department of Housing and Urban Development, Office of Policy Development and Research, *Current Housing Unit Damage Estimates*, February 12, 2006 (revised April 7, 2006), p. 23.

5. Interview with Arnold Hirsch, May 23, 2006.

6. Interview with Michael Centineo, director, New Orleans Department of Safety and Permits, August 1, 2006.

7. Interview with John Lockwood, May 30, 2006.

8. Gwen Filosa, "HUD Chief Approves Road Home Grants," New Orleans *Times-Picayune*, May 31, 2006.

9. U.S. Department of Housing and Urban Development, "Jackson Approves Louisiana's $4.6 Billion 'Road Home Program,'" news release, May 30, 2006.

10. Bring New Orleans Back, Urban Planning Committee, "Action Plan for New Orleans," January 11, 2006, p. 61, fig. 43.

11. HUD news release, May 30, 2006.

12. U.S. Sen. Mary L. Landrieu, "Landrieu Statement on Passage of Hurricane Relief Bill," June 15, 2006, http://landrieu.senate.gov/~landrieu/releases/06/2006616630.html.

13. Interview with Natalie Wyeth, press secretary, Louisiana Recovery Authority, July 18, 2006.

14. Interview with LaToya Cantrell, May 12, 2006.

15. U.S. Army Corps of Engineers, "Corps Revising N.O. Flood-Insurance Maps for FEMA," news release, June 7, 2004.

16. U.S. Department of Homeland Security, "Powell, FEMA Release New Orleans Advisory Flood Data: U.S. Army Corps of Engineers Revises Cost Estimates for Levees," news release, April 12, 2006.

17. Ibid.

18. Federal Emergency Management Agency, "Advisory Base Flood Elevations for Orleans Parish, Louisiana," April 12, 2006.

19. Interview with Walter Leger, May 31, 2006.

20. Interview with Butch Kinerney, FEMA mitigation specialist, July 10, 2006.

21. John Schwartz and Adam Nossiter, "Complex Equation Determined Rules for Rebuilding in New Orleans, Federal Officials Say," *New York Times*, April 14, 2006.

22. Kinerney interview, July 10, 2006.

23. Interview with Nathan Kuhle, May 31, 2006.

24. Interview with David Winkler-Schmit, May 31, 2006.

25. William Niskanen, chairman, Cato Institute, *The Commerce and Consumer Protection Implications of Hurricane Katrina*, testimony before the House Energy and Commerce Committee's Commerce, Trade, and Consumer Protection Subcommittee, September 22, 2005.

26. Insurance Information Institute, "Facts and Statistics: National Flood Insurance Program," http://www.iii.org/media/facts/statsbyissue/flood/.

27. William O. Jenkins Jr., director, Homeland Security and Justice Issues, U.S. Government Accountability Office, *National Flood Insurance Program: Oversight of Policy Issuance and Claims*, testimony before the House Committee on Financial Services' Subcommittee on Housing and Community Opportunity, GAO-05-532T, April 14, 2005, p. 1.

28. David M. Walker, comptroller general of the United States, U.S. Government Accountability Office, *Federal Emergency Management Agency: Challenges for the National Flood Insurance Program*, testimony before the Senate Committee on Banking, Housing and Urban Affairs, GAO-06-335T, January 25, 2006, p. 7.

29. Federal Emergency Management Agency, *National Flood Insurance Program, Program Description*, August 1, 2002, p. 6.

30. E-mail from Butch Kinerney to author, July 18, 2006.

31. Jenkins testimony, April 14, 2005, p. 2.

32. Federal Emergency Management Agency, *NFIP Losses by State*, "Totals for Fiscal Year 2005," May 31, 2006.

33. Jenkins testimony, April 14, 2005, p. 2.

34. Walker testimony, January 25, 2006, p. 4.

35. Emile J. Brinkmann and Wade D. Ragas, *An Estimate of the Cost of Hurricane Katrina Flood Damage to Single Family Residential Structures in Orleans Parish*, Mortgage Bankers Association, February 2006, p. 9.

36. Interview with Cora Charles, May 29, 2006.

37. *An Estimate of the Cost of Hurricane Katrina Flood Damage*, p. 9.

38. Interview with J. Robert Hunter, May 24, 2006.

39. JayEtta Hecker, director, Physical Infrastructure Issues, U.S. General Accounting Office, *Flood Insurance: Emerging Opportunity to Better Measure Certain Results of the National Flood Insurance Program*, testimony before the U.S. Senate Committee on Appropriations' Subcommittee on VA, HUD and Independent Agencies, GAO-01-736T, May 15, 2001, p. 4.

40. U.S. Army Corps of Engineers news release, June 7, 2004.

41. Kinerney interview, July 10, 2006.

42. *An Estimate of the Cost of Hurricane Katrina Flood Damage*, p. 9.

43. Interview with Jerry Schulin, May 30, 2006.

44. City of New Orleans, "FEMA Flood Zones" map of Lakeview area, https://secure.cityofno.com/Resources/Portal37/zip_70124.pdf.

45. Hecker testimony, May 15, 2001, p. 3.

46. U.S. Government Accountability Office, *Federal Emergency Management Agency: Improvements Needed to Enhance Oversight and Management of the National Flood Insurance Program*, GAO-06-119, October 18, 2005, p 16.

47. Walker testimony, January 25, 2006, p. 12.

48. Jenkins testimony, April 14, 2005, pp. 6–7.

49. Interview with Pat Hemard, May 26, 2006.

50. Interview with Julie Quinn, May 26, 2006.

51. Louisiana House of Representatives, Insurance Committee and member personal pages, http://house.louisiana.gov/H-Reps/Cmte_IN.asp and http://house.legis.state.la.us/.

52. Monroe (La.) *News-Star*, "La.'s Chief Insurance Regulator to Step Down," February 3, 2006; and Federal Bureau of Prisons Inmate Locator, http://www.bop.gov/iloc2/LocateInmate.jsp.

53. Interview with Loretta Worters, July 7, 2006.

54. Louisiana Department of Insurance, "Consumer Complaints Lead to Probe of Hurricane Claims Handling Practices of Two Insurers," news release, May 17, 2006.

55. E-mail from Orice M. Williams, director, Financial Markets and Community Investment, U.S. Government Accountability Office, to author, August 9, 2006.

56. Quinn interview, May 26, 2006.

57. James J. Donelon, commissioner, Louisiana Department of Insurance, "Advisory Letter 06-04," June 5, 2006.

58. Louisiana Department of Insurance, "Over 99% of Homeowners Insurance Market Signs on to Extend Prescription," news release, August 14, 2006.

59. Congressional Research Service, *Hurricanes and Disaster Risk Financing through Insurance: Challenges and Policy Options*, March 25, 2005, p. 12.

60. Greater New Orleans Community Data Center, "Metro New Orleans Fair Market Rent History," 2006, http://www.gnocdc.org/reports/fair_market_rents.html.

61. Leger interview, May 31, 2006.

62. Wyeth interview, July 18, 2006.

63. Housing Authority of New Orleans, "Post-Katrina Frequently Asked Questions," May 11, 2006, p. 1.

64. U.S. Department of Housing and Urban Development, "HUD Names New Recovery Advisor and Receiver to Advance Current HANO Hurricane Recovery Efforts," news release, April 14, 2006.

65. Gwen Filosa, "Brick Housing Biting the Dust," New Orleans *Times-Picayune*, November 3, 2005.

66. Interview with William Quigley, May 29, 2006.

67. Interview with Beth Butler, head organizer Louisiana ACORN (Association of Community Organizations for Reform Now), May 11, 2006.

68. Interview with Adonis Exposé, January 19, 2007.

69. Susan Saulny, "5,000 Public Housing Units in New Orleans Are to Be Raised," *New York Times,* June 15, 2006.

70. HUD news release, June 4, 2006.

## EPILOGUE

1. Max Mayfield, director, Tropical Prediction Center / National Hurricane Center, *The Lifesaving Role of Accurate Hurricane Prediction,* testimony before the Senate Commerce, Science and Transportation Committee's Subcommittee on Disaster Prevention and Prediction, September 20, 2005, http://www.legislative.noaa.gov/Testimony/mayfieldfinal09 2005.pdf.

2. U.S. Army Corps of Engineers, "Hurricane Protection System," http://www.mvn .usace.army.mil/hps/.

3. Interview with Dan Hitchings, director, U.S. Army Corps of Engineers, Task Force Hope, June 14, 2006.

4. U.S. Army Corps of Engineers, "Hurricane Protection System: New Orleans East Bank Depth Inundation Maps (Metro Area)," http://www.mvn.usace.army.mil/hps/depth _inundation_maps_07_26.htm.

5. U.S. Army Corps of Engineers, "Hurricane Protection System: Questions and Answers, Hurricane Recovery and Levee Issues," http://www.mvn.usace.army.mil/tfh/Q%20 and%20As/Hurricane%20Recovery%20and%20Levee%20Issues.htm.

6. E-mail from Hassan Mashriqui to author, July 18, 2006.

7. Lt. Gen. Carl Strock, commander and chief engineer, U.S. Army Corps of Engineers, "Press Briefing on Gulf Coast Rebuilding" transcript, White House news release, August 22, 2006, http://www.whitehouse.gov/news/releases/2006/08/20060822-5.html.

8. Mark Schleifstein, "Corps' Coastal Report Short on Specifics," New Orleans *Times-Picayune,* July 11, 2006.

9. Ibid.

10. Interview with Al Naomi, senior project manager, U.S. Army Corps of Engineers, June 13, 2006.

11. U.S. Sen. David Vitter, "Senator Vitter Comments on Corps Report," news release, July 10, 2006, http://vitter.senate.gov.

12. Interview with Tim Searchinger, attorney, Environmental Defense, July 12, 2006; and Environmental Defense and National Wildlife Federation, "Hurricane Protection Plan by Corps of Engineers Doomed to Fail, Flood Experts Charge," news release, July 11, 2006.

13. Interview with Ivor van Heerden, June 22, 2006.

14. Amendment 4861 to S-728, *The Water Resources Development Act of 2006,* 109th Cong., 2nd sess., *Congressional Record,* vol. 152, pp. S7814-15.

15. Interview with Michael Brown, July 7, 2006.

16. Interview with Terry Ebbert, May 25, 2006.

17. Interviews with Joseph Allbaugh, June 29, 2006, and Brown, July 7, 2006.

18. Eileen Sullivan, "FEMA Reform Prompts Major Departmentwide Restructuring," *Congressional Quarterly*, January 17, 2007.

19. Mark Schleifstein, "Nagin Lays Out Evacuation Plan; Category 2 Means Everybody Out, He Says," New Orleans *Times-Picayune*, May 3, 2006.

20. City of New Orleans, "2006 Emergency Preparedness Plan," http://secure.cityofno.com/portal.aspx?portal=46&tabid=38.

21. "Nagin Lays Out Evacuation Plan."

22. Interview with Phyllis Parun, July 13, 2006.

23. Louisiana State Legislature, Act 43, 2006 First Extraordinary Session, http://www.legis.state.la.us/billdata/streamdocument.asp?did=363755.

24. Susan Saulny, "New Orleans Details Steps to Repair Its Legal System," *New York Times*, August 8, 2006.

25. Interview with Lawrence Powell, July 11, 2006.

26. Louisiana Department of Health and Hospitals, "Health Standards: Hurricanes Katrina and Rita," http://www.dhh.louisiana.gov/offices/page.asp?id=112&detail=5185.

27. U.S. Government Accountability Office, "Hurricane Katrina: Status of the Health Care System in New Orleans and Difficult Decisions Related to Efforts to Rebuild It Approximately 6 Months after Hurricane Katrina," GAO-06-576R, March 28, 2006, pp. 4–5, http://www.gao.gov/new.items/d06576r.pdf.

28. Interview with James Moises, May 9, 2006.

29. Bring New Orleans Back, Health and Social Services Committee, *Report and Recommendations to the Commission*, p. 42.

30. Interview with Peter DeBlieux, May 8, 2006.

31. Ronette King, "Flood-Ravaged Hospitals Are Diagnosing Their Needs for This Hurricane Season," New Orleans *Times-Picayune*, May 27, 2006.

32. Jaime Guillet, "New Orleans–Area Hospitals Set Forth Plans for Hurricane Evacuations," *New Orleans CityBusiness*, May 22, 2006.

33. U.S. Government Accountability Office, *Disaster Preparedness: Limitations in Federal Evacuation Assistance for Health Facilities Should Be Addressed*, GAO-06-826, July 20, 2006, pp. 4–5, http://www.gao.gov/new.items/d06826.pdf.

34. Interview with Cynthia Bascetta, June 29, 2006.

35. Interview with Michael Hopmeier, April 24, 2006.

36. Joe Becker, senior vice president of preparedness and response, American Red Cross, *Hearing to Review the Response by Charities to Hurricane Katrina*, testimony before the House Committee on Ways and Means' Subcommittee on Oversight, December 13, 2005.

37. John Davies, president and chief executive officer, Baton Rouge Area Foundation, *Hearing to Review the Response by Charities to Hurricane Katrina*, testimony before the House Committee on Ways and Means' Subcommittee on Oversight, December 13, 2005, http://waysandmeans.house.gov/hearings.asp?formmode=view&id=4572.

38. Ibid.

39. *The Chronicle of Philanthropy,* "How the Red Cross Should Move Forward: Advice from Experts," January 12, 2006.

40. Interview with Daniel Borochoff, June 9, 2006.

41. American Red Cross, "American Red Cross, NAACP and Faith-Based Groups Partner to Provide Disaster Training Nationwide," news release, May 2, 2006, http://www.redcross.org/pressrelease/0,1077,0_314_5355,00.html.

42. Interview with Joe Leonard Jr., July 12, 2006.

43. Interview with Christy Carl, director, Galveston County Historical Museum, August 29, 2006.

44. Federal Emergency Management Agency, "Three Years after the Flood Grand Forks Surviving and Rebuilding Better," news release, April 19, 2000, http://www.fema.gov/news/newsrelease.fema?id=7439.

45. Federal Emergency Management Agency, "Advisory Base Flood Elevations for Orleans Parish, LA," April 12, 2006.

46. Ibid.

47. Interview with J. Robert Hunter, former administrator, National Flood Insurance Program, May 24, 2006.

48. Federal Emergency Management Agency, *Flood Recovery Guidance: Frequently Asked Questions,* April 12, 2006, http://www.fema.gov/pdf/hazard/flood/recoverydata/abfe_faqs_la_leveeparishes.pdf.

49. Hunter interview, May 24, 2006.

50. Interview with Greg Miller, April 24, 2006.

51. Interview with Al Naomi, senior project manager, U.S. Army Corps of Engineers, May 9, 2006.

52. Naomi interview, May 9, 2006.

53. U.S. Army Corps of Engineers, "Freshwater Diversion" brochure, www.mvn.usace.army.mil/pao/bro/FreshwaterDiversion.pdf.

54. America's WETLAND, "Resource Center: America's WETLAND Official Numbers (last updated April 18, 2006)," http://www.americaswetlandresources.com/background_facts/thenumbers/index.html.

55. Searchinger interview, January 22, 2007.

56. Interview with Roy Dokka, May 5, 2006.

# About the Authors

JENNI BERGAL is a project manager at the Center for Public Integrity. She worked as a reporter for nearly two decades at the South Florida *Sun-Sentinel* in Fort Lauderdale, where she specialized in investigative reporting about health care, social services and economic crime. She has won dozens of state and national awards, including the Worth Bingham Prize for Distinguished Reporting, the Gerald Loeb Award for Distinguished Business and Financial Journalism, the National Press Club Consumer Journalism Award and the Clark Mollenhoff Award for Excellence in Investigative Reporting. She has been a Pulitzer Prize finalist twice—once in 1996 for beat reporting and again in 1999 for investigative reporting. She lives in the Washington, D.C., area.

SARA SHIPLEY HILES is a freelance journalist who specializes in environmental issues. Her work has taken her from Louisiana's chemical corridor to the Douglas fir forests of the Northwest and the mining towns of Peru. She graduated from Loyola University in New Orleans and worked for six years for *The Times-Picayune* as a reporter and copy editor. During her 14-year journalism career, she also has worked as a staff reporter for *The Statesman Journal* of Salem, Ore., *The Courier-Journal* of Louisville, Ky., and the *St. Louis Post-Dispatch*. She has won awards for news and feature writing from organizations that include the Press Club of New Orleans and the Oregon Newspaper Publishers Association. She lives in Bowling Green, Ky.

FRANK KOUGHAN is a freelance journalist who spent eight years as an associate producer for CBS News' *60 Minutes*. He began his career as a researcher and fact-checker at a number of magazines, including *Harper's*, *Vanity Fair* and *Spy*, before going on to produce documentaries for Bill Moyers and PBS' *FRONTLINE*. He has won numerous awards, including the George Polk Award, two Investigative Reporters and Editors Awards, a duPont–Columbia Silver Baton, two Emmys and a Gerald Loeb Award for Distinguished Business and Financial Journalism. He was raised in Swampscott, Mass., and graduated from Boston College. He lives in Querétaro, Mexico.

JOHN MCQUAID is co-author of *Path of Destruction: The Devastation of New Orleans and the Coming Age of Superstorms*. He is an Open Society Institute Katrina Media Fellow who spent 22 years at *The Times-Picayune*, where he co-wrote a 2002 series on hurricane risk that anticipated a storm like Hurricane Katrina. His writing about the levee failures after Katrina was part of the newspaper's package of stories that won a 2006 Pulitzer Prize. He has worked as an investigative reporter, foreign correspondent and national political reporter and his work has won many national awards, including a Pulitzer Prize for Public Service in 1997 for a series about the global fisheries crisis and the John B. Oakes Award for Distinguished Environmental Journalism for a series about how pollution disproportionately impacts the poor. He lives in the Washington, D.C., area.

JIM MORRIS is a project manager at the Center for Public Integrity. He is a veteran journalist who specializes in coverage of industries and government agencies. He has won numerous journalism awards, including the George Polk Award, the National Association of Science Writers Award, the Sidney Hillman Foundation Award and five Headliners Foundation of Texas awards. He has worked as a deputy editor at *Congressional Quarterly*, supervising a team of five homeland security reporters, and as an investigative reporter at publications including the *Houston Chronicle*, *The Dallas Morning News*, *The Sacramento Bee* and *U.S. News and World Report*. He lives in the Washington, D.C., area.

KATY RECKDAHL is a freelance writer who has written about New Orleans since 1999. Her stories have tackled topics ranging from homelessness and HIV-positive women to Mardi Gras Indians and jazz musicians. She has won numerous awards, including a Casey Journalism Center Medal, the James Aronson Award for Social Justice Journalism and the

Press Club of New Orleans' Alex Waller Memorial Award for three years running. She also has been awarded a Katrina Media Fellowship from the Open Society Institute. On August 28, 2005, the day before Hurricane Katrina struck, she gave birth to her son, Hector, in a New Orleans hospital. She left the city two days after the storm, then lived as an evacuee in Phoenix before returning to New Orleans in July 2005.

CURTIS WILKIE grew up in southern Mississippi, near New Orleans. He graduated from the University of Mississippi in 1963 and worked as a reporter in the Mississippi Delta for six years during the height of the civil rights movement. After receiving a Congressional Fellowship in Washington, D.C., in 1969, he reported for the *News Journal* papers in Wilmington, Del. In 1975, he joined the staff of *The Boston Globe*, where he was a national and foreign correspondent for 26 years. In 2005, he received a special award for excellence in nonfiction from the Fellowship of Southern Writers. He is the author of *Dixie: A Personal Odyssey through Events that Shaped the Modern South*. He has lived in New Orleans' French Quarter since 1993 and also has a home in Oxford, Miss., where he holds the Kelly G. Cook Chair in Journalism at the University of Mississippi.

# About the Center for Public Integrity

The Center for Public Integrity is a nonprofit, nonpartisan, independent journalism organization that produces investigative reports on significant public issues to make institutional power more transparent and accountable.

Based in Washington, D.C., the Center does not develop or advocate public policy, nor does it endorse candidates or parties. It is supported by grants from major foundations and individual donors, and does not accept contributions from corporations, labor unions, governments or anonymous donors. It has a national Board of Directors and a national Advisory Board.

Since 1990, the Center has published hundreds of investigative reports and 15 books. It has received the prestigious George Polk Award and more than 20 other journalism awards from national organizations, including PEN USA and Investigative Reporters and Editors. In April 2006, the Society of Professional Journalists recognized the Center with a national award for excellence in online public service journalism for the fifth consecutive year.

The views expressed in this book do not necessarily reflect those of individual members of the Center's Board of Directors or Advisory Board.

For more information about the Center and its publications, contact:

The Center for Public Integrity
910 Seventeenth Street, N.W.
Seventh Floor
Washington, D.C. 20006

E-mail: contact@publicintegrity.org
Internet: www.publicintegrity.org
Telephone: (202) 466-1300
Fax: (202) 466-1101

# Index

| DATE | | | |
| --- | --- | --- | --- |
| | | | |
| | | | |
| | | | |
| | | | |
| | | | |
| | | | |
| | | | |
| | | | |
| | | | |
| | | | |
| | | | |
| | | | |

# THE AMERICAN SHORT STORY

# THE AMERICAN SHORT STORY

## A Critical Survey

by ARTHUR VOSS

UNIVERSITY OF OKLAHOMA PRESS : NORMAN

Library of Congress Cataloging in Publication Data

Voss, Arthur, 1910–
The American short story.
1. Bibliography: p. 357
1. Short stories, American—History and criticism.
I. Title.
PS374.S5V6 813'.01 72–9264


First edition, 1973; second printing, 1975.

**For Isabel**

# FOREWORD

AMERICAN LITERATURE is distinguished as much for its short fiction as for its novels, poetry, and plays, and it would appear that as much attention is given to the short story as to the other literary forms by teachers and critics of American writing. Yet, although there is no lack of anthologies containing American short stories or of published scholarship and criticism devoted to the work of individual American short-story writers, there exists no up-to-date, comprehensive critical survey of the American short story.

I have endeavored to fill this need in addressing this book to the student, the teacher, or the general reader seeking to increase his knowledge and appreciation of those writers who can be regarded as having contributed significantly to the development of the American short story. My chief concerns have been to provide information and critical opinion about the stories of a particular writer and how they can best be read; the literary and other influences that affected their composition; and the narrative techniques the writers employed—and how successfully—to achieve their artistic effects and make meaningful their themes. Although I have naturally given most space to the major writers, I have tried to do justice also to the writers of lesser significance who deserve to be included in a survey of this kind, writers whose work has both historical and aesthetic interest and who make manifest, along with the major figures, how truly impressive the achievement of our American short-story writers has been.

The Bibliography is included to give added reference value to the book. The entries, to many of which I am indebted for stimulating my thinking and enlarging my insights, are extensive enough that the reader should find them serviceable guides in finding particular stories he may wish to read, as well as critical commentary relating to them.

ARTHUR VOSS

*Lake Bluff, Illinois*

# CONTENTS

# THE AMERICAN SHORT STORY

The Americans have handled the short story so wonderfully that one can say that it is a national art form.

Frank O'Connor
*The Lonely Voice:*
*A Study of the Short Story*

# 1 The Beginnings of the American Short Story

## Washington Irving

ALTHOUGH SHORT PROSE NARRATIVES date back to ancient times, the short story has existed as a distinct literary form for only about a century and a half. Other countries, among them Russia, France, and England, have produced short-story writers of great distinction, but it is in America particularly that the modern short story has flourished, with writers from Irving, Hawthorne, and Poe down to the present playing a leading part in establishing and developing the form and in so doing giving us a body of literature of great richness and variety.

During the seventeenth and eighteenth centuries most American prose writing was utilitarian rather than belletristic. Authors published accounts of travel and exploration, histories, biographies and autobiographies, sermons and other religious writings, and political tracts. Toward the end of the eighteenth century Philip Freneau emerged as the first American poet (with the exception of the Puritan Edward Taylor) to write verse of undoubted literary quality as well as historical interest. During the same time the first plays by American authors were produced and the first American novels were published. The latter, most of which were influenced by English models, include *Modern Chivalry* (1792–1815), by Hugh Henry Brackenridge, an imitation of *Don Quixote* which comments humorously and satirically on the evils and excesses of the day; sentimental novels of domestic life, such as *Charlotte Temple* (1791), by Mrs. Susanna Rowson; and the Gothic novels of Charles Brockden Brown.

Brown's *Wieland* (1798), *Edgar Huntly* (1799), and several other novels belong to a school of fiction which flourished during the eighteenth century, prominent examples by English writers being Horace Walpole's *The Castle of Otranto* (1764), Mrs. Ann Radcliffe's *The Mysteries of Udolpho* (1794), and Matthew Gregory Lewis' *The Monk* (1795). Although sometimes didactic and moral in tone, these novels sought to evoke sensations of terror and horror in the reader by depicting supernatural and other sensational occurrences and actions, usually taking place in medieval castles or in other strange and wild settings which contributed importantly to the atmosphere of mystery and foreboding. Brown's novels compare favorably with those of his English contemporaries, but it was not until the 1820's, when *The Spy*, *The Pioneers*, and the other earlier novels of James Fenimore Cooper appeared, that America was to produce its first novelist of significance.

The earliest published pieces of short fiction in America also appeared late in the eighteenth century. Some of them were Gothic narratives, while others were tales and sketches in imitation of Addison, Goldsmith, and other English periodical writers. The moralizing Oriental tale so popular in England during the eighteenth century, an example of which is Addison's "The Vision of Mirzah," was copied by American writers, as were the English character sketches written to inculcate good manners and morals. A typical early American tale of this kind, "Chariessa; or A Pattern for the Sex," published in 1788 in the *Columbian Magazine,* portrays an admirable spinster whose virtues are heightened by the device of contrasting her with her mean and avaricious relatives. Other sketches of this sort bore such titles as "Constantia, or Unexampled Magnanimity," "The Benefits of Charity," and "The Triumphs of Friendship." At least one short narrative, however, was published before 1800 with some of the characteristics which have come to be commonly associated with the short story. An account of mob violence in a city of the Roman Empire in the fourth century A.D., "Thessalonica: A Roman Story" maintains a swift narrative pace and has considerable suspense and dramatic quality. One of Charles Brockden Brown's few pieces of short fiction, it appeared in May 1799, in the *Monthly Magazine*, a periodical edited by the author.

The first notable advance in the development of the short narrative in America came with the publication of *The Sketch Book* (1820) by Washington Irving (1783–1859). Irving, America's first distinguished man of letters, was born in New York City. Trained to be a lawyer, he turned instead to writing and collaborated with his brother, William, and James Kirke Paulding in bringing out a series of English *Spectator*-like essays and sketches entitled *Salmagundi* (1807–1808). In 1809 the publication of a burlesque account of the early Dutch inhabitants of his native city, entitled *Diedrich Knickerbocker's History of New York*, added considerably to his literary reputation. *The Sketch Book* came after a period of several years of nonliterary activity and was written in England, where Irving had taken up residence in 1815. It was an immediate success there as well as in America. It consists mainly of essays and sketches dealing with English subjects and themes: manners and customs, both present and past ("Rural Funerals," "Rural Life in England"), historic places ("Westminster Abbey," "Stratford on Avon"), books and authorship ("The Mutability of Literature," "The Art of Bookmaking"), and a plea that British and Americans endeavor to overcome the animosity and prejudice with which they too often regarded each other ("British Writers on America"). There are also some highly sentimental sketches ("The Broken Heart," "The Widow and Her Son"), a few miscellaneous pieces, and three tales: "Rip Van Winkle," "The Legend of Sleepy Hollow," and "The Spectre Bridegroom." The first two of these are the chief reason why Irving remains a classic, if minor, American author. Like much of Irving's other work, both are derivative, being adaptations of German legends, but they have American settings and are told in a manner which is Irving's alone.

There is perhaps no more famous story in all American literature than that of the lazy but likable Rip Van Winkle, who goes hunting one day in the Catskill Mountains to escape his termagant wife, encounters a band of strange-looking little men in old-fashioned Dutch dress playing at ninepins, drinks overmuch from a flagon of their liquor, and falls asleep, to awaken twenty years later, puzzled by his experience and confused by a much-changed world, to which, however, in time he is able very satisfactorily to adjust himself.

Deceptively simple on the surface, "Rip Van Winkle" is actually a rich and subtle story, made so by its reflections of folklore, legends, and myths and by its psychological implications. One of the most important of these has to do with Rip's sense of having lost his identity. He tells himself that "everything's changed, and I'm changed, and I can't tell what's my name, or who I am!" Might not something like this also happen to us (if it hasn't already), a sudden shocked awareness that without our realizing it time has swiftly passed and made for great change, with the result that we are not sure who or where we are? But we can also be reassured by the fact that the outcome for Rip is a happy one. His wife has died and can no longer harass him, he is old enough to be idle with impunity, and he has achieved a kind of fame through his wondrous experience, which he relates to strangers at every opportunity. Irving seems to be suggesting, in a story which charms us with its good humor, lightness of touch, and wonderfully vivid pictorial quality, that time and change should treat us all as gently as Rip.

"The Legend of Sleepy Hollow," longer and more discursive than "Rip Van Winkle," is told in the leisurely manner which is characteristic of many of Irving's tales. It is some time before the reader arrives at the main action of the tale, what with the extended but vivid portrait of Ichabod Crane (which, good-humored though it is, reflects the somewhat contemptuous attitude of the more urbane New Yorkers toward their Connecticut neighbors), the description of the tranquil, lonely Sleepy Hollow region, the account of Dutch manners and customs, and Irving's occasional digressions. In no sense, however, are Irving's description of the setting and his characterization of Ichabod padded; rather they serve to enrich the narrative and to prepare us for its outcome. But the chief triumph of the story is the way in which its main action is developed in a series of three vignettes.

First there is the picture of Ichabod making his preparations to go to the Van Tassel party and his journey there on his workhorse Gunpowder. Then comes the party, which, though it has a disastrous sequel, begins auspiciously for Ichabod, who eats largely from the abundant board, dances with his beloved Katrina Van Tassel, and listens to tales of ghosts and other apparitions—including that of the headless horseman, told by his redoubtable rival for Katrina's hand, Brom Bones.

Ichabod, in turn, regales the company with extracts from Cotton Mather's account of witchcraft in New England. Irving has already told us of Ichabod's extraordinary appetite for the marvelous, as well as his interest in the buxom Katrina and even more in the possessions she will inherit, and the fact that Brom Bones has made Ichabod the object of his practical jokes. Finally there is the climactic scene, full of suspense and terror for Ichabod, when he leaves the party late at night and is pursued by the headless horseman.

No summary can adequately convey the pictorial quality of these scenes and the skill with which they are managed. In its blending of the elements which Irving said were among those he particularly aimed at in writing a story—"the play of thought, and sentiments and language; the weaving in of characters, lightly yet expressively delineated; the familiar and faithful exhibition of scenes in common life; and the half-concealed vein of humor that is often playing through the whole"—"The Legend" achieves to a high degree the "nicety of execution" which Irving believed to be the prime requisite of the short tale.

In *The Sketch Book*, Geoffrey Crayon (Irving's *nom de plume*) had told of a stay at the mansion of an English family, the Bracebridges. Another visit to this family provided the framework for the fifty or so sketches, anecdotes, and character studies which make up *Bracebridge Hall* (1822). Interpolated among many short, *Spectator*-like pieces devoted to the Bracebridges and their way of life are several tales told by various persons at the hall. In "The Stout Gentleman," the narrator, having nothing better to do in an inn on a rainy day, speculates on the identity and character of a gentleman who keeps to his room and departs the next day, the narrator having failed to gain so much as a glimpse of him.

Of this story Irving declared, with tongue in cheek, that it had "in it all the elements of that mysterious and romantic narrative so greedily sought after" in his day. Nevertheless, the three other tales in the volume obviously were written to appeal to this popular taste. "The Student of Salamanca" is a romantic story about a young alchemist's assistant who is less interested in occult researches than in the beautiful daughter of his master and who not only manages to save the alchemist from the Inquisition and the daughter from the villain but also turns out to be the heir of a powerful Spanish grandee. "Annette Delarbre" is a sentimental tale of a girl whose mind becomes deranged after her lover is reported lost during a storm at sea and who, like Ophelia, decks herself with flowers and sings plaintive ballads. The story tells how her physician is able to cure her after her lover, who has been miraculously saved, returns, but the principal emphasis is on the pathos of the girl's situation. The final tale in the volume, "Dolph Heyliger," is a long and somewhat rambling piece with an American setting. Dolph, a shiftless but well-meaning lad who, according to his own account (he was renowned as "a drawer of the longbow"), became well-to-do after discovering the buried treasure of an ancestor.

There are four unrelated groups of stories in Irving's next volume, *Tales of a Traveller* (1824). According to the Preface, he wrote these tales in Mainz, Germany, while he was prevented from traveling by an indisposition and had no better way of occupying his time. Vague about his sources, he said merely that the stories came from reading and observation, and he took occasion again, as he had in the Postscript to "The Legend of Sleepy Hollow," to ridicule lightly the desire of readers to be taught by apologue. Asserting that his tales had sound morals, even if they did not appear on the surface, he added, "On the contrary, I have often hid my moral from sight, and disguised it as much as possible by sweets and spices, so that while the simple reader is listening with open mouth to a ghost or a love story, he may have a bolus of sound morality popped down his throat, and be never the wiser for the fraud."

Certainly there are no very apparent morals in the tales in Part I, "Strange Stories by a Nervous Gentleman," which were heard at a hunting dinner by the same nervous gentleman who related the story of the Stout Gentleman in *Bracebridge Hall.* Obliged to stay overnight by a snowstorm, the guests tell tales, most of which deal with haunted bedchambers. The longest and most substantial piece is "The Story of the Young Italian," which employs the conventional plot of separated lovers led to believe each other faithless by the machinations of the villain. But the best of these tales is the mock-Gothic "Adventure of the German Student," which is Poe-like in its economy of telling and in the vividness with which the setting is portrayed but is typically Irving in its ending. The narrator reveals that he had his incredible tale from the best authority: "The student told it to me himself. I saw him in a mad-house in Paris."

Part II, "Buckthorne and His Friends," which Irving tried unsuccessfully to make into a novel, to be called *History of an Author*, has little novelty or imaginative quality. The story of a literary hack, entitled "The Poor-Devil Author," recalls Tobias Smollett's picture of London literary men in *The Expedition of Humphrey Clinker*, while the lengthy and rather tedious account of Buckthorne and his "great expectations," though less moralistic and less emphatic in portraying man's folly, is reminiscent, respectively, of the "Old Man of the

Hill" episode and of the story of Mr. Wilson in Henry Fielding's *Tom Jones* and *Joseph Andrews*.

In the tales in Part III, "The Italian Banditti," a group of travelers in an inn at Terracina relate stories they have heard about the brigands who infest the mountainous regions of Italy. Despite their romantic subject matter, the tales are unexciting, sometimes even dull, and are of less interest than the two stories with American backgrounds included in "The Money Diggers," Part IV, the last part of *Tales of a Traveller*. "The Devil and Tom Walker" is a variation on the legend of the man who sells his soul to the devil, while "Wolfert Webber, or Golden Dreams" tells of a Dutchman who digs without success for buried pirate treasure in his cabbage garden but finally becomes wealthy when his farm becomes valuable real estate. Both stories good-naturedly ridicule avarice, and the first is one of the best stories in the entire volume.

More of Irving's tales are to be found in *The Alhambra* (1832). A product of his residence in Spain from 1826 to 1829, it treats of the history, legends, and superstitions associated with the old fortress of the medieval Moorish kings of Granada. Domiciled in the Alhambra, Irving discovered in the manners, customs, and traditions of the Andalusian peasants who inhabited it and in its romantic atmosphere and "haunted ground" just the sort of material to stimulate his fancy and narrative gifts. Interspersed among the descriptive and historical sketches are a number of legends, several of them having to do with treasure buried by the Moors. Irving himself stated the formula for the legends of this nature: "The hidden wealth is always laid under magic spell, and secured by charm and talisman. Sometimes it is guarded by uncouth monsters or fiery dragons, sometimes by enchanted Moors, who sit by it in armour, with drawn swords, but motionless as statues, maintaining a sleepless watch for ages."

This is the matter of "The Legend of the Moor's Legacy," in which the humble little water carrier, Peregil, befriends a dying Moor and as a result is bequeathed a magic scroll and taper which unlocks Moorish treasure. What makes this story superior in charm and interest to the other legends is Irving's characterization of Peregil, who loves his children dearly, patiently endures his shrewish wife (indeed, a familiar type in Irving's tales), and must contend with both a prying barber who discovers his secret and a corrupt and avaricious mayor before he is justly and amply rewarded for his acts of kindness.

Very similar, though less effective, is "The Legend of the Two Discreet Statues," while in the other legends, if there is no buried treasure, its lack is compensated for by magic steeds and carpets, Arabian astrologers, beautiful captive Christian princesses, enchanted armies, and subterranean palaces. Sometimes Irving makes light of his romantic materials, as in "The Legend of the Rose of the Alhambra" and in the stories about Governor March, the redoubtable one-armed governor of the Alhambra.

After 1832, Irving devoted himself mainly to writing history and biography; thus *The Sketch Book* and the other volumes discussed contain virtually his entire output of tales. The artistry which enabled Irving to give enduring interest and appeal to "Rip Van Winkle" and "The Legend of Sleepy Hollow," and which is present also, if in a lesser degree, elsewhere in his short fiction and the influence he exerted on Hawthorne, Poe, and other writers, justify calling him, as historians of American literature have done, the "father of the American short story."

UNTIL THE MIDDLE 1830's, when Nathaniel Hawthorne's first tales of literary significance began to appear in magazines, no stories of much literary merit aside from Irving's were published, with the exception of "Peter Rugg, The Missing Man" (1824), by William Austin, a tale which has something of a Rip Van Winkle motif but anticipates Hawthorne in manner and method. Of the many who wrote in the vein of Irving, Nathaniel Parker Willis was the most successful. He was prolific and very popular, but his urbanity and wit tended toward the superficial and trivial. Other writers confined themselves mainly to following Irving's lead in depicting the customs and manners of a region. James Hall, who wrote of frontier life in Illinois, said in the Preface to his *Legends of the West* (1832): "The sole intention of the tales comprised in the following pages is to convey accurate descriptions of the scenery and population of the country in which the author resides." Much the same intention is reflected in the sketches of Caroline Kirkland, which describe the settlers on the Michigan frontier; in J. P. Kennedy's *Swallow Barn* (1832), with its scenes of Virginia plantation life; in A. B. Longstreet's *Georgia Scenes, Characters, Incidents, etc. in the First Half Century of the Republic* (1835); and in Albert Pike's *Prose Sketches and Poems Written in the Western Country* (1834), which depicts the Indians and the American settlers in New Mexico. Thenceforth the peculiar and distinctive characteristics of a particular locality or region were to be represented frequently in tales and sketches, although it was not until after the Civil War that this kind of story attained its full development.

Aspiring writers of short narratives, besides being stimulated by Irving's great success, were also greatly encouraged after 1830 by the steadily increasing number of possible outlets for their work. By that time the annual or gift book had become popular with readers, and publishers found it profitable to issue many such volumes. Ornate and showy in binding and format, they contained much light and sentimental poetry and prose but also published some serious short fiction, for example that of Hawthorne, a number of whose early tales and sketches appeared in *The Token.* The widely read *Godey's Lady's Book,* a monthly founded in 1830 on the model of the annuals, but directed especially at a feminine audience, published writers like Poe and Longfellow. By the 1830's, also, the number of literary periodicals had markedly increased. They were published not only in the East but also in the South and West, and although many of them were short-lived, some, like *The Knickerbocker* (founded in 1833), the *Southern Literary Messenger* (1834), and *Graham's* (1841), continued their existence for extended periods, thus encouraging the increased production of literature and helping to raise the level of literary taste.

As a result of all this, between 1830 and 1860, the quantity of short stories written and published in America was considerable. As for quality, it was present in no small measure in the work of two of America's most distinguished short-story writers—Nathaniel Hawthorne and Edgar Allan Poe.

# 2
# Romance, Allegory, and Morality
# Nathaniel Hawthorne and Herman Melville

THE TALES AND SKETCHES of Nathaniel Hawthorne (1804–64), like those of Washington Irving, reflect to a considerable degree the literary tastes and conventions of their time. One finds in Hawthorne, who spoke of himself as a writer of romances (albeit of a kind he called psychological), as in Irving, Poe, and many earlier as well as contemporary European writers an emphasis on the picturesque externals of nature, a predilection for the supernatural and other Gothic elements, and, on occasion, overmuch pathos and sentiment. But, unlike many of the other romancers, Hawthorne, especially in his best work, transcended his use of stock fictional devices and formulas, and, unlike Irving, he did not write to amuse and entertain by commenting in a lightly satirical vein on man's foibles and follies. Hawthorne's work usually had a much more serious intent. It penetrated into the moral and spiritual nature of man, and its most significant effects were achieved through symbolism, allegory, and other means of which Irving made little or no use.

Born in Salem, Massachusetts, Hawthorne came of an old

New England Puritan family. One ancestor, William, is still remembered as a stern judge in the Salem witchcraft trials of 1692–93. Others, like Hawthorne's father, who died when his son was four, were sea captains. After graduating from Bowdoin College in 1825, Hawthorne lived in Salem in comparative seclusion until 1837, doing a great deal of reading in New England history and writing even more. Not a little of what he wrote during this period, according to the Preface to the third edition of *Twice-Told Tales* (1851), he burned "without mercy or remorse and, moreover, without subsequent regret," and he attempted unsuccessfully to withdraw from circulation the immature novel *Fanshawe* (1828), which he had written during his college days and had published at his own expense. His feelings about literature as a profession are also reflected in an early tale, "The Devil in Manuscript," which portrays a discouraged and disillusioned young author who throws his tales in the fire because their composition has led him "into a strange sort of solitude,—a solitude in the midst of men,—where nobody wishes for what I do, nor thinks nor feels as I do," and who complains that "no American publisher will meddle with an American work—seldom if by a known writer, and never if by a new one,—unless at the writer's risk."

By 1837, however, Hawthorne had managed to publish over thirty tales and sketches, most of them anonymously, in annuals and magazines and in that year collected eighteen of them in *Twice-Told Tales*. The work was reviewed favorably by Longfellow, and when an enlarged second edition containing an additional twenty-one tales and sketches appeared in 1842, it was praised by Poe, who, although he devoted much of his review to his own conception of what a short story should be, termed Hawthorne "original in *all* points" and said that his tales belonged to "the highest region of art." Later, in reviewing *Mosses from an Old Manse* (1846), Poe reversed himself, terming Hawthorne "peculiar" rather than original and calling him "infinitely too fond of allegory." Hawthorne himself was inclined to speak disparagingly of his tales and to see in them deficiencies which accounted for their relatively limited reception by the reading public. Although *Mosses* contained such outstanding stories as "Young Goodman Brown," "Rappaccini's Daughter," "The Birthmark," and "The Artist of the Beautiful," he said in his Preface that he did not regard the collection at all highly. Of the *Twice-Told Tales*, which he reissued in 1851, he said, "They have the pale tint of flowers that blossomed in too retired a shade. . . . Instead of passion there is sentiment; and, even in what purport to be pictures of actual life, we have allegory, not always so warmly dressed in its habiliments of flesh and blood as to be taken into the reader's mind without a shiver."

*The Snow Image and Other Twice-Told Tales* (1851) was Hawthorne's final volume of tales and sketches. Made up mostly of early pieces previously uncollected, it does not come up to his other collections, but it is notable for the inclusion of "My Kinsman, Major Molineux," one of the most impressive of Hawthorne's early tales, and of his last short narrative, "Ethan Brand," written late in 1848, not long before he began *The Scarlet Letter* (1850). Continuing to devote himself to the writing of longer fiction, Hawthorne soon produced *The House of the Seven Gables* (1851), *The Blithedale Romance* (1852), and, at the end of a seven-year stay abroad in England and Italy, *The Marble Faun* (1860), his final work except for several later romances left unfinished at his death.

Hawthorne's shorter writings are, generally speaking, of three kinds: sketches, tales of New England history, and allegories and moral tales. But this is only an approximate classification, since the sketches sometimes have narrative elements as well as moral applications; certain of the allegories, especially those Hawthorne called "fantasies," should probably be called sketches; and some of the historical tales can be termed, as Hawthorne himself called them, allegories or parables. What might be called the sketches proper are descriptive rather than narrative and usually take the form of discursive personal essays. Although they reflect a good deal of Hawthorne's personality and sometimes express attitudes and themes present in the tales, the conventions of the gift books and annuals for which most of them were written are all too apparent in them. Even in the more substantial ones, such as "Sights from a Steeple" and "The Toll-Gatherer's Day," in which Hawthorne observes and comments on the human scene, one finds little of the interest and charm of the very fine personal, reflective, and quietly humorous essays which serve as prefaces to *Mosses* and *The Scarlet Letter.*

About a dozen of Hawthorne's tales deal more or less explicitly with early New England history or legend. Almost all of them have a dramatic conflict which goes deeper than the external action in which the characters are engaged and have as well a symbolic character and moral import. The historical facts, for example, on which "The Maypole of Merrymount" was based, according to Hawthorne, "wrought themselves, almost spontaneously, into a sort of allegory." How Hawthorne used his sources in this instance can be seen by consulting the conflicting accounts in William Bradford's *History of Plymouth Plantation* and Thomas Morton's *New English Canaan*, which tells how the gay colonists, led by Morton at Mount Wallaston, or Merrymount, near Plymouth, were conquered by the Puritans. To dramatize the conflict between Cavalier and Puritan and to reinforce his moral, Hawthorne made the Puritans attack the Merrymount settlers during o his sensitive spirit, caused by the blows inflicted on y the Puritan boy he thought to be his friend is more us. Lacking the love of his own mother to restore him, pines away and dies. Moving though the story is, it is ther as unified nor as artistically effective as it might be. here is undue emphasis on pathos and sentiment, and alhough Ilbrahim is the chief character, there are times when the story seems to be more Catherine's than his. Furthermore, Hawthorne engages our interest in Tobias, who is a secondary character, by portraying at some length the conflict in him occasioned by his conversion to Quakerism, only to ignore him at the end of the story, when, in the course of years, the spirit typified by the gentle boy ameliorates both the harsh severity of the Puritans and the fanaticism of the Quakers.

Less noteworthy than the foregoing historical stories are the four tales collected under the heading "Legends of the Province House," all of which deal with incidents more or less supernatural or "marvellous," supposed to have been associated with the Boston mansion which was inhabited in colonial times by the royal governors. Writing in a tone and supplying his tales with a framework after the manner of Irving, Hawthorne purported to have heard these legends from two elderly frequenters of the mansion, which in his own day had been turned into a hostelry. The action in each tale is given an emblematic significance. In "Howe's Masquerade" the ghosts of the former governors of New England pass before the guests at a masked ball given by Sir William Howe, the last royal governor, while Boston is being besieged by General Washington; they form, in the words of one of the spectators, "the funeral procession of royal authority in New England." In "Edward Randolph's Portrait" the faded picture of Randolph's face, made horrible by the misery of the curse laid on him by the colonists for depriving them of



their "immoral" ceremonies around the Maypole. He also created the two young lovers, Edgar and Edith, centering the action around their nuptials and thereby providing added interest and pathos. Endicott, the stern Puritan leader, performs two symbolical acts. He cuts down the Maypole, with the result that New England becomes "a land of clouded visages, of hard toil, sermon, and psalm"; but, moved by compassion, he also performs "a deed of prophecy" in throwing a wreath of roses over the heads of the lovers. The moral gloom of the world triumphs over gaiety, and life is a difficult path, but the lovers have courage to tread it and think no more of the vanities of Merrymount.

The story is one of Hawthorne's most artistic in its structure and tone, consisting of a single episode in which the stage is vividly set at the outset, after which Hawthorne interpolates the backgrounds of his story and contrasts the Merrymount settlers and the Puritans, the latter being termed "most dismal wretches," their Maypole being the whipping post. Effective, too, is the manner in which Hawthorne gives premonitory hints of the outcome; the approach of the Puritans is paralleled by the sun going down and the gloom of evening falling over the Maypole festivities just as the Puritans attack.

"Endicott and the Red Cross" deals with another episode which Hawthorne makes a turning point in colonial history. Endicott is drilling the Salem trainband when Roger Williams brings the news that King Charles I and Archbishop William Laud plan to send to the colony a governor-general who will take away the Puritans' freedom of worship. Endicott rends the cross from the English flag, an act symbolizing the defiance which ultimately resulted in the colonies winning their independence. Although this defiance is praised by Hawthorne, the story is also an ironic indictment of Puritan narrowness and bigotry. The picture of a stern Puritan society which harshly punished dissenters is even more graphic than that in "The Maypole of Merrymount," although it does not seem as essential to the point of the story.

Indicative of the kind of grim humor Hawthorne employs in both stories is the following passage from "Endicott and the Red Cross": "At one corner of the meeting-house was the pillory, and at the other the stocks; and, by a singular good fortune for our sketch, the head of an Episcopalian and suspected Catholic was grotesquely incased in the former machine; while a fellow-criminal, who had boisterously quaffed a health to the king, was confined by the legs in the latter." There is also a lighter humor in the description of the woman who "wore a cleft stick on her tongue, in appropriate retribution for having wagged that unruly member against the elders of the church; and her countenance and gestures gave much cause to apprehend that, the moment the stick should be removed, a repetition of the offense would demand new ingenuity in chastising it." Another of the offenders is the young woman doomed to wear the letter A on the breast of her gown, the prototype of Hester Prynne in *The Scarlet Letter*.

"The Gray Champion," although first published two years earlier than "Endicott and the Red Cross," reads like a sequel to that story. The theme is the same—that of Puritan resistance to English domination—and although the time is fifty years later, when James II abdicated, the descendants of the early Puritans have changed neither their sober mien or garb nor their opposition to the threat of curtailment of their religious freedom, symbolized by their royal governor-general, Sir Edmund Andros. Yet the crisis does not seem as meaningful as in the other story, for we do not know why Sir Edmund has chosen to parade his soldiers on the day the story takes place; perhaps it is only as a show of strength. An old Puriton dressed in the garb of fifty years earlier mysteriously ap-



pears to outface the governor
King James has been deposed.
as he came, but his prophecy is
arrives that William III is Englan
thorne found the suggestion for his
legendary New England figure who
saved the inhabitants of the town of Had
attack, he gave his legendary champion
cance, making him "the type of New Engla
spirit" which manifested itself again in the res
English at Lexington and Bunker Hill.

The longest, as well as one of the best, of Hawth
torical tales is "The Gentle Boy," which deals with th
cution of the Quakers by the early New England Pu
It begins with a historical foreword, culminating in a re
ence to the execution of two Quakers in the Massachuse
Bay Colony in 1659. But the story proper is concerned mainly with six-year-old Ilbrahim, the son of one of the victims and himself the innocent victim both of the fanaticism of his parents, who deliberately sought martyrdom, and of the harsh intolerance and narrow bigotry of the Puritans. His father dead and his mother, Catherine, driven into the wilderness, the gentle boy would have perished had not the compassion of the Puritan couple, Tobias and Dorothy, led them to brave the religious prejudice and persecution of their neighbors to care for the child. Not long after, Catherine returns to denounce the Puritans as they are gathered in their meetinghouse and to express her remorse for having abandoned her son. Religious zeal, however, is still stronger in her than mother love, and she leaves Ilbrahim with Dorothy and Tobias, who continue to give him the love he needs. But Ilbrahim is subsequently beaten severely by the neighbor children, and though he recovers from the bodily injury, the in-

their provincial character, is mysteriously restored as a warning to the royal governor who is later responsible for the Boston Massacre. In "Lady Eleanor's Mantle," which has more than a little of the macabre character of Poe's "The Masque of the Red Death," Lady Eleanor wraps herself in pride, as in a mantle, and scorns all human sympathies. In "Old Esther Dudley," an aged woman is a symbol of the loyalists who wait in vain for the royal governor to return to the province house.

Also legendary, but much more artistically symbolical, is "My Kinsman, Major Molineux," one of the earliest and certainly one of the finest of Hawthorne's tales. It tells how a young countryman, who thinks himself shrewd but is really naïve, arrives in Boston one evening seeking the wealthy uncle who had offered to help him rise in the world, only to discover after much fruitless inquiry his kinsman tarred and feathered and being ridden out of town by a mob. As circumstantial in its detail as a story by Kafka, it is Kafkaesque also in its dream-fable and nightmarelike quality and in the multilevel readings it has evoked. Some view the story as a historical allegory, dramatizing, through young Robin's joining the crowd in laughing at his hapless uncle, young America's impulse to resist and ultimately overthrow British authority. Inevitably, too, the story has been interpreted in Freudian terms, Robin's actions illustrating the necessity of overthrowing paternal authority (the uncle filling the role of a father figure) if he is to achieve maturity. The tale also anticipates the "initiation" story as written by Hemingway and Sherwood Anderson—although what kind of knowledge and exactly how much Robin has gained from his sobering experience is made none too clear. In any event, he has achieved the beginning of wisdom in realizing that he is not as shrewd as he had supposed.

More obviously allegorical, but nevertheless another of Hawthorne's masterpieces, is "Young Goodman Brown," probably suggested by the descriptions of witches' meetings in Cotton Mather's *Wonders of the Invisible World*. Brown, whose surname suggests that he was not only a simple husbandman but also a "good man," takes leave at sunset to make a journey, despite the protestations of his pretty wife, Faith. Tempted, as all men are, to know what sin is, Brown has a covenant with the Devil. When they meet in the forest, Brown hesitates to accompany him farther, despite the latter's assurances that Brown's father and grandfather had been his good friends and that many in New England, including the governor and his council, are "firm supporters of my interest." But, learning that the old woman who taught him his catechism, the deacon, and the minister, all of whom he has held in high respect, are on their way to the Devil's meeting, and overwhelmed by the conviction that his "Faith is gone," he rushes madly into the depths of the forest. There he finds a vast and awful assemblage of his neighbors of both good and bad repute in communion with the Evil One, and, along with himself, to be welcomed into this universal communion is Faith. As he shouts to her to resist, the vision disappears. "Had Goodman Brown fallen asleep in the forest and only dreamed a wild dream of a witch-meeting?" Hawthorne does not say; but whatever the case, what Goodman has seen or dreamed brings him disillusionment and a sense of estrangement from which he never recovers.

Hawthorne's superb narrative artistry, employing vivid pictorial description, Gothicism, symbolism, indirection, and indefiniteness gives his story a dramatic power unequaled elsewhere in his fiction except in *The Scarlet Letter*. The ambiguity, raising as it does certain questions of interpretation, has evoked much debate among the story's many critics, but it seems clear that Brown's experience causes him to lose his faith in the essential goodness of human beings, including himself, and, further, there is the implication that Brown is to be condemned for retaining his one-sided view of human nature. Brown comes to realize through his vision in the forest that evil is very real; unlike Brown, we should accommodate ourselves to this revelation by still believing there is also good in man.

The theme that man is naturally depraved and that every human heart conceals iniquity of thought or deed is given almost equally effective, although much more restrained, expression in "The Minister's Black Veil." Subtitled "A Parable," the story was suggested by an account of an eccentric New England clergyman named Moody, who, after accidentally killing a friend (or, according to another version, after the death of his wife), kept his face covered with a handkerchief. In Hawthorne's story, however, instead of wearing a handkerchief, the Reverend Mr. Hooper appears before his congregation with his features, except for mouth and chin, concealed by a far more awe-inspiring black crepe veil. He does not cover his face because of any specific sin or sorrow, but his sermon on secret sin gives a clue to why he has assumed the veil. It symbolizes his awareness of guilt in the human heart. To Elizabeth, his betrothed, who asks him what grievous affliction has caused him to wear the veil and who tells him it is rumored that he hides his face because of a secret sin, he answers, "If I hide my face for sorrow there is cause enough, and if I cover it for secret sin, what mortal might not do the same?" Like Dimmesdale in *The Scarlet Letter*, whose prototype he is in certain respects, his sense of guilt results in his becoming a more effective clergyman, and yet, like Goodman Brown, his revelation of sin in the human heart dooms him throughout a long and irreproachable life to be a man apart from other men. On his deathbed, smiling the faint, sad smile so often seen on his lips, he turns his veiled face on those gathered there, and says, "I look around me and, lo! on every visage a Black Veil!"

"Roger Malvin's Burial" also deals with secret sin and its consequences. Reuben Bourne, a wounded survivor of an Indian fight, succeeds in making his way back to the settlement but leaves his companion, the fatally wounded Roger Malvin, to die in the wilderness. Malvin had pointed out that to stay would only result in Bourne's own death and be a useless sacrifice, but, unable to rid himself of a sense of guilt and fearing the condemnation of Malvin's daughter and others, Bourne allows them to believe that he remained with the dying man and gave him burial. Oppressed by his secret and by his failure to keep his promise to return later and inter Malvin's bones, he becomes moody and misanthropic. His marriage to Dorcas Malvin does not bring him happiness, and the farm she inherits from her father is finally lost through Bourne's neglect, so that some years later they are forced to seek a new home in the wilderness. Compulsively drawn in some mysterious way to the place where Malvin died, Bourne, while hunting, accidentally shoots and kills his son. As a result of this act he feels a sense of having expiated his own sin, an outcome which, although it may seem melodramatic and largely inexplicable as far as its motivation and religious implications are concerned, Hawthorne somehow succeeds in convincing us has psychological truth.

At one point in the story Bourne's secret is likened to a serpent gnawing into his heart. In "Egotism, or the Bosom Serpent," the guilt of Roderick Elliston, namely "a monstrous egotism," is typified by a snake which, to Elliston's diseased imagination, actually gnaws within his bosom. Every man's breast, Elliston believes, has its own serpent, "the type of each man's fatal error, or hoarded sin, or unquiet conscience." But Hawthorne's use of the serpent as a central symbol in the story is not convincing. Furthermore, Elliston's egotism is said to be manifested in the form of jealousy, but of what exactly is never specified, and he is cured of his obsession suddenly and unexpectedly at the end of the story when he learns, through the love and forgiveness of his wife Rosina, to forget himself "in the idea of another."

Unlike Elliston, most of Hawthorne's other characters who are preoccupied with self—and there are many of them—are incapable of redemption. Pride, vanity, or some engrossing purpose destroys their human sympathies, infects them with a kind of madness, and isolates them from other men. The painter in "The Prophetic Pictures" lives only for his art and pries into the souls of his subjects in order to represent on canvas what they will become, almost as if he were the creator of their destinies. Richard Digby, the religious bigot in "The Man of Adamant," sure that only he has found the way to salvation, eschews the love and friendship of others, and after his death his corpse turns to stone like that of the cave which he has inhabited in the forest. Each of the adventurers—the "Seeker" motivated by vain ambition, the doctor, the merchant, the poet, the nobleman, and the cynic—who seek the precious gem in "The Great Carbuncle," except for the young husband and wife, is impelled by "his own selfish and solitary longing." To the young couple the stone is simply the object of natural desire, an emblem of happiness. The search does not bring them frustration and defeat as it does the others because they learn that to be happy they must not desire what other men cannot share with them. The vanity of ambition is also the theme of "The Ambitious Guest." Its unity of effect and powerful irony of situation make it deservedly one of Hawthorne's most famous tales.

In three of Hawthorne's most notable stories, intellectual curiosity or excessive preoccupation with scientific endeavor makes for tragic consequences. Aylmer, the brilliant scientist in "The Birthmark," loves his beautiful young wife, Georgiana, but he loves science more. The birthmark shaped like a tiny hand on her left cheek is for him a symbol of imperfection, and he persuades her to allow him to attempt to remove it. Although very close to being an alchemist, even a sorcerer, Aylmer is a symbol of man's desire for perfection, which impels him to try to transcend his physical and mortal part, typified by Aminadab, the Caliban-like laboratory assistant, who during Alymer's experiment mutters, "If she were my wife, I'd never part with that birthmark." Alymer is destined to fail because he aspires to the infinite. His aim had been a noble one, Georgiana tells him as she is dying; but true wisdom on Alymer's part, Hawthorne seems to say in the somewhat ambiguous concluding paragraph, would have been for him to perceive that his happiness lay in accepting as near perfect a being as the earth had to offer. For Freudian critics the story has a second meaning, the birthmark symbolizing sexuality and Aylmer being obsessed with the need to remove it because he shares with other of Hawthorne's heroes—such as Brown, Hooper, and Giovanni in "Rappaccini's Daughter"—a fear of and revulsion toward sex.

Far more reprehensible in his scientific zeal is Dr. Rappaccini in "Rappaccini's Daughter." Unlike Alymer, there is nothing holy or lofty about his scientific experimentation. It is diabolic in that he has experimented with his daughter, Beatrice, nourishing her with poisons from her birth, and making her the human sister of the beautiful but deadly flower he has grown in his garden. Giovanni Guasconti, the young student who falls in love with her, might save her by his love and sympathy. But because he cannot overcome his fear of her apparently poisonous influence, he urges her to take a powerful antidote, which results in her death. The greater part of this, for Hawthorne, very long story is a highly imaginative elaboration of a sentence which he copied in 1839 into his notebook from Sir Thomas Browne's *Vulgar Errors*: "A story there passeth of an Indian King that sent unto Alexander a fair woman, fed with aconite and other poisons, with this intent complexionally to destroy him!" Despite a certain power and richness, and the poignant figure of the innocent Beatrice, infected with evil through no fault of her own, Hawthorne's story has some glaring inconsistencies and absurdities and an obvious air of contrivance and artificiality.

"Ethan Brand" is Hawthorne's most powerful treatment of the baleful effects of an obsessive scientific or intellectual curiosity. The specific idea on which the story is based is indicated in two of Hawthorne's notebook entries of 1844:

> The search of an investigator for the Unpardonable Sin;—he at last finds it in his own heart and practice.
>
> The Unpardonable Sin might consist in a want of love and reverence for the Human Soul; in consequence of which, the investigator pried into its dark depths, not with a hope or purpose of making it better, but from a cold philosophical curiosity,—content that it should be wicked in whatever kind or degree, and only desiring to study it out. Would not this, in other words, be the separation of the intellect from the heart?

This is the fate of Ethan Brand, who has lived a solitary and meditative life as a lime burner and has conceived the idea of searching for the Unpardonable Sin. At the beginning of the tale he has returned, after wandering for eighteen years, to the lime kiln he formerly tended. He has looked into many a human heart in an effort to discover what he sought, only to realize at last that in doing so he has committed the

Unpardonable Sin. Brand, the reader is told, became a fiend through conducting psychological experiments on others, "converting man and woman to be his puppets and pulling the wires that moved them to such degrees of crime as were demanded for his study." Hawthorne is so vague, however, about what Brand has done that the story is unsatisfying. Yet this man, who casts himself into the kiln—where his heart, as if it were marble, is turned to lime—not in remorse or despair, but because he has cast off his humanity, is a terrifying and awful figure.

Whether or not, as some readers have believed, Brand is the kind of man Hawthorne feared he might become, Hawthorne very likely had in mind his own experience in artistic creation when he wrote "The Artist of the Beautiful," which deals with the situation of the artist in conflict with the world of hard reality and common sense. Owen Warland neglects his watchmaking trade in his endeavor to create a delicate mechanism that will contain the beauty and spiritual essence of a butterfly. His genius and skill are finally equal to the task, but only after a long period of frustration, discouragement, and struggle. In the eyes of the practical Peter Hovenden, to whom Warland was formerly apprenticed, Warland's aspirations are nonsensical, and his skepticism and scorn inhibit Warland, as does the "hard, brute force" of Warland's well-meaning old school friend Robert Danforth, the blacksmith. Furthermore, Hovenden's daughter Annie, whom Warland loves and looks to for inspiration, is incapable of manifesting the sympathy and understanding which would have meant help and strength to him, and he loses her to Danforth. Warland is regarded as little better than a madman by his fellow townsmen. In one of several auctorial comments on the situation of the artist Hawthorne says:

> It is requisite for the ideal artist to possess a force of character that seems hardly compatible with its delicacy; he must keep his faith in himself while the incredulous world assails him with its utter disbelief; he must stand up against mankind and be his own sole disciple, both as respects his genius and the objects to which it is directed.

When his handiwork is destroyed by Annie's child, Warland does not regard his labor as in vain. "He had caught," says Hawthorne, "a far other butterfly than this. When the artist rose high enough to achieve the beautiful, the symbol by which he made it perceptible to mortal senses became of little value in his eyes while his spirit possessed itself in the enjoyment of the reality." The psychological critics, it should be added, do not concede Warland the victory he appears here to have won. For them, rather than being simply an aspiring artist, he seeks in art a refuge from the world because his lack of manhood prevents him from finding a satisfactory place in it.

"The Artist of the Beautiful" has a dramatic quality and a realism in its character portrayal which is lacking in a number of Hawthorne's other parables and allegories. These rather wooden and abstract pieces, some of which Hawthorne called fantasies because they were developments of an unusual or farfetched idea that struck his fancy, are usually characterized by some kind of procession or cataloguing of persons or actions, a favorite narrative device with Hawthorne. In "The New Adam and Eve" replicas of the first man and woman wander through a deserted city and view with wonder and amazement all the works which perverted, corrupt, supposedly civilized man—who has been swept from the face of the earth—has left behind him. In "The Hall of Fantasy" various classes of men are seen: poets, businessmen, inventors, reformers. To dwell here always instead of in the real world unfits a man for the real employments of life, but Hawthorne concludes, "There is but half a life—the meaner and earthier half—for those who never find their way into the hall." In "The Procession of Life" men are grouped together not by artificial distinctions, such as wealth, social position, and religious beliefs, but by truer and more fundamental distinctions, such as their intellectual power, their suffering, and their good and evil actions. Death is the chief marshal of the procession, and, though he does not know where it will end, "God, who made us," Hawthorne asserts in conclusion, "knows, and will not leave us on our toilsome and doubtful march, either to wander in infinite uncertainty, or perish by the way!"

More noteworthy, chiefly because of its irony and satire, is "Earth's Holocaust," in which zealous reformers and revolutionaries cast into a great bonfire the badges and symbols of rank, the implements of war, the instruments of execution, books, spirituous liquors, and the appurtenances of religion, including the Bible. Yet these efforts to reform the world by destroying those things which purportedly have corrupted men is doomed to failure until the heart, which is the root of all evil, is first of all purified. Similar in tone is "The Celestial Railroad," an ironical parody of *Pilgrim's Progress*, in which the narrator discovers that Bunyan's route to heaven can now be traversed much more comfortably in a railroad coach. Apollyon, who fought so fierce a battle with Christian, is now the chief engineer; a tunnel has been bored through the Hill of Difficulty; the Valley of the Shadow of Death is illuminated by gas lamps; the giants Pope and Pagan have been supplanted by the giant Transcendentalist, who shouts at the travelers in so strange a phraseology that they cannot understand him and do not know whether to be frightened or not;

and Vanity Fair with its many churches and highly esteemed clergymen has many advantages as a permanent residence. The narrator ultimately discovers to his horror that his guide is the Devil, but to his relief he wakes to find that the journey has been a dream.

There remain a few other tales which should be mentioned to provide a reasonably comprehensive survey of Hawthorne's work in the short story. Some of them, like "The Great Stone Face," "David Swan," "Feathertop: A Moralized Legend," and "The Snow-Image," are, as Mark Van Doren has observed, "more moralizing than moral." Others, like "Alice Doane's Appeal," "The Wedding Knell," and "The White Old Maid," are weak because Hawthorne's craftsmanship and imagination failed him. A few, like "Mrs. Bullfrog" and "Mr. Higginbotham's Catastrophe" are not at all characteristic. The former is a humorous trick-ending story in not very good taste; the latter is also humorous, and Hawthorne's handling of mystery and suspense makes it easy to understand why Poe called the story "vividly original and managed most dexterously." "Wakefield," halfway between sketch and tale, is a provocative study of fate and perversity as the causes of human isolation. "The Canterbury Pilgrims" and "The Shaker Bridal," despite the somewhat artificial situation of the former and the sentimental ending of the latter, are interesting treatments of the communistic and celibate religious sect known as the Shakers. Both "Peter Goldthwaite's Treasure" and "Dr. Heidegger's Experiment" are undistinguished tales; they are of interest, however, because Hawthorne later returned to their themes: concealed family wealth in the first tale appears in *The House of the Seven Gables*, and the elixir of life in the second tale is treated in the unfinished novels *Septimius Felton* and *The Dolliver Romance*.

Unlike Irving, Hawthorne does not seem to have given

much thought to problems of form and style, or to how he might best achieve his purpose in telling a story. But if he wrote with what appears to be less concern for technique and less conscious artistry than Irving, he could achieve on occasion, as in "My Kinsman, Major Molineux," "Young Goodman Brown," and "The Ambitious Guest," a compactness of structure and a unity of effect worthy of Poe. And there are technically less admirable stories which have somehow as much, or almost as much, dramatic impact and significance of meaning.

Academic scholarship and criticism dealing with Hawthorne has been a thriving industry for a number of years. His character and personality and the religious, moral, psychological, and aesthetic aspects of his work have been exhaustively analyzed, interpreted, and debated. Critics who have preoccupied themselves with his religious and moral symbolism and imagery have been particularly attracted to Hawthorne, with the result that they have tended both to slight his psychological implications and to overrate his total achievement. More recently, however, it is being increasingly recognized that in a good many instances there seems to be much more to a Hawthorne story than its overt moralizing or allegorizing and that he is perhaps somewhat less congenial to us today than we once thought him. Hawthorne wrote a small number of stories that are superb, or fall only a little short of being so, but they, along with *The Scarlet Letter*, are sufficient to ensure him a lasting place in the first rank of American fiction writers.

HERMAN MELVILLE (1819–91) wrote only a small body of short fiction; yet this work, including as it does such stories as "Bartleby," *Benito Cereno*, "The Encantadas," and

*Billy Budd*, is impressive in its quality. The son of a once-prosperous merchant and importer, Melville grew up in New York City and in Albany and its vicinity. His formal schooling was irregular and none too extensive, although he later supplemented it by wide reading in Shakespeare, Spenser, other Elizabethans, Dante, Rabelais, Robert Burton, Sir Thomas Browne, and others, whose influence is often reflected in his writing. His experiences between 1839 and 1844 as a sailor on a merchant ship, on whaling vessels, and on an American warship, were important in providing him with the materials and inspiration for writing.

Melville's first two novels, *Typee* (1846) and *Omoo* (1847), based on his adventures in the South Seas, though much embellished by his reading of travel books and his own imagination, made him a popular author. *Mardi* (1848), also a story of Pacific adventures but more serious in theme, was too heavily allegorical and lacking in unity to be successful, either artistically or commercially. *Redburn* (1849), reflecting Melville's experiences on a voyage to Liverpool, and *White Jacket* (1850), based on his naval experiences, were more effective in fusing exciting adventure and significant themes; but when *Moby-Dick* was published in 1851, it had few appreciative readers, and the reception of *Pierre: or, The Ambiguities* (1852), an unsuccessful attempt at blending romance and psychological analysis, was almost completely unfavorable.

Discouraged by the loss of his reading public, and given the opportunity to publish short fiction in *Putnam's Monthly* and *Harper's New Monthly Magazine*, Melville produced fifteen tales and sketches between 1852 and 1856, all but two of them published in one or the other of these magazines. *The Piazza Tales* (1856), Melville's only volume of short fiction, includes six of these pieces: "The Piazza," "Bartleby," "The Encantadas," "The Lightning-Rod Man," *Benito*

*Cereno*, and "The Bell-Tower." Melville wrote no other short fiction except for the novelette *Billy Budd*. Composed toward the end of his life, after many years of literary inactivity, except for the occasional writing of poetry, it remained unpublished until 1924.

Several of Melville's sketches are of interest more for what they seem to suggest about his state of mind at the time of their writing than for their artistry and substance. They are more nearly personal essays than stories and usually culminate in the author's experience teaching him a moral lesson. In "The Piazza" the narrator goes for a walk up a mountain which he can view from his porch. Surely, he thinks, to live in this region must be like living in fairyland. But he is disabused of this notion by a girl who lives in an old house on the mountain. She is discontented with her lot and yearns for the happiness which the inhabitant of the house far below—the narrator's—must surely enjoy. He resolves henceforth to stay on his piazza and to make no further attempts to find fairyland. Other sketches of this sort, among them "The Happy Failure" and "Jimmy Rose," point the moral that a man, even though he may be poor and unsuccessful, can maintain his faith and hope and be happy.

Less moralizing is "I and My Chimney," which recounts with much of the charm and good humor of an essay by Lamb, the narrator's inordinate fondness for the huge central chimney in his house and his efforts to resist his wife's efforts to have it removed so that she may have a grand entrance hall in its place. The chimney can be regarded as a symbol of man's fundamental, private self, and it may well have, as some of Melville's critics have suggested, specific autobiographical implications.

It seems probable that in "The Lightning-Rod Man," Melville was condemning the harsher and more bigoted aspects of Calvinist doctrine. During a thunderstorm the narrator is visited by a seller of lightning rods, who gloomily and frighteningly describes the innumerable ways that lightning may strike. The narrator casts him out of the house, declaring, "The hairs of our head are numbered, and the days of our lives. In thunder as in sunshine, I stand at ease in the hands of my God." But despite this rude treatment and the narrator's efforts to discredit him in the neighborhood, the lightning-rod man "still dwells in the land; still travels in storm time, and drives a brave trade with the fears of man."

There is equally strong social criticism in three pairs of contrasting sketches. In "The Paradise of Bachelors and the Tartarus of Maids," a description of the comfortable life enjoyed by affluent lawyers residing in the Temple in London is paired with an account of the hard lot of women workers in a New England paper mill. The first part of "Poor Man's Pudding and Rich Man's Crumbs" attacks with irony the attitude of the wealthy who profess to believe that the poor are much better off than they really are. The second part describes the distribution of leftover food to beggars after a great official banquet in London, regarded as a most generous charity by the English, but not by the American narrator. Charity is also the subject of "The Two Temples," wherein Melville indicated that a wealthy and exclusive church in New York City did not want poor people in its congregation. *Putnam's Monthly* declined to print the sketch on the grounds that it might offend the religious feelings of readers.

Some of Melville's best descriptive writing is to be found in "The Encantadas, or Enchanted Isles," a series of ten sketches dealing with the physical features, history, and legend of the Galápagos Islands, situated in the South Pacific about six hundred miles west of Ecuador. Based on visits he had made to the archipelago and on narratives of Pacific voyages,

Melville's account is a vivid picture of barren and desolate islands, their volcanic ash and sparse vegetation providing a fit habitation only for the tortoises with which they abound, yet affording a refuge of sorts for buccaneers, castaways, and deserting seamen. Not until the seventh sketch does any narrative material appear, and it and the following two sketches tell the stories of persons associated with the islands: a Creole soldier of fortune who reigned briefly over one of the islands before his subjects, consisting of Peruvian immigrants and sailor-deserters, deposed him; a half-blood Indian woman who went to procure precious tortoise oil with her husband and brother, both of whom accidentally drowned, and who lived alone on one of the islands for many months before her rescue; and the misanthropic Oberlus, more like Caliban than a human being, who made brutes of the seamen whom he captured and enslaved.

Although in "The Encantadas" Melville wrote of a faraway place unfamiliar to most of his readers, in only one tale did he employ a romantic, remote setting and situation of the kind to be found in many short stories of the period, including more than a few by Hawthorne and Poe. Written in a somewhat ornate, archaic style and laid in Italy in a far-off time, "The Bell-Tower" tells of Bannadonna, a cunning artificer who builds a huge bell tower complete with clockwork and a robot to strike the hours. Determined to go beyond the natural philosophers, alchemists, and others who seek the secrets of nature in order to enslave her, he considers his automaton as only the first step in his eventual development of a mechanical being in which "all excellences of all God-made creatures which served man were here to receive advancement and they to be combiined in one." But the presumptuous and fallible creator is accidentally killed by his creation. "And so," ends this Hawthorne-like parable (which

Melville may have meant to be a warning against undue worship of the machine), "pride went before the fall."

In "Bartleby," Melville pointed no moral, leaving the reader to draw his own conclusions concerning the nature of the eccentric Bartleby and the consequences of his behavior. When first employed as a scrivener, or copyist, in a New York City law office, Bartleby proves capable and industrious, but he will perform no other tasks than his copying, and after a time he will not do even that, saying only in explanation, "I would prefer not to." Bartleby refuses to leave the office when he is discharged, and his employer, who has on the whole exercised remarkable forbearance, finally moves, leaving Bartleby behind to plague the next tenant, whose complaints cause the landlord to have him arrested for vagrancy and imprisoned in the Tombs, where he soon dies.

This enigmatic yet richly suggestive and greatly moving story has been regarded as a portrait of a real individual, based perhaps on Thoreau or on one of Melville's friends, who withdraws from society; as a study of the mental disorder now called schizophrenia; as a demonstration of the thesis that predestination is a stronger force in determining men's actions than free will; as a satirical indictment of the barren and monotonous activity associated with the business world of Wall Street; and as an allegory of Melville's own unhappy situation as a writer, and, by extension, the conflict between the nineteenth-century American artist and his environment.

Whether or not the critics who have identified Bartleby with Melville himself are correct—a theory that would seem to require at certain points in the story a somewhat forced and overly ingenious reading—the details in the story which can be construed as supporting the various other interpretations cannot be ignored. Bartleby's nonconformity, passive resistance, and withdrawal from society can be attributed to either deliberate choice or some psychological compulsion, or even a combination of both. The puzzled and frustrated lawyer-narrator, seeking for some explanation, comes to believe that "these troubles of mine touching the scrivener had been all predestinated from eternity, and Bartleby was billeted upon me for some mysterious purpose of an all-wise Providence, which it was not for a mere mortal like me to fathom." The satire of business society is apparent in the walled-in lawyer's office being likened to a prison, in the peculiar actions of the two other scriveners, Turkey and Nippers, whose adjustment to their Wall Street existence is only partial, and in the person of the lawyer, who smugly congratulates himself on being known as "an eminently *safe* man," who does "a snug business among rich men's bonds, and mortgages, and title deeds." The lawyer's generally positive and sympathetic response to Bartleby may indicate, as one critic has suggested, that Bartleby is a kind of alter ego, or double, embodying the lawyer's unconscious reaction against his prudent, acquisitive, essentially sterile way of life.

Intriguing though the figure of Bartleby is, the story is not so much his as the lawyer's, with the emphasis throughout mainly on the latter's personality and character and on the conflict engendered in him by the lawyer-scrivener relationship. The lawyer is like other men, although better perhaps than most. If he is somewhat too self-centered and self-satisfied, he is also aware that he has these weaknesses, and he is remarkably honest with himself about his motives in his dealings with Bartleby. When Bartleby first shows signs of passive resistance, the lawyer decides to be charitable. He will regard Bartleby's conduct as involuntary and will not discharge him, since that might be driving him out to starve. Yet the real reason for this decision, the lawyer admits, is that he believes Bartleby will still be useful to him, and thus the lawyer "can cheaply purchase a delicious self-approval." Later, when Bartleby gives up copying but refuses to be discharged, the lawyer, candidly admitting that it will be to his self-interest, again resolves to practice charity in allowing Bartleby to remain. The uncharitable remarks of his professional friends, however, make the lawyer fearful of Bartleby's "scandalizing my professional reputation," and he finally, although not without considerable compunction, moves his office. It is significant that even then, since it is Bartleby's terrible isolation more than his perverseness that weighs on the lawyer, he cannot completely desert Bartleby. He returns and tries to persuade Bartleby to take up some other kind of work which the lawyer will gladly use his influence to obtain for him. Later, as soon as the lawyer learns that Bartleby has been imprisoned, he visits him in the Tombs, where, as before, Bartleby rejects his proffered assistance. Bartleby's history is concluded when the lawyer reports a rumor he hears some months after Bartleby's death: Bartleby had once worked in the Dead Letter Office in Washington. Whether or not he is justified in believing that this explains why Bartleby had become obsessed with human loneliness does not seem as important as the lawyer's final words: "Ah, Bartleby! Ah, humanity!" They point up not only the lawyer's natural compassion but also the insight he has gained: isolation and loneliness are an essential fact of the human condition.

Although Melville found his main action and principal characters for *Benito Cereno* in the seventeenth chapter of Amasa Delano's *A Narrative of Voyages and Travels in the Northern and Southern Hemispheres* (1817), both the story's narrative and the theme were essentially his own creation. The action takes place in 1799, off Santa Maria, an uninhabited Pacific island, where Captain Delano has anchored his ship, the *Bachelor's Delight*, to take on water. Seeing a strange ship attempting to enter the harbor and apparently in distress, Captain Delano, a benevolent man and "a person of a singularly undistrustful good nature," sets off in his whaleboat to board the ship and give assistance in piloting it in. He discovers the vessel to be the *San Dominick*, a Spanish merchantman, whose cargo of Negro slaves as well as passengers and crew have undergone much privation and suffering, many having died of scurvy and fever. Dispatching his boat crew to bring back food and water, Captain Delano remains on board to pilot the unhappy vessel into the harbor. During the period of several hours which elapses before the wind grows strong enough to move the ship, Captain Delano observes and hears much that is mystifying and arouses in him from time to time not only suspicions of wrongdoing but also fears for his safety: the slovenly neglect of the ship and the general lack of discipline, especially in regard to the Negroes; the seemingly furtive actions on the part of several of the crew; and the moody, reserved, hypochondriac behavior of the young captain, Don Benito, whose recital of the calamities endured by his ship Captain Delano finds difficult to credit in all its particulars. Good-natured and charitable, Captain Delano is capable of explaining away the incidents and actions which give rise to his recurring, but surely, he believes, fanciful and unwarranted, misgivings.

Suspense and tension build to the highly dramatic scene which begins when Don Benito, not with treacherous intent, as Captain Delano first surmises, but to save the captain and himself, leaps into the captain's whaleboat as the latter is about to put off from the ship, and which ends when Captain Delano's men capture the fleeing Spanish vessel after overcoming the mutinous Negroes in a hand-to-hand battle. The remainder of the story is revelation and explanation, given mainly in the form of extracts from desposition by Don Benito to the court which tries his supposedly devoted Negro bodyservant, Babo, and the latter's fellow ringleaders. These men had incited the slaves to overcome and put to death most of the white passengers and crew, including Don Benito's dearest friend, and had compelled Don Benito on pain of death to be a party to a plot calculated to effect the capture of Captain Delano's ship. When Captain Delano learns from Don Benito how far removed appearance had been from reality, he explains that the suffering he had seen on the ship, more apparent than real, had increased his good nature, compassion, and charity, and "enabled me to get the better of momentary distrust, at times when acuteness might have cost me my life without saving another's." But the less sanguine Don Benito suggests that not all men, even the best of them like the captain, who are deceived by appearance have the good fortune to be undeceived. Broken in health and unable to overcome the baleful effects of the evil embodied in Babo, Don Benito does not long survive.

In general, Melville's critics have praised *Benito Cereno* highly, some regarding it as more than an exciting mystery story on the theme of appearance versus reality. Melville, they have variously suggested, was writing disguised autobiography or condemning Negro slavery or indicting mid-nineteenth-century civilization for its spiritual hollowness or dramatically illustrating the theme that evil triumphs over good. What Melville seems particularly to be pointing up at the end of the story is the necessity of recognizing fully the reality of evil. Admirable though he is because of his benevolence and courage, Captain Delano reveals, as Richard Chase has put it, "a limited grasp on life" in his inability to understand why Don Benito does not recover from his experience. Like Delano, there were good men in Melville's own day who seemed incapable of seeing the profound influence of evil in the world.

Good versus evil, innocence and trustfulness versus conspiracy and guile, are also the themes of *Billy Budd.* But in developing them, Melville attempted a much more comprehensive and penetrating inquiry into the metaphysical, moral, and social questions concerning the nature and destiny of man than he did in *Benito Cereno.* Begun in 1886 and developed through several stages of revision, *Billy Budd, Sailor (An Inside Narrative),* as it is titled in the most recent, textually reliable edition, was probably suggested in part by an incident in 1842 in which a midshipman and two seamen of the American brig *Somers* were hanged at sea for mutiny. Melville changed his setting to a British man-of-war and the time to just after the British naval mutinies at Spithead and Nore in 1797. In Billy Budd, Claggart, and Captain Vere he created characters suggested to him by other sources.

Billy, whose impressment aboard H.M.S. *Bellipotent (Indomitable* in earlier editions) from the English merchantman *Rights-of-Man* takes place early in the story, is a young seaman of twenty-one who is uncommonly endowed with physical strength, beauty, and good nature. On the warship, as he had been on the merchant vessel, he is well liked by his fellows, except for Claggart, the master-at-arms, whose antipathy for Billy eventually leads him to accuse Billy to Captain Vere of plotting mutiny. When Claggart is called upon by Vere to repeat his charge to Billy's face, Billy, afflicted by a stammer that prevents him from speaking when excited, instinctively strikes out at Claggart, dealing him a blow on the forehead that kills him instantly. "Struck dead by an angel of God! Yet the angel must hang!" exclaims Vere, and the drumhead court convened to try Billy, influenced by Captain Vere's arguments that martial law and their duty demand Billy's conviction, confirms, although not without misgivings, that prophecy.

Although Melville characterizes Billy, Claggart, and Captain Vere fully, some of the implications of their natures and actions remain ambiguous. All three, Melville emphasizes, are rare and unusual beings, Claggart's evil nature being "born with him and innate," and Billy and Vere being "two of great Nature's nobler order." To some of Melville's critics Billy has seemed unsatisfactory as a tragic hero—if that is what Melville intended him to be—as well as a less convincingly drawn character than the other two. Like Adam before the Fall, Billy is absolute innocence. Incapable of understanding his fate and therefore of suffering because of it, he accepts as would a child or animal what happens to him. Claggart is like Satan in his pride; he personifies the evil that hates good and must ever try to destroy it. Vere, significantly, is endowed with those qualities which make him the well-nigh perfect embodiment of the just and impartial judge. Modest, brave, an experienced sailor, he is fair and kind toward his subordinates without brooking any laxity in discipline. Intellectually inclined and a wide reader, he is fond of those writers "who, free from cant and convention, like Montaigne, honestly and in the spirit of common sense philosophize on realities." Conservative in his social and political beliefs in an age of revolutionary unrest, he is so on grounds of reason rather than emotion or prejudice. Although he stands for authority, he is no lover of authority for its own sake. His dilemma and theirs, as he points out to the drumhead court, is evoked by the "clash of military duty with moral scruple—scruple vitalized by compassion."

In Billy's Christian meekness and humility, his acceptance of his fate, his "God bless Captain Vere!" just before his execution, and his ascension in "the full rose of the dawn," some of Melville's critics see an affirmation that goodness persists and triumphs over evil and injustice. For other critics, Melville, in portraying the inevitable triumph of expedient over absolute justice in society, shows no more faith than despair; and for yet others, there is a bitter irony in such details as Billy's blessing of Vere and the report in the official naval publication which makes Billy a villain and Claggart a hero. The "ironist" interpretations ask the reader to believe that Melville was saying that Billy's life is without meaning and that Vere is to be condemned. This the reader may find difficult to do, particularly when he remembers the private meeting in which Vere communicates the sentence of the court to Billy, culminating, it is suggested, in a kind of sacrament. Later Melville makes it explicit that neither Billy's agony, "mainly proceeding from a generous young heart's virgin experience of the diabolical incarnate and effective in some men," nor Vere's, which is greater, survives the "something healing" that occurred there.

Melville's shorter fiction, as Newton Arvin has said, "is never, or almost never, wholly lacking in interest and character, and sometimes it has the character of greatness." Most of Melville's tales are lesser ones, it is true, but the achievements in "Bartleby" and *Benito Cereno* give him the distinction of being the only American writer before the Civil War, except for Hawthorne, to produce noteworthy stories with profound significant themes. And *Billy Budd* is assuredly one of the finest of short novels.

## 3
## Terror, Mystery, and Ratiocination

## Edgar Allan Poe

In 1840, three years after Hawthorne published the first edition of *Twice-Told Tales*, there appeared a two-volume work entitled *Tales of the Grotesque and Arabesque*, by a Southern writer named Edgar Allan Poe (1809–49). The orphaned child of traveling actors, Poe was brought up by John Allan, a prosperous Richmond tobacco importer, who became his guardian but never legally adopted him. Poe attended schools in England, where the Allans lived for five years, and in Richmond, Virginia. He spent a year at the University of Virginia, and after quarreling with Allan went to Boston, where he published *Tamerlane and Other Poems* (1827) and began a two-year enlistment in the army. In 1829 he published *Al Aaraaf, Tamerlane, and Minor Poems*, and in 1831 was dismissed from West Point after a brief period as a cadet. In the same year he published *Poems* and soon afterward began to write fiction. Altogether Poe published about seventy tales between 1832 and 1849, the year he died under mysterious circumstances in Baltimore. During those years he lived successively in Richmond, Philadelphia, and New

York City, and was a prolific contributor to periodicals, several of which he also edited. Only about half of the tales which Poe published in the magazines were collected during his lifetime. *Tales of the Grotesque and Arabesque* contained twenty-five, and *Tales* (1845) included twelve, three of which were reprinted from the earlier volume.

In the Preface to the former collection Poe said, "The epithets 'Grotesque' and 'Arabesque' will be found to indicate with sufficient precision the prevalent tenor of the tales here published." Although he did not define these terms, the grotesque tales and sketches, of which there are fourteen, are clearly those in which Poe attempted, although without conspicuous success, to be humorous and satirical. In several early tales he burlesqued or otherwise satirized the extravagances and sensationalism found in certain kinds of contemporary popular tales. Tall-tale exaggeration is parodied in "Loss of Breath," in which a Mr. Lacko'breath loses his breath in the process of denouncing his wife, and since he continues to live, his peculiarity causes him to suffer broken arms, a fractured skull, hanging, and burial before he gets his breath back from one Mr. Windenough. "The Assignation" makes light of the Byronic hero and the tale of passion, and in "How to Write a Blackwood Article" a lady writer learns the formula from the magazine's editor for writing those tales known as "Intensities," an example of which was "The Dead Alive," "the record of a gentleman's sensations when entombed before the breath was out of his body; full of taste, terror, sentiment, metaphysics, and erudition." Poe also parodied this kind of tale in "A Predicament," in which the same lady writer records her sensations when she is trapped in a clock tower and the scimitarlike minute hand of the huge clock moves inexorably toward her neck, ultimately decapitating her. Despite his ridicule Poe exploited much the same situation in "The Pit and the Pendulum," and in other tales, also written with serious intent and often with consummate skill, he recorded the sensations evoked by premature burial and other terrifying experiences.

Little need be said concerning Poe's other grotesques. Predominantly satirical, the objects of their ridicule, usually carried to an extreme, include the practice of making social lions out of persons possessing eccentric or other unusual characteristics (in "Lionizing" the hero's abnormally large nose makes him a social lion until he shoots off the nose of a nobleman in a duel, who thereupon supplants him as the darling of society). Modern progress and reformers ("Some Words with a Mummy") and the not so admirable practices of some magazine editors of the day ("The Literary Life of Thingum Bob, Esq.") were also satirized by Poe, whose forte, though he seems never to have realized it since he continued to publish such pieces from time to time until the end of his career, was clearly not humor and satire.

It was undoubtedly to his arabesque tales that Poe referred, presumably in answer to the criticism that he had been influenced by E. T. A. Hoffmann and other German romancers, when he said in the Preface to *Tales of the Grotesque and Arabesque*: "If in many of my productions terror has been the thesis, I maintain that terror is not of Germany, but of the soul—that I have deduced this terror only from its legitimate sources, and urge it only to its legitimate results." This disclaimer is belied, however, by "Metzengerstein," first published in 1832, to which Poe added the subtitle "A Tale in Imitation of the German" when it was reprinted in 1836. Beginning with the words, "Horror and fatality have been stalking abroad in all ages," it tells how the enemy of a German baron achieves vengeance after death by means of metempsychosis, his soul inhabiting the body of a horse which is the cause of an accident fatal to the baron. Whether Poe was parodying the German tale of terror or whether he was seriously trying to outdo the genuine article is not easy to tell. At any rate, there is no suggestion of the burlesque in "Ms. Found in a Bottle," which in 1833 was awarded a $100 prize offered by a Baltimore periodical. Hurtled through a dark and mysterious ocean by a storm which has swept the crew of his ship overboard, the narrator is thrown aboard another vessel, with which it collides. The vessel is like no other, and its ghostly crew gives no sign of the narrator's existence. As the ship thunders on toward what must surely end in destruction, the narrator records his sensations, consigning them to a bottle just before the ship is engulfed in a whirlpool. Although Poe may have intended it to be read as a parable of human life, the tale is of interest mainly because he employed in it for the first time the narrative devices characteristic of some of his best later tales. It is told in the first person; every effort is made to give the effect of verisimilitude to an incredible occurrence; an atmosphere of terror and horror is created; and the sensations of the narrator as he undergoes the terrifying experience are emphasized.

In another early tale, "Berenice," Poe developed a formula which he followed in a half-dozen stories, in all of which the principal character is a madman or is afflicted with a morbid state of mind verging on madness, a trait which is usually explained as an inherited tendency. Poe admitted that he made "Berenice" excessively horrible, but justified doing so on the grounds that it was the kind of story readers wanted. Egaeus, the monomaniacal hero, aware that his reason is failing, becomes obsessed with the idea that to regain it he must have the teeth of his loved one. After she is seized by an epileptic trance and buried prematurely, without being conscious of his act he disinters her and removes the teeth from the still-living body.

Concerning "Berenice," Poe said that he would "not sin quite so egregiously again," and "Morella" and "Ligeia," although they follow the same general pattern, are less gory. Both have as their ostensible theme the idea that if the will of the human spirit is strong enough it will survive the death of the body, but Poe emphasizes in both tales the horrifying way in which this survival is achieved. In the former the dead wife, whom the narrator has come to abhor, is reincarnated in their daughter. As he sees the child, whom he loves fervently, grow more and more like her mother, his agony increases until he is compelled to call her Morella too. When she dies and is carried to the tomb, the horror of the father is complete when he finds no trace of the body of the first Morella. In the latter story, which Poe thought his best tale, the soul of the narrator's first wife, Ligeia, takes possession of the body of her successor. Like the bereaved lovers in Poe's poems dealing with the death of a beautiful woman, the husband suffers intense grief when Ligeia falls ill and dies. After a period of aimless wandering he chooses a remotely situated dwelling, a gloomy abbey, becomes a slave to opium, and marries the Lady Rowena, who is soon attacked by a strange malady which, one fearful night, results in her death. Again and again she appears to revive momentarily, and each time the narrator, although his mind is filled with thoughts of Ligeia, vainly attempts to aid her. In the final struggle to revive, the corpse manages to rise from the bed, and its appearance, so unlike that of Rowena, causes a horrible doubt to assail him, one which is confirmed when the shroud falls from the figure revealing the wild eyes of the Lady Ligeia. Even if all this is read as an obsessed narrator's hallucination with psychological and/or allegorical meanings, the tale nevertheless remains pure Gothic, and as an example of the type is masterful in its technique. Poe thought it the "loftiest kind" of tale because it made use of the "highest imagination," but he might better have said "highest artifice."

Rivaling "Ligeia" in its use of this same "imagination" is "The Fall of the House of Usher." Here again are all the stock trappings of Gothic romance: a gloomy, decayed mansion; a hero, Roderick Usher, who is afflicted with a strange mental disorder which makes him morbidly sensitive and who lives in terror of losing his reason; the death from a mysterious disease of Usher's twin sister, who rises from the tomb one wild, stormy night; and the simultaneous death of Usher and dissolution of his mansion, which sinks into the dark tarn beside it. Despite Usher's split personality and morbid sensibility, some critics have suggested that Poe's depiction of him would seem to have too little psychological validity to make him significant as a prototype of those fictional heroes of our own time who cannot make a satisfactory adjustment to the real world. Usher is in every respect the preposterous Gothic hero, albeit a very impressive specimen. Such respected critics as Cleanth Brooks and Robert Penn Warren believe that the horror in his story is generated for its own sake. Others insist that it should be taken more seriously, suggesting, variously, that the obsessed Usher and his narrator friend share a hallucination that Madeline rises from the tomb and that the event did not actually take place; that the story develops a vampire motif; and that it should be read allegorically, as others of Poe's fictions can also be read, in the light of Poe's philosophical views—as expressed in *Eureka* and elsewhere—or as a reflection of Poe's romantic desire to escape from reality. But however one reads the story, he must admire its effective use of symbolism, its vivid evocation of atmosphere

(the opening and closing sections of the story being particularly notable in this respect), and its sustained tension and dramatic power.

In several other tales Poe depicted the actions and sensations of madmen. Obsessed by the look in his victim's eye, the murderer in "The Tell-Tale Heart" perpetrates his deed with a cunning that conceals all evidence of his guilt but confesses to the unsuspecting police where he has hidden the body when he becomes convinced that they too can hear the beating of the victim's heart. In "The Black Cat" another murderer is afflicted with a mania which he calls "perverseness." A companion piece to this tale, but without its extravagant situation and excessive horror, is "The Imp of the Perverse," in which Poe analyzed in detail that quality in man which causes him to act against his will and better judgment. Still another tale of obsession is "William Wilson," a vivid if somewhat fantastic allegory in which Wilson's conscience is personified as an alter ego who exercises an unbearable authority over him. Driven finally to desperation, he kills his tormentor and thereby destroys himself. Wilson's early education, described at length, corresponds with Poe's, and the dissolute life Wilson leads, his unstable temperament, and his lack of self-control, although they are far more extreme, resemble certain aspects of Poe's life and character.

Poe's ability to give the appearance of truth to the implausible or incredible is manifested in many of his tales, but it is perhaps most evident in those stories which purport to be accounts of actual experiences, as in *The Narrative of Arthur Gordon Pym* and "The Journal of Julius Rodman," narratives of a voyage to the Antarctic and of an expedition to the banks of the Missouri River, respectively. They are only partly imaginary; Poe draws heavily on sources referred to in the text. The former story, Poe's longest, has a preface designed

to persuade the reader of the authenticity of the narrative, and throughout Poe is as matter-of-fact and circumstantial as either Daniel Defoe or Jonathan Swift. The first thirteen chapters, however, constitute primarily an adventure story filled with excitement and suspense, in which a lad of eighteen stows away on a whaling vessel and experiences such hazards as mutiny, storms, hunger, thirst, and even cannibalism, practiced by the survivors of the disabled vessel. But after Pym and one of the crew members are rescued by a trading ship whose captain is bent on exploring Antarctic waters, the character of the narrative changes radically. A great deal of data concerning previous voyages of exploration and discovery is given, as well as a detailed description of the explorations of Pym's ship, which culminate in an encounter with savages, who destroy the vessel and kill the ship's company, Pym and his companion being the only survivors. The narrative ends abruptly and inconclusively as the two are carried in a frail canoe into a strange and terrifying sea which threatens their destruction. Despite its lack of unity and character development, the story has been much discussed by Poe's critics, who have justifiably found considerable power and intensity in it but who have also tended to see in it perhaps more symbolical significance than the tale, which Poe once called "a very silly book," is able to support.

Poe also employed the method of the Defoe-like circumstantial narrative in several burlesques and hoaxes. One such work is "Hans Phall." If it were not for its framework, the tone of which is reminiscent of Irving in his Diedrich Knickerbocker manner and which gives away the fact that it is imaginary, this account of the construction of a balloon and a nineteen-day trip to the moon would be one of Poe's most effective attempts at making the improbable seem probable. Poe also exploited the balloon-voyage theme in two other short pieces

of science fiction, one an account of a three-day flight across the Atlantic so convincing that it duped the readers of a New York newspaper in which it was first printed. It was subsequently reprinted under the title "The Balloon Hoax." The other tale, "Mellonta Tauta," is ostensibly a journal of a balloon journey in 2848, but Poe was less concerned with picturing the world as it might be a thousand years after he was writing than he was in satirizing those aspects of his own age which he disliked, among them republican government, the social ideas of men like François Fourier, and the philosophy of John Stuart Mill.

The most convincing and artistically effective of Poe's circumstantial narratives is undoubtedly "A Descent into the Maelstrom," a detailed and vivid account by a survivor of a fishing vessel that is sucked into the Moskoe-ström, a tidal whirlpool off the northwest coast of Norway. He escapes by throwing himself overboard lashed to a cask, having been astute enough to observe that cylindrical objects are not drawn to the bottom of the whirlpool and engulfed.

A much less credible tale of a narrow escape from death, but one of Poe's most masterly depictions of the agonies of mind evoked by a terrifying experience, is "The Pit and the Pendulum." Imprisoned by the Spanish Inquisition in a dungeon, the narrator has not one but three narrow escapes: by lucky accident he avoids plunging into a deep pit to an awful death; by his wits he avoids being sliced in two by a razor-sharp pendulum descending toward him; and finally he is forced toward the pit by the contracting fiery walls of his dungeon. As he totters on the brink, they retreat. He has been saved through the defeat of the Inquisition forces by the French army.

At the beginning of this tale the narrator fears his dungeon may be a tomb. Premature burial was too good a subject not to exploit more fully than Poe had done here and in an occasional other tale. "The Premature Burial" begins with the citation of several instances indicating that persons thought dead had suffered the hideous fate of reviving in the grave; the tale strives to convey the impression that the experience of the narrator was an actual one. Obsessed by his fear that during one of his cataleptic fits he will be thought dead and be buried alive, the narrator believes that the elaborate precautions he has taken to avoid such a fate have been of no avail when he recovers from one of his trances in what he is certain is a coffin but which is instead a wooden bunk in a boat. The outcome of this agonizing experience is that it has a therapeutic value. He forgets his fears and is cured of his cataleptic disorder.

Poe concentrated on atmosphere in much of his fiction, and in some of his tales he relied on it almost entirely for effect. The most impressive of these tales is "The Masque of the Red Death," a single striking episode in which Prince Prospero defies the plague devastating his kingdom. Secluding himself and his retainers in a seemingly impregnable castle, he gives a masked ball in a suite of seven chambers richly and fantastically decorated in various colors. One of the rooms is shrouded in black velvet with window panes of a blood color. Here, as a gigantic ebony clock strikes midnight, the Red Death, in the awful guise of a stiffened, plague-stricken corpse, interrupts the revelry. Accosted by the prince, who falls dead before it, the figure is seized by the other revelers and is discovered to have no tangible form under the deathlike mask and graveclothes. For some critics atmosphere is everything here, for others the story is a moral parable on the *memento mori* theme, and others find even more portentous meanings.

Also perhaps implicit in this tale is the idea of nemesis; but the theme of revenge, a staple one in the tale of terror and horror, Poe rather surprisingly did not utilize again after the very early "Metzengerstein" until almost the end of his career, in "The Cask of Amontillado." A classic tale of its kind, one of its notable features is the economy with which it is told. Instead of the discursive opening which characterizes a good many of Poe's tales, the theme and plot are indicated succinctly at the outset. The first sentence states that it was an insult—the nature of which is not explained since in itself it is not important—that finally moved Montresor to take revenge on Fortunato. It is also made clear that the revenge, to be worthy of the name, had to satisfy certain requirements. "I must not only punish," says Montresor, "but punish with impunity. A wrong is unredressed when retribution overtakes its redresser. It is equally unredressed when the avenger fails to make himself felt as such to him who has done the wrong." The conclusion makes it clear that the requirements were satisfied. Although intoxicated when lured into the catacombs, Fortunato is sober when he is walled up by his enemy and thus is aware of what is happening to him. Furthermore, that Montresor has gone unpunished for his crime is made evident by his statement that fifty years have passed without Fortunato's tomb being disturbed.

The tale is replete with irony. There is irony in Fortunato's name, the evil action takes place during the carnival season, and Fortunato, whose vanity makes him Montresor's victim, is garbed in a fool's costume. The more Montresor feigns reluctance to impose on Fortunato, the more the latter insists that he must taste Montresor's newly purchased wine to determine whether it is genuine Amontillado. As they enter the damp wine vaults, Fortunato is seized with a cough and Montresor is most solicitous: these catacombs will endanger his friend's health. But Fortunato will not go back. "I shall not die of a cough," he says, to which Montresor replies,

"True—true." They pause for a draught of wine, and Montresor drinks to Forunato's long life. When the tipsy Fortunato inquires whether Montresor belongs to the masonic brotherhood, the latter produces a trowel from the folds of his cloak. More meaningful than these ironic touches, which call a little too much attention to themselves, is the irony that Montresor does not confess his crime because of a guilty conscience. His account, although some critical opinion views Montresor as another of Poe's obsessed narrators or finds moral meaning in it, is simply a matter-of-fact recital by a man who personifies the spirit of cold-blooded revenge. The extent to which its unity of effect contributes to this tale becomes all the more apparent when it is compared with "Hop-Frog: or, The Eight Chained Ourang-Outangs," Poe's one other tale of revenge, a readable enough, but far less skillfully told, story of how a cruel king who loves to play practical jokes is the victim of one played on him by the court jester he has persecuted.

Although Poe is usually credited with being the father of modern detective fiction, there is no evidence that in "The Murders in the Rue Morgue," "The Mystery of Marie Roget," and "The Purloined Letter" he was aware that he had written what was, in at least some respects, a new kind of story. In the first of these tales, Poe created his detective, C. Auguste Dupin, and gave him a mystery to solve, illustrating how what Poe called the "faculty of ratiocination" works. The story proper is preceded by several pages of exposition, in which a distinction is made between mere concentration and imaginative insight and between mere ingenuity and true analytical power. At the conclusion Poe says, "The narrative which follows will appear to the reader somewhat in the light of a commentary upon the propositions just advanced." Even at this point Poe is not ready to begin his story; several more pages are devoted to introducing Dupin and to giving an example of Dupin's "peculiar analytic ability," in which Dupin reads the narrator's mind and then explains how, through observation and inference, he has been able to do so.

The circumstances surrounding the murders are presented in a series of newspaper accounts perused by Dupin and his narrator friend. A Madame L'Espanaye and her grown daughter, who occupied the fourth story of a house in Paris, are found brutally murdered, the corpse of the daughter forced up the fireplace into the chimney and that of the mother, her head severed from her body, discovered in a yard in the rear of the building. Depositions are taken by the police from persons who heard screams and two voices, one of which, it was generally agreed, was that of a Frenchman, whereas there is no agreement about the language spoken by the other voice. The doors and windows of the women's apartment were found locked or fastened, leaving no apparent means of egress. According to the physician who examined the bodies, they were mutilated in such a way that the murderer must have been someone of very great strength. A bank clerk, who accompanied Madame L'Espanaye to her home on the day of her death with a sum of money she had drawn from the bank, is arrested, but beyond that there is nothing to incriminate him. It is at this point, when the narrator—and all Paris for that matter—regards the murders as "an insoluble mystery," that Dupin, because it will "afford us amusement" and because the bank clerk had once done him a favor, applies himself to the problem.

What has confounded the police, Dupin points out, are the apparently unexplainable aspects of the murders, but, paradoxically, it is those aspects that make it easy for Dupin, after he has analyzed the depositions and examined the scene of the murders, to construct a hypothesis, the validity of which is ultimately demonstrated. Dupin discovers that the means of both ingress and egress must have been a window which was apparently, but not actually, fastened with a broken nail. To enter and leave through the window would have required unusual agility, and from this fact, from the seeming absence of motive, from the fact that the killings required superhuman strength, and from the testimony that the voice not belonging to the Frenchman had sounded foreign to those who heard it, Dupin deduces that the killer was not human. Dupin finds a small tuft of hair, presumably overlooked by the police, clutched in Madame L'Espanaye's fingers, and the imprint of fingers too large to be those of a human being on the daughter's throat indicate that the killer must have been an orangutan that had escaped from its owner, a Maltese sailor. The sailor is tracked down by Dupin, and he confesses that he witnessed, without being able to prevent, the horrible deeds of the animal, giving an account of the murders which corroborates Dupin's hypothesis.

In "The Mystery of Marie Roget," Poe reintroduced Dupin, referring to his exploits in solving the Rue Morgue murders, and set him to work again, exercising his analytical faculty on a mystery which, as Dupin indicates, is almost the exact opposite of the Rue Morgue case, since it involves an ordinary rather than an extraordinary crime. The story closely follows an actual case, that of a Mary Rogers, murdered in New York City in 1841. Poe changed her name and placed the murder in Paris under similar circumstances. At the beginning of the story Poe pointed out that readers would recognize details which corresponded to those of the Mary Rogers case, and when the story was republished in the *Tales*, Poe revised it to eliminate certain contradictions and illogical arguments and appended footnotes indicating the correspondences, identifying the names of the characters, places, newspaper accounts, and so on with the actual names.

> Herein [wrote Poe], under pretense of relating the fate of a Parisian *grisette*, the author has followed, in minute detail, the essential, while merely paralleling the inessential, facts of the real murder of Mary Rogers. Thus all argument founded upon the fiction is applicable to the truth: and the investigation of the truth was the object.

With his keen journalistic sense, Poe no doubt counted on his story having a wide appeal because of the considerable publicity given the mystery in the newspapers. It is suggested at the end of the story that the police, by following Dupin's reasoning, were able to apprehend the murderer, a naval officer in whose company she had been when she disappeared mysteriously for a week some time earlier, and Poe himself asserted that confessions obtained after the story was published confirmed Dupin's conclusion and the hypothetical details on which it was based. The case of Mary Rogers, however, was never solved, and how sound Poe's deductions were will probably never be known.

"The Mystery of Marie Roget" is the least successful of Poe's detective stories, partly because of its inconclusive ending but chiefly because it is less a story than a somewhat tedious essay in deductive reasoning. "The Murders in the Rue Morgue" has more story quality, but it too lacks the unity, compactness, and dramatic character of "The Purloined Letter." In this story there is no preliminary treatise on analysis or on coincidences and probabilities as in the other two stories. At the beginning the narrator, in company with Dupin in the latter's study, is meditating on the Rue Morgue affair and the Marie Roget mystery, the subject of their conversation earlier in the evening, when the prefect of the Paris police enters. He has been unable to solve a case which baffles him because of its very simplicity. An unscrupulous Minister D—— has stolen a letter from a member of the royal family, who knows that he has taken it. The contents of the letter will somehow embarrass or incriminate her if they are made known to another eminent person. Possession of the letter thus gives the minister a power over her which he can use for political purposes. The prefect has been unable to get back the letter, which he is certain is still in the minister's possession, despite a rigorous search of the minister's premises and of the person of the minister under the ruse of having him waylaid by footpads.

"For three months," says the prefect, "a night has not passed, during the greater part of which I have not been engaged, personally, in ransacking the D—— Hotel." The narrator can only conclude, along with the prefect, that the letter is not on the premises, but Dupin suggests that another search be made. The unhappy prefect returns a month later, after having conducted another unsuccessful search. He would give, he says, fifty thousand francs to anyone who could help him. Dupin tells him to write out a check for that amount and in return for it hands over the letter to the dumbfounded prefect while the narrator looks on astounded. In Dupin's subsequent explanation to the narrator Poe elaborated perhaps more than he needed to on the principles of deduction, but the contrast between the unimaginative prefect and the astute Dupin is effectively brought out. The prefect, in short, underestimated the intellect of the minister. Regarding him as little more than a fool because he is a poet, the prefect could not utilize the principle, even if he were aware of it, that makes for Dupin's success, namely, "an identification of the reasoner's intellect with that of his opponent." Dupin, knowing the minister's cleverness and ingenuity and aware that he must always have the letter near at hand for his purposes, concludes that "the Minister had resorted to the comprehensive and sagacious expedient of not attempting to conceal it at all."

Dupin's account of how he tests his hypothesis and obtains the letter is succinct and highly dramatic. Calling on the minister one morning, Dupin carefully scrutinizes the apartment during their conversation under cover of a pair of green spectacles necessitated, he tells the minister, by weak eyes. He sees a letter, apparently carelessly thrust in a card rack, but bearing a different seal and address from the letter described by the prefect. Dupin concludes that it is the same letter turned inside out, redirected and resealed. (The careful reader will observe here that Dupin is not only studying the letter from the other end of the room through dark glasses but also sees both the seal and the address, which would be on opposite sides of the letter.) Calling on the minister the next day, Dupin substitutes a facsimile for the letter during a moment when the minister's attention is attracted to a Dupin-staged disturbance in the street. Dupin's ruse to possess himself of the letter is especially appropriate since it employs essentially the same method the minister had used to obtain it. The outcome is satisfying not only because it will bring about the downfall of an unprincipled man but also because it enables Dupin to even the score for an ill turn the minister once did him. What is most impressive about Dupin's feat, however, is that, in addition to demonstrating his superiority to the police, as in the other stories, he outwits an opponent worthy of his mettle.

What are the elements in the Dupin stories which have often been reflected in detective fiction since Poe's day? First, there is the familiar figure of the brilliant, often eccentric amateur or private detective. Why Poe made Dupin eccentric may perhaps be explained by the fact that Poe's heroes are almost invariably unusual men. Poe was less concerned with characterizing Dupin, however, than with demonstrating his method of solving a mystery, and as a result Dupin is less memorable than Sherlock Holmes and some of his other literary descendents. Poe adds little to his initial portrait of Dupin in "The Murders in the Rue Morgue"—young man of good family but meager income who loves books and seclusion and is strangely lethargic insofar as the ordinary activities of men are concerned. In "The Mystery of Marie Roget," Poe's obvious use of Dupin as a mouthpiece results in his being almost completely dehumanized; but in "The Purloined Letter," his explanation of how he outwitted both the prefect and the minister suggests that, like most other human beings, he is not entirely above vanity and malice.

Dupin's narrator friend remains anonymous throughout; we know almost nothing about him except that he met Dupin by accident and that the two live in an old mansion rented by the narrator. Nor is there any need to know more. He serves a useful, if obvious, function, and, like Dupin, is an all too familiar character in detective fiction since Poe. He enhances the impression of the detective's brilliance and his presence is flattering to the reader, who may be just as much in the dark, but who can feel persuaded, when apprised of the solution, that he has not been quite so undiscerning. Furthermore, the use of such a character as narrator enables the writer to mask the existence of clues necessary to solving the mystery without seeming to do so, since, unlike the brilliant detective, the narrator is unaware of their significance. Devices like Poe's apparently locked room and concealment in the most obvious place have also been staple ingredients of the detective story, and his police are the precursors of a host of methodical, persistent, but unimaginative police inspectors and "dumb cops" in the often uninspired, mechanical stories which have subsequently employed Poe's formula and make it now seem hackneyed.

Poe wrote two other tales of mystery and analysis which, it is sometimes asserted, are not true detective stories because they withhold evidence essential to solving the mystery until after the solution has been revealed. "Thou Art the Man" is written in a tone of levity scarcely appropriate to a murder mystery, which suggests that it may have been intended as a burlesque of the story of ratiocination. Poe's attempt at humor falls very flat. However, Poe seems here to have once more pioneered in the use of now commonplace devices to keep the reader in the dark by having his murderer plant false clues and otherwise contrive to cause suspicion to fall on an innocent person. The psychological device of unexpectedly confronting the criminal with the corpse to achieve a confession also appears to have been new to the mystery story.

"The Gold Bug" fully deserves its reputation as a classic mystery story. Its hero is a man whose analytical faculty rivals that of Dupin and who markedly resembles him in other respects. Once well-to-do but now reduced to living with his Negro servant, Jupiter, in impoverished circumstances in a hut on an island in Charleston Harbor, Legrand has read widely and is somewhat eccentric. The Dupin-narrator relationship is repeated in the narrator's friendship with Legrand, the narrator coming from Charleston to visit him from time to time. During one such visit Legrand has just discovered an unusual scarabaeus—a beetle of a brilliant gold color—and, not having it in his possession at the moment, he makes a sketch which the narrator observes looks more like a skull than a bug. Legrand examines the drawing and acts strangely. A month later Jupiter comes to the narrator in Charleston with a note saying that Legrand wants to see him at once. Jupiter reports that his master has been writing queer figures on a slate and talking about gold in his sleep. When Legrand asserts that the gold bug will make his fortune if used properly, the narrator is almost ready to believe, along with the superstitious Jupiter, that the bite of the bug has made Legrand mad. Against his better judgment, the narrator, to humor his friend, consents to accompany Legrand and Jupiter on an expedition, concerning the nature of which the narrator is entirely ignorant until it results in the discovery of a buried treasure of immense value.

Legrand's explanation, although it takes up almost half the story, is as full of interest and suspense as what has gone before. The paper—actually a piece of parchment on which Legrand had sketched the gold bug—had been picked up by Jupiter near where the scarabaeus had been found and had been used to wrap up the bug. The narrator had sat near the fire while examining the sketch, and the heat had caused the skull to appear. Various reasons caused Legrand to associate the parchment with Captain Kidd and his buried treasure, and further heating of the parchment brought out a cipher, the decoding of which Legrand explains, as well as how he was required to exercise his powers of analysis further before directions for finding the treasure acquired meaning. He also confesses that he deliberately set about to mystify the narrator because he was annoyed by the latter's evident suspicions of his sanity. Poe could thus keep the reader in the dark without seeming to withhold information unfairly. Another touch, appropriate to a tale of buried treasure, is that Legrand admits that he needed good luck along with his powers of analysis to make the discovery.

The virtuosity displayed in "The Gold Bug" and in many of Poe's other tales is impressive. The popular misconception (perhaps not as common now as it once was) that his tales and poems were the products of a genius given to liquor and drugs to induce inspiration for literary creation is disproved both by the facts of Poe's life and by the writings themselves. Poe developed and enunciated precise, if somewhat limited, theories of what a poem and a prose tale should be. He wrote in a review of Hawthorne's *Twice-Told Tales*:

> A skillful literary artist has constructed a tale. If wise, he has not fashioned his thoughts to accommodate his incidents; but having conceived, with deliberate care, a certain unique or single *effect* to be wrought out, he then invents such incidents—he then combines such events as may best aid him in establishing this preconceived effect. If his very initial sentence tend not to the outbringing of this effect, then he has failed in his first step. In the whole composition there should be no word written, of which the tendency, direct or indirect, is not to the one pre-established design. And by such means with such care and skill, a picture is at length painted which leaves in the mind of him who contemplates it with a kindred art, a sense of the fullest satisfaction.

This subsequently influential and now famous doctrine was propounded by one of the most professional of writers, who knew exactly what he was about and who, to a remarkable degree, made his practice accord with his theory. But there is another passage in the Hawthorne review which should also be noted because it bears significantly on the problem of evaluating Poe's tales. Terror, passion, horror, and many other such effects, said Poe, could be treated to advantage in the prose tale, whereas Beauty was the province of the poem. He continued:

> And here it will be seen how full of prejudice are the usual animadversions against those *tales of effect*, many fine examples of which were found in the earlier numbers of *Blackwood*. The impressions produced were wrought in

> a legitimate sphere of action, and constituted a legitimate although sometimes an exaggerated interest. They were relished by every man of genius: although there were found many men of genius who condemned them without just ground. The true critic will but demand that the design intended be accomplished, to the fullest extent, by the means most advantageously applicable.

If we can be content with Poe's definition here of "the true critic," we can rate his tales very high indeed. But what of the "design" itself? Literature of a high order results from the application of technique to significant themes, and there is by no means general agreement about the extent that these qualities are present in Poe's work. Poe certainly has his ardent advocates, one going so far as to call him the greatest American writer and the one of most significance in world literature. At the other extreme there is Henry James's famous statement that "an enthusiasm for Poe is the mark of a decidedly primitive stage of reflection." And somewhere in between are Allen Tate, who credits Poe with discovering the great subject of the disintegration of the modern personality, but at the same time says that Poe is primarily of historical interest to us; and T. S. Eliot, who reminds us of the adulation accorded Poe by Baudelaire, Valéry, and other French poets and critics, but who, like Tate, cannot quite give Poe his unqualified approval.

Despite his faults, Poe has many virtues. He is eminently readable; he does write with power and intensity; he is a master of effects; he is the originator of the detective story; and although his greatest achievement is undoubtedly in the short story, he is also impressive as a poet and critic. We can appreciate and admire him whether or not we find profound moral, psychological, and philosophical meanings in his stories. He is the forerunner of such very different writers as

Dostoevski, Sir Arthur Conan Doyle, H. G. Wells, and Robert Louis Stevenson, and he will surely retain the appeal he seems always to have had for all classes of readers. He is, to borrow one of his favorite words, unique.

# 4
# Local Color and Western Humor
## Bret Harte and Mark Twain

ALTHOUGH Washington Irving gave "The Legend of Sleepy Hollow" and some of his other tales with American settings a pronounced regional flavor, there were only a few other writers before the Civil War who, in anything more than incidental fashion, followed him in his primary concern with depicting those aspects of the speech, dress, mannerisms, customs, and geographical setting that were peculiar to a particular locality or region of the country. In what was to be the phenomenal development after the war of this kind of story —often called the local-color story—a major event was the publication in 1868 of "The Luck of Roaring Camp" by Bret Harte in a new California magazine, the *Overland Monthly*. This story, with its picturesque California mining-camp setting and its artful blending of paradox, incongruity, and sentiment in the manner of Dickens, was enthusiastically acclaimed in the East, and Harte soon followed it with similar stories in the *Overland*, among them being the two others for which he is best remembered, "The Outcasts of Poker Flat" and "Tennessee's Partner."

Born in Albany, New York, Francis Bret Harte (1836–1902) went to California as a boy of seventeen. After working at various occupations, including schoolteaching, mining, and typesetting, he began to contribute to various California newspapers and magazines, eventually becoming the first editor of the *Overland Monthly. Condensed Novels and Other Papers* (1867) is a representative selection of the writing he did during the ten years before the publication of "The Luck of Roaring Camp." It contains amusing parodies of James Fenimore Cooper, Charlotte Brontë, Alexandre Dumas père, Charles Dickens, and other novelists, descriptive essays and narrative sketches on miscellaneous subjects, and several Irvingesque tales dealing with the legends of early Spanish California. Included also are two California mining-camp stories published in 1860, "A Night at Wingdam" and "The Work on Red Mountain" (later rewritten and entitled "M'liss"), in which Harte first employed the subject matter and narrative method that were to make him famous.

That subject matter and the way Harte treated it are nowhere better exemplified than in "The Luck of Roaring Camp." The time is 1850, and the story opens with a mining-camp prostitute named Cherokee Sal dying in childbirth. Except for Sal, the population of the camp has been entirely male, and there follows the ludicrous spectacle of the crude, rough miners attempting to rear the infant, whom they have christened Thomas Luck. His presence appeals to their better natures and is a regenerative influence on them. There is less shouting and swearing, and even the mining claims yield enormously. But tragedy strikes when winter floods inundate the camp and the child is drowned.

Paradox and incongruity are everywhere in the story. Concerning the inhabitants of Roaring Camp, Harte wrote:

> The greatest scamp had a Raphael face, with a profusion of blond hair; Oakhurst, a gambler, had the melancholy air and intellectual abstraction of a Hamlet; the coolest and most courageous man was scarcely over five feet in height, with a soft voice and an embarrassed timid manner. . . . The strongest man had but three fingers on his right hand; the best shot had but one eye.

The little man known as Stumpy serves as midwife, godfather, father, and mother to the child, and the rough Kentuck, who can't forget how "the d——d little cuss" had "rastled with" his finger, performs the almost impossible feat, for him, of washing and donning a clean shirt every afternoon so that he can hold the child. Harte also wrote in an elevated literary English, a striking contrast to the actions and the speech of his characters, who express themselves in the picturesque lingo of the mining camp and gambling table. Equally obvious is the exploitation of sentiment, as in the attempt at the beginning of the story to arouse pity and compassion for Cherokee Sal: "Dissolute, abandoned, and irreclaimable, she was yet suffering a martyrdom hard enough to bear even when veiled by sympathizing womanhood, but now terrible in her loneliness." When she dies, the reader is told that "she had climbed, as it were, that rugged road that led to the stars, and so passed out of Roaring Camp, its sin and shame, forever." The story ends, as it began, with a sentimentalized death: " 'Tell the boys I've got The Luck with me now,' " says the dying Kentuck, "and the strong man, clinging to the frail babe as a drowning man is said to cling to a straw, drifted away into the shadowy river that flows forever to the unknown sea."

Like the miners of Roaring Camp, the characters in "The Outcasts of Poker Flat," whose presence even a rough mining community decides it can no longer tolerate, turn out to have hearts of gold. The sluice robber and drunkard known as Uncle Billy is an exception, since the exigencies of Harte's plot require him to be the villain, but the gambler Oakhurst and the two women known as The Duchess and Mother Shipton —whose "impropriety was professional," to quote Harte's euphemistic description—more than rise to the occasion when their better instincts are appealed to. Snowed in, after Uncle Billy has run off with the mules, in a rude mountain cabin in company with a young couple on their way to Poker Flat to be married, the party passes the time singing to the accompaniment of the girl Piney's accordion and listening to her companion, Tom Simson (better known as The Innocent of Sandy Bar), narrate in Sandy Bar vernacular the exploits of the heroes of the *Iliad*. Food and firewood become scarce after a week, but no one complains. Oakhurst, who could have saved himself by deserting his weaker companions on the way to Sandy Bar, accepts "the losing game" coolly, and on the tenth day Mother Shipton dies, having starved herself for a week so that Piney could have her rations. When rescuers arrive several days later, brought by The Innocent, who has managed to get to Poker Flat on a pair of snowshoes made by Oakhurst from a packsaddle, they find Piney and the Duchess frozen to death in each other's arms and Oakhurst, "who struck a streak of bad luck," a suicide with a bullet in his heart. Contrived and melodramatic though it is, the story is one of the most artful developments of the Harte formula and the best of his many variations on the theme that immoral and even criminal men are capable of generous and virtuous actions.

The loyalty of a miner to his partner, another of Harte's favorite themes, is given characteristically exaggerated development in "Tennessee's Partner." In a lecture on the gold miners as he saw them, published as the introduction to *Tales of the Argonauts* (1875), Harte said of the partner relationship: "To be a man's 'partner' signified something more than a common pecuniary or business interest; it was to be his friend through good or ill report, in adversity or fortune, to cleave to him and none other." Tennessee's partner more than measures up to these requirements. He forgives Tennessee for running off with his wife. He compromises himself by continuing his association with Tennessee after the latter turns highwayman, and when Tennessee is apprehended and brought to trial, he offers his savings to the judge as restitution for Tennessee's act, not realizing that the offer constitutes a bribe. After Tennessee is hanged, his partner provides him with a funeral and proper burial and then pines away and dies. In this altogether one-sided relationship the dumb, dogged devotion of the partner resembles that of a faithful dog to its master, and it is this quality, if we can refrain from viewing him too realistically, that makes him both a comic and a pathetically appealing figure.

In most of Harte's other early California stories the situations turn on acts of devotion and sacrifice. In "Miggles" a beautiful girl gives up the gay, loose life she has led to care for a man who was kind and generous to her before he was rendered mentally and physically helpless by a stroke, and, of course, she is ennobled by her act of renunciation. The New England schoolmistress Miss Mary, in "The Idyl of Red Gulch," renounces the handsome ne'er-do-well drunkard she loves to save the illegitimate child he has had by a woman of dubious reputation from the presumably baneful influence of his parents. In "Brown of Calaveras," when Jack Hamlin, the gentleman gambler, learns how much the woman he has planned to run away with means to her husband, Jack does the only decent thing under the circumstances, advising the husband to take his wife away and make a new start, and on his part fading out of the picture.

Hamlin, who figures in a number of Harte's stories, is cynical, ruthless, and without conscience in his dealings with those of his own kind but invariably compassionate and generous in his relations with those who are innocent, weak, or helpless, as in "A Protégée of Jack Hamlin's," in which he befriends a young girl who has been duped and deserted by another gambler. Although one of his chief trademarks is his association with women of doubtful reputation, Hamlin acts with a scrupulous morality toward the girl, and although she falls in love with him and he becomes increasingly attracted to her, they cannot, in keeping with the Harte formula, be permitted happiness together.

Two other characters who make many appearances in Harte's stories, usually in minor roles, are Colonel Starbottle, a version of the stock Southern Gentleman and a lawyer-politician noted for his eloquent oratory, and the stagecoach driver known as Yuba Bill. The latter performs his hazardous occupation—dangerous and difficult because of highwaymen and bad roads—with calmness and courage, and he has the characteristics of the western "tall man" in his capacity for liquor, his strength, and his tendency to take matters in his own hands when he deems it necessary. An example of this last trait—and incidentally of Harte's use of the western humorist's device of understatement—appears in Yuba Bill's account, in "Captain Jim's Friend," of what he did to a newspaper publisher who had presumed to criticize him for his lack of civility to passengers:

> "I jest meandered into that shop over there, and I sez, 'I want ter see the man ez runs this yer mill o' literatoor and progress. . . .' Bimeby the door opens again, and a fluffy

> coyote-lookin' feller comes in and allows that *he* is responsible for that yer paper. . . . Then I sez, 'You allowed in your paper that I oughter hev a little sevility knocked inter me, and I'm here to hev it done. You ken begin it now.' With that I reached for him, and we waltzed oncet or twicet around the room, and then I put him up on the mantelpiece and on them desks and little boxes, and took him down again, and kinder wiped the floor with him gin'rally, until the first thing I knowed he was outside the winder on the sidewalk. On'y blamed if I didn't forget to open the winder. If it hadn't been for that, it would hev been all quiet and peaceful-like, and nobody hev knowed it. But the sash being in the way, it sorter created a disturbance and unpleasantness *outside*."

In "An Ingénue of the Sierras," one of Harte's best later stories, Bill has as a passenger a young girl who is running away from home to meet a road agent, or highwayman, who has promised to marry her and reform. Bill is suspicious of her lover's intentions and takes it upon himself to see that justice is done by making sure that he marries the girl. Unlike the early California tales, which treat such situations sentimentally and melodramatically, this story has a humorous twist, the girl turning out to be a confederate of the road agents and Bill being duped into allowing his coach to be used by them to transport contraband in the girl's luggage.

In 1871, Harte collected his *Overland* stories in *The Luck of Roaring Camp and Other Sketches* and in the same year left California for the East, where he signed a ten-thousand-dollar contract for a year with the *Atlantic* for the exclusive publication of his sketches and poems, of which he was to contribute at least twelve. The contract was not renewed, and after a period during which he lectured on California and on American humor, wrote further stories, and tried novel and playwriting with indifferent success, he left America in 1878, living first in Germany and then in Scotland, in which countries he served as a United States consular agent, and after 1885 in England. If, as the years passed, Americans found Harte's stories less novel and entertaining, English readers continued to be appreciative, and Harte wrote on until his death in 1902. In almost all the more than 125 stories he published after leaving California, he followed the pattern of the *Overland* stories which won him fame, but in only a few did he attain the same level.

Although Harte frequently asserted that he could vouch for the realism of his characters and situations, he romanticized his materials to a greater degree than many other local colorists, especially those who identified themselves more closely with the regions they portrayed and who wrote less to exploit the eccentric and picturesque for its own sake than for some other purpose. From Harte's example, however, many of them learned that their stories must have a tone and form if they were to be more than sketches or loosely developed tales, and only a few of them were able to match the technical skill which makes "The Luck of Roaring Camp" and "The Outcasts of Poker Flat," whatever they may lack in verisimilitude, important landmarks in the development of the American short story.

LIKE BRET HARTE, Mark Twain (Samuel Langhorne Clemens, 1835–1910) first became widely known to readers through a short narrative with a California setting, "The Celebrated Jumping Frog of Calaveras County." First published in a New York newspaper in 1865 and reprinted in many periodicals throughout the country before it became the title story in 1867 of Mark Twain's first book, this famous yarn had behind it a long tradition of frontier humor. Sketches and anecdotes reflecting the humor and folklore of the frontier society of the South and Southwest, and later the Far West, first began to find their way into print early in the nineteenth century, and after the appearance of *Sketches of Davy Crockett* (1833) and A. B. Longstreet's *Georgia Scenes* (1835), the publication of such pieces increased by leaps and bounds. Besides reflecting native idiom and manners, they employed exaggeration of all kinds, abounded in absurdities and incongruities, portrayed the marvelous and fantastic in the most matter-of-fact and straight-faced manner, and at their best displayed individuality and vitality, if not literary art. Often their authors were satirists as well as humorists, and in time some of them developed into well-known professional funnymen who, besides delighting readers, entertained audiences from the lecture platform. Among the humorists whose pseudonyms became household words were John Phoenix, Artemus Ward, Petroleum V. Nasby, Orpheus C. Kerr, and Mark Twain himself.

Born in the small village of Florida, Missouri, Mark Twain grew up in nearby Hannibal, on the Mississippi River, learned the trade of printer, and contributed to his brother's newspaper. As a journeyman printer in half a dozen or so cities all the way from Keokuk, Iowa, to New York, as well as during his subsequent activity as a pilot on the Mississippi, from time to time he tried his hand at writing newspaper humor. Finally he became a newspaperman in earnest in Nevada Territory, after taking a fling at mining and speculating in mining claims. There for the Virginia City paper, as well as later in California, he wrote hoaxes, burlesques, and other humorous sketches. When these pieces had a satirical point, as they frequently did, his tendency to carry the literary devices of the frontier humorist to extreme lengths was likely to obscure the point, as Mark Twain himself pointed out in "The Petrified Man" and "My Bloody Massacre," which recount his experiences with two of his Virginia City hoaxes, each being what he called a "reformatory satire." Although his petrified-man story consisted of "a string of roaring absurdities," he had made them so convincing, he declared, that readers missed his point completely. Similarly, he had made the altogether preposterous "horrible details" of his other piece so engrossing that readers "never got down to where the satire part of it began." Some of Mark Twain's early satire, however, is more restrained, as in "The Story of the Bad Little Boy" and "The Story of the Good Little Boy," both written about 1865, which anticipate *Tom Sawyer* in their burlesque of the children depicted in the Sunday School books.

Mark Twain's jumping-frog yarn is the high point of his early writing, and not many of his later humorous narratives come near to equaling it. In "Private History of the 'Jumping Frog' Story," "How To Tell a Story," and elsewhere, he had a good deal to say—not all of it entirely serious—about its backgrounds and the technique of its telling. At the California mining camp called Angel's, he had heard how one Jim Smiley had been taken in by a crafty stranger, who, by surreptitiously filling Jim's frog with quail shot, had brought about its ignominious defeat in a jumping contest with the stranger's frog. Preoccupied with the "facts" of the story, neither the narrator nor his audience, according to Mark Twain, "was aware that a first-rate story had been told in a first-rate way, and that it was brimful of a quality whose presence they never suspected—humor." Mark Twain's version is similar in that neither Simon Wheeler, who gets so much satisfaction out of telling about Jim Smiley, nor the listener, on whom he inflicts his narrative, sees anything funny about it. Perhaps because Mark Twain emphasized that a humorous story of the kind exemplified by "The Jumping Frog" depends for its effect upon a grave manner of telling rather than on its matter, critics have usually been content to say that manner of telling is all-important in the story, neglecting to add that its situation and characterization are also responsible for its being a humorous sketch of a very high order. In creating Simon Wheeler, Mark Twain did much more than employ the stock device of dead-pan narration. The bored listener who provides the framework for Wheeler's story may profess to find it monotonous and interminable, but as soon as Wheeler opens his mouth, we hear an idiom and perceive an interest in and an attitude toward human and animal nature which remind us of *Huckleberry Finn*. Few episodes in that novel are more vivid or have finer comic effects.

"Baker's Bluejay Yarn," found in the last part of Chapter II and in Chapter III of *A Tramp Abroad* (1880), is another masterpiece of humorous storytelling. If anything, it is more artistically wrought than its more famous predecessor, being more unified and better pointed up and relying less on the stock devices of the professional humorist to achieve its effects. The narrator is somewhat akin to Simon Wheeler and is just as serious about his story, which he refers to as "a perfectly true fact," but he is not as obviously the garrulous yarn spinner, and his speech, more consistently than Wheeler's, approximates the vernacular of Huck Finn. There is no painfully bored listener, only Mark Twain himself, who briefly introduces Jim Baker and then lets him tell his story. As in Wheeler's description of Smiley and the various animals he bets on, which precedes the account of Smiley's encounter with the stranger, there is much humor in Baker's preliminary explanation of why a bluejay is just as human as a man, which, in addition, is a telling piece of satire on human nature. Baker maintains that he can understand the speech of birds and beasts, and the "little incident" he then relates of the bluejay, who without knowing it was trying to fill up a house with acorns by dropping them in a hole in the roof, is, if taken on its own terms, a convincing illustration of Baker's thesis—and wonderfully comic as well.

Interpolated in Mark Twain's books of travel and reminiscence are other yarns and anecdotes which are independent short narrative entities. *Roughing It* (1872) has "The Story of the Old Ram" and "Tom Quartz," both on the order of "The Jumping Frog," and "Buck Fanshaw's Funeral," which could very well pass for the work of Bret Harte. There is more tall-tale exaggeration in this anecdote than there is in Harte generally, especially in the description of the late Buck, who "could run faster, jump higher, hit harder, and hold more tanglefoot whisky without spilling it than any man in seventeen counties." But Scotty Briggs, the "stalwart rough" with a heart of gold and Buck's bosom friend, is a character right out of "The Luck of Roaring Camp" or "Tennessee's Partner." Like these stories, Mark Twain's sketch depends largely for its effects upon the incongruous elements in its characters and situation. No Harte character expresses himself more copiously or vividly in the lingo of the mining camp and gambling table than does Scotty when he calls on the young minister, newly arrived from the East, to arrange with him to conduct the funeral service for Buck. The excessively erudite speech of the minister may be overdone, but the resulting difficulty the two have in understanding each other is highly amusing.

Mark Twain's writing may have benefited, as he declared it did, from Harte's patient schooling of him in the mysteries of literary craftsmanship, but it also suffered on those occasions when he followed Harte's example in infusing it overmuch with sentiment. A few pieces like "A True Story," in which a Negro woman who had been a slave recounts her history, have a genuine pathos, but "The Californian's Tale" almost outdoes Harte in its sentimentalized account of a miner who believes his wife, who has died nineteen years before, is still alive, and of the efforts of his fellow miners to humor him in his delusion. Sentimentality also spoils some of the stories in which Mark Twain indicted man for his ingratitude, cruelty, and hypocrisy, among these being "A Dog's Tale," "A Horse's Tale," and "Was It Heaven or Hell?" Probably as mawkishly sentimental as anything Twain wrote is "The Death Disk," suggested by a "touching incident" he had come across in Carlyle's *Letters and Speeches of Oliver Cromwell* concerning the little daughter of an officer in Cromwell's army. Sentenced to death for having exceeded his authority in battle, the father is pardoned by the stern Puritan commander, whose heart is stirred by the little girl's plea for mercy.

For the most part, however, Mark Twain's shorter narratives reveal the humorist and satirist at work rather than an emulator of Harte-like sentiment and story form. Yarns, anecdotes, and satirical sketches and fables, most of them savoring of the tall tale, the hoax or joke, or the otherwise extravagantly exaggerated, flowed from his pen in profusion throughout his long career, as "Cannibalism in the Cars," "How I Edited an Agricultural Newspaper," "Journalism in Tennessee," "The Invalid's Story," "The Facts in the Great Beef Contract," "The Belated Russian Passport," "The Canvasser's Tale," and many other such pieces testify. Relying often on the patently absurd and implausible for his humorous and satirical effects, he had a sharp eye for manifestations of the artificial and improbable in serious fiction, as his attack on the creator of Deerslayer in "Fenimore Cooper's Literary

Offenses" and several burlesques of certain conventional kinds of tales attest. "A Medieval Romance" and "A Ghost Story" parody the kinds of stories their titles suggest; "The Stolen White Elephant" and "A Double-Barreled Detective Story" make sport of Sherlock Holmes detective fiction; and "The Loves of Alonzo Fitz Clarence and Rosannah Ethelton" and "The Esquimau Maiden's Romance" burlesque the romantic love story, which is also treated in mock-serious fashion in "The £1,000,000 Bank-Note," a lengthy yarn liberally sprinkled with broad satire at the expense of the English, about an impecunious San Francisco mining broker's clerk who wins fame and fortune in London.

Mark Twain's later writings include several stories of novelette or near-novelette length, two of them sequels to *The Adventures of Tom Sawyer* and *Adventures of Huckleberry Finn*. In *Tom Sawyer Abroad*, Tom, Huck, and the Negro slave Jim go on a balloon voyage which takes them as far as Africa, where they explore the Sahara Desert and the wonders of the pyramids. There is very little action in what is ostensibly an adventure story, the three characters being engaged in talk a great deal of the time. Huck and Jim are still very literal-minded and not nearly as well informed as Tom, who has a difficult time disabusing them of their many mistaken notions. This kind of thing is played up much more than in *Huckleberry Finn* and done less effectively. The Tom, Huck, and Jim of the longer work are recognizable, but they do not live for us in the same way, and their relationship to one another and the experiences they have are much less meaningful. In *Tom Sawyer, Detective*, Mark Twain again sent Tom and Huck down to Arkansas to visit Uncle Silas and Aunt Sally. There Tom becomes a hero when he saves Uncle Silas from being convicted of murder by deducing who the real murderers are, the story being related by an admiring Huck, who is dazzled by the brilliance of Tom's intellect. Along with some rather heavy-handed humor, the story has much the same kind of melodrama, violence, and other sensational elements which figure importantly in *Tom Sawyer*, but it is lacking in the beauty, charm, and moral quality which the earlier novel so clearly possesses.

Mark Twain's other longer stories reflect in different ways and in varying degrees the profound misanthropy and pessimism of his later years. "Extract from Captain Stormfield's Visit to Heaven" is a satire in his earlier yarn-spinning manner on the traditional notion of heaven as a place where harp playing, hymn singing, and hosannaing are the principal activities. But this generally good-humored fantasy—in which Captain Stormfield races a comet on his way to heaven and after getting there learns what it is really like from his observation and experiences, as well as from long talks with "an old bald-headed angel" named Sandy McWilliams—also takes a number of hits at man's pride, vanity, and folly.

Very different in tone and spirit is the bitterly sardonic representation of how men fall prey to greed in "The Man That Corrupted Hadleyburg." Although it is to Twain's credit that he expressed his theme almost entirely through character and dramatic action instead of relying overmuch, as was sometimes his habit, on explicit statement to make his point, the story has an air of contrivance, and its characters are more wooden than real. The stranger is bent on taking revenge for an unspecified offense done him by the supposedly incorruptible town of Hadleyburg, and of the group of citizens whom he is successful in corrupting with his sack of gold it seems necessary that at least the central figures of old Mr. and Mrs. Richards, whose fall is of the stuff of tragedy, be convincing. Regarded as symbols of honor and honesty because they escape the public exposure the others suffer, their situation is highly ironic, but the conflicts engendered in them by temptation and their plight after they succumb do not really engage the sympathies of the reader. In somewhat less caustic fashion, Clemens also satirized man's greed in "The $30,000 Bequest," in which a husband and wife fantasize about speculating on the stock market, amassing a huge imaginary fortune before a crash wipes them out. Although it has some humorous touches, the story is inferior to those parts of the novel, *The Gilded Age*, in which Clemens, in collaboration with Charles Dudley Warner, dealt with speculation and other get-rich-quick schemes.

Mark Twain's strongest and most comprehensive indictment of what he called "the damned human race" is to be found in *The Mysterious Stranger* (sometimes characterized as a short novel). This posthumously published narrative, which has as its setting a sixteenth-century Austrian village, has a plot of sorts involving an admirable old priest, Father Peter, and his niece. Both are victims of machinations of an evil astrologer. But the story is primarily a vehicle for the expression of a pessimistic and deterministic philosophy, which in most of its points Clemens also argued in *What Is Man?* published anonymously in 1906. Replacing the dogmatic Old Man of that essay-dialogue, Mark Twain's spokesman in *The Mysterious Stranger* is an angel who visits the village and reveals himself to three boys as a nephew and namesake of Satan, although he goes by the name of Philip Traum among men. Satan shows the boys visions of the past and of the future filled with war, murder, and other violence; calls their attention to instances in their village of man's pride, folly, and inhumanity; and transports Theodor, the narrator and Satan's special confidant, to various parts of the world to show him further examples. Theodor also learns from Satan that no life is worth living. Satan spares Theodor's friend

Nickolaus and the girl Lisa long lives of unhappiness and suffering by bringing about their deaths by drowning, and although he thwarts the astrologer's designs, he does Father Peter a far greater kindness by making him mad, since no sane man can be happy. Satan's greatest service, however, is reserved for Theodor in the final revelation that "there is no God, no universe, no human race, no earthly life, no heaven, no hell. It is all a dream—a grotesque and foolish dream."

Since *The Mysterious Stranger* is in the tradition of the philosophical tale written by Voltaire and others, it can perhaps be excused for leaving something to be desired as fiction, but the ambiguities and contradictions in its philosophy are a different matter. In spite of his avowed contempt for man, Satan's attitude toward him is markedly ambivalent. Some of Satan's pronouncements give the reader pause, as for instance when he expounds a determinism which relieves man of moral responsibility, yet also condemns him for abusing that moral sense which enables "him to distinguish between right and wrong, with liberty to choose which of them he will do . . . and in nine cases out of ten he prefers the wrong." Clemens also speaks through Theodor, however, and, although the story's artistic weaknesses and philosophical illogicalities keep the reader from sympathizing as fully as he should with Theodor's reactions to Satan's revelations, the boy's progress from innocence to knowledge has at least a certain measure of imaginative intensity and emotional impact.

In the yarn and comic anecdote Mark Twain was supreme. In his more formal stories he could be brilliantly funny and satirical as well as compelling in venting moral indignation, but he was usually too much the raconteur and improviser to be able to give his tales appropriate form and artistic unity. Sherwood Anderson, Hemingway, and other twentieth-century story writers are indebted to him, but their debt is to the Mark Twain of *Huckleberry Finn*, who spoke in the language of real speech, with high art and truth, on the great theme of the encounter of innocence with experience.

# 5
# The Regional Story in New England, the South, and the Middle West

## Sarah Orne Jewett
## Mary E. Wilkins Freeman
## George Washington Cable
## Joel Chandler Harris
## Thomas Nelson Page
## Hamlin Garland
## and Others

ALTHOUGH New England afforded less novel and colorful materials for local-color fiction than those available to Bret Harte, that did not deter writers of the region, for several decades beginning with the 1850's, from producing many stories with characters and settings taken from New England rural and village life. The customs, manners, and dialect of the rustic New Englander had already found expression in a considerable body of writing, some of it dating back to colonial times. Often the Yankee rustic was portrayed as no more than the stock figure of comic anecdote, but he was also represented more realistically in such writings as Sylvester Judd's novel *Margaret* (1845), in the Yankee dialect verses of James Russell Lowell's first series of *The Biglow Papers* (1848), and in Seba Smith's sketches in *Way Down East* (1854).

NOW REMEMBERED almost solely as the author of *Uncle Tom's Cabin*, Harriet Beecher Stowe (1811–96) also has the distinction of being one of the early writers of New England regional fiction. Her first published story, "A New England Sketch," appeared in 1834, and although for a number of years thereafter she was mainly a propagandist for abolition in her novels and short stories, she turned again to the portrayal of New England life and character in such novels as *The Minister's Wooing* (1859), *The Pearl of Orr's Island* (1862), and *Old Town Folks* (1869) and in tales, the best of which were written for the *Atlantic Monthly* and collected in *Oldtown Fireside Stories* (1871). Their narrator, Sam Lawson, the village blacksmith and raconteur, pausing often to quote Scripture, to make a shrewd comment on human nature, and to point an edifying moral, tells stories to an audience of small boys, who hang on his every word and who often have to urge him to get on with his story. Most of the stories have legendary elements and look back to the New England of colonial and revolutionary days, but some also deal with the more immediate past. One of the latter, "The Minister's Housekeeper," is a heartwarming little idyl in which Mrs. Stowe is at her best in the handling of dialect and in picturing, albeit somewhat romantically, human nature and rural folkways.

ANOTHER pioneer writer of New England local-color short fiction was Rose Terry Cooke (1827–92). Beginning with eight stories which were published in the *Atlantic* in 1857–58, the first year of that magazine's existence, she contributed almost one hundred tales to it and other periodicals over a period of three decades. Usually credited with being more realistic than Mrs. Stowe, Mrs. Cooke does give the impression of faithfully representing farm and village life in Connecticut, her native state. In some of her stories the events are vouched for as having actually happened. Of the miserly old widower who cruelly treats his new wife in "Mrs. Flint's Married Experience," the author says: "If this story were not absolutely true, I should scarce dare to invent such a character as Deacon Flint." "Miss Lucinda" begins by warning the reader that it contains nothing of the elevated or tragic, being "only a little story about a woman who could not be a heroine." In "Uncle Josh," the effect of his mother's death on a little boy is thus described: "When Josh woke up and knew his mother was dead, he did not behave in the least like good little boys in books, but dressed himself without a tear or a sob, and ran for the nearest neighbor."

At times, however, Mrs. Cooke succumbed to a tendency to emphasize unduly the peculiarities of her characters or to draw types rather than individuals. A character she was especially fond of portraying was the wife in the small village or on the farm who drudged unremittingly, but usually uncomplainingly, and whose life was often one long struggle with poverty and illness. Though Mrs. Cooke was justifiably indignant over the plight of such women, the very strength of her feeling usually kept her from employing the restraint necessary to make them convincing characters. Occasionally, too, she could lapse into the sentimental and even the melodramatic, as in "Eben Jackson," a lachrymose tale of a sailor who dies of fever and of the woman who is faithful to him unto death, and in "Clary's Trial," in which a beautiful indentured girl is rescued from the villain by a noble hero.

Certain stereotyped story situations also recur in Mrs. Cooke's work, one being that in which the rustic young lovers are kept apart by feuding parents or other circumstances but in which love finds a way to bring them together by the end of the story. Another conspicuous weakness in many of her stories is their extremely loose construction. "Grit," in which a stubborn father is opposed by an equally determined and resolute daughter, is a notable exception, and if Mrs. Cooke could have given more of her stories its unity, proportion, and concision, she might, with her often sympathetic and penetrating insight into New England character and her skill at reproducing the vernacular, have produced a body of short fiction of literary excellence as well as historical significance.

MUCH SUPERIOR to Mrs. Cooke as a literary artist was Sarah Orne Jewett (1849–1909). A native of the Maine village of South Berwick, near Portsmouth, she was only nineteen when the first of her more than one hundred stories dealing with the inhabitants of Maine coastal villages and farms was published in the *Atlantic* in 1869. During the next few years there also appeared in the *Atlantic* several of the sketches of life in a decaying little Maine seaport town which constitute the chapters of her first book, *Deephaven* (1877).

> It is wonderful [says the author], the romance and tragedy and adventure one may find in a quiet old-fashioned country town, though to heartily enjoy the every-day life one must care to study life and character and must find pleasure in thought and observation of simple things, and have an instinctive, delicious interest in what to other eyes is unflavored dullness.

This interest informs the book, as it does almost all of Miss Jewett's work, giving poignancy and charm to its characters and scenes. These qualities are realized with superior artistry in the later *The Country of the Pointed Firs* (1896), assuredly a minor American literary masterpiece. In this account of a summer sojourn in the maritime village of Dunnet Landing, the inhabitants, including such memorable characters as Mrs. Todd, her bachelor brother, William, and her old mother, come vividly alive, and over them the author sheds a warm glow without ever becoming sentimental or patronizing.

A good many of Miss Jewett's stories, including most of the best ones, are similar in narrative method to those sketches in *Deephaven* and *The Country of the Pointed Firs* in which a character is first portrayed and then relates a story to the narrator. Miss Debby's account in "Miss Debby's Neighbors" of how the Ashby's fought with each other is less interesting than the portrait of Miss Debby herself, who represents a kind of woman of whom Miss Jewett was especially fond and of whom she says in the opening paragraph:

> There is a class of elderly New England women which is fast dying out:—those good souls who have sprung from a soil full of the true New England instincts; who were used to the old-fashioned ways, and whose minds were stored with quaint country lore and tradition. The fashions of the newer generation do not reach them; they are quite unconscious of the western spirit and enterprise, and belong to the older days, and to a fast-disappearing order of things.

Another very fine portrait of this kind is old Mrs. Goodsoe in "The Courting of Sister Wisby," who tells how Lizy Wisby lived with Deacon Brimblecom for a time before she married him to determine whether he would make a suitable husband.

The pathos in Miss Jewett's many stories of middle-aged and elderly women is genuinely moving. "A Lost Lover" is a variation on the Enoch Arden theme, telling of an old maid whose lover was thought to have been lost at sea but who returns a drunken tramp. In 'Marsh Rosemary" a middle-aged wife is deserted by her younger husband. Usually happier in tone but also poignant are the stories of the elderly spinsters or widows who live alone, often with only the most meager resources: the destitute elderly maiden sisters in "The Town Poor," restored finally to their old home through the generosity of friends; the eighty-five-year-old woman in "Aunt Cynthy Dallet," who is sturdily independent, living by herself in an isolated mountain cabin, but who at the end of the story is to have the company of her middle-aged niece; the widow in the brief but beautifully written "Going to Shrewsbury," who loses her farm to unscrupulous relatives but is determined not to be dependent on her town kinfolk and somehow to make her own way; and the inmate of the poorhouse, the elderly Betsey of "The Flight of Betsey Lane," who has always wanted to travel and, when she comes into a bequest of one hundred dollars, sets out not telling anyone, to visit the Philadelphia Centennial and has a wonderful time.

Though her range was not broad, Miss Jewett's achievement was substantial. To her, writing was an art, and she strove always for perfection. Of a story on which she was working she said, "I am so afraid I can't give it breadth and largeness enough, and that it will have a dull kind of excellence and not real life and vitality." A wide and perceptive reader, she knew well Jane Austen, Thackeray, James, Tolstoi, Flaubert, and other great writers of fiction and sought to learn from them. It was Flaubert who confirmed her feeling that the trivial and commonplace can be significant and important. She knew that, as she herself said, "one must have one's own method: it is the personal contribution that makes true value in any form of art or work of any sort." That method in *The Country of the Pointed Firs* and in a sizable number of her short stories is indeed personal. It resulted in a style of great precision and beauty, an avoidance of the overly romantic and sentimental, and portrayals of characters and their environment full of sympathy and insight.

NOT QUITE AS distinctive a writer as Miss Jewett, but after her the best of many authors, nearly all of them women, who contributed to the very large output of New England local-color stories, was Mary E. Wilkins Freeman (1852–1930), who during her long writing career produced some 230 short stories and a number of novels.. Beginning with stories of rural and village life in eastern Massachusetts, a region well known to her through long residence in the town of Randolph, she turned later, especially after she married and moved to New Jersey in 1902, to other kinds of fiction, including ghost stories, animal stories, a novel of social protest, and a historical romance. Nearly all her best stories are to be found among the twenty-eight included in *A Humble Romance and Other Stories* (1887) and the twenty-four in *A New England Nun and Other Stories* (1891).

Many of Mrs. Freeman's New England heroines are uncompromisingly resolute in following a course of action which they believe to be their duty. The girl in "A Taste of Honey" loses her lover because she does not feel she can marry him until she has carried out her promise to her late father to pay off the mortgage on the house he had struggled to build. The need these women feel to follow a certain code of conduct is exemplified by the heroine of "Robins and Hammers," who "had been brought up on the rigid New England plan, and had a guilty feeling that it was a waste of time if she stopped a minute to be happy." She also refuses to marry until she can earn enough money to buy the two silk dresses and furniture which she believes a bride should bring with her to her new home.

Independence, resoluteness, and even downright stubbornness are traits often possessed by the many older women in Mrs. Freeman's stories. In "An Independent Thinker" an elderly woman believes she can use her time to better advantage by earning money from knitting on Sunday rather than going to church, since she is deaf and cannot hear the preacher. Her conduct causes her to be censured by the townspeople, including the mother of her grandaughter's suitor, but the old lady is astute enough to be able to resolve this and other complications without losing her independence.

Also independent and determined is the penniless old woman in "A Church Mouse," who through her persistence gets the job of sexton of the village church and, having no other place to live, moves in with her stove and bed, much to the consternation of the congregation. Equally determined is the farmer's wife in "The Revolt of Mother," who has waited in vain for forty years for the new house her husband had promised her when they were married. Tired of seeing him provide far better buildings for his livestock than for his family, she decides that the handsome new barn he is building for his cows will do very well for a house, and when he goes away on a trip just before its completion she moves in and refuses to budge.

Some of Mrs. Freeman's character portraits are finely perceptive. Very real is the old woman in "The Bar Lighthouse," who has lost her religious faith, having become an unbeliever for pragmatic reasons. "An' thar's been things I've wanted different," she tells the minister, "but I ain't never had 'em—things that I've cried an' groaned an' prayed to the Lord for—big things an' little things—but I've never got one." And there is Louisa Ellis in "A New England Nun," who prefers her placid, orderly, uneventful existence as an old maid to marriage. Engaged to Joe Daggett, who returns prepared to marry her after fourteen years in Australia, where he had gone to make his fortune, her difficulty is resolved when she learns by chance that Joe, although he is too honorable to go back on his word to her, loves Lily Dyer, and thus it is possible for Louisa to release him and remain an uncloistered nun.

One remembers also the two proud middle-aged maiden sisters, whose harmless duplicity is the subject of "A Gala Dress." They have only one good dress between them and take turns going out to social functions in it, changing the trimming each time so that it will not be recognized as the same dress. But their story would be less significant if it did not have the prying sharp-eyed neighbor with whom they are contrasted and who confesses that she had tried to expose them out of envy, having no good dress of her own.

Certain qualities in Mrs. Freeman's stories give her the appearance of being more predominantly a realist than most of the other New England local colorists. Her method of telling a story is usually objective and detached; her style is plain and at times even lacking in grace; her descriptive details are abundant and sharply limned; her representation of the New England vernacular seems authentic and never obtrusive; her characters' traits and the drab austere lives they lead are far from idealized; and there is a somberness, if not grimness, of tone, with only infrequent touches of humor. Yet there are other aspects of her stories which are less realistic and convincing: heroines who often come close to dying of unrequited love and sometimes even do so; abrupt and seemingly unmotivated reversal of attitude which a character may experience; characters in difficulty who find kindness and generosity from others where none existed before; happy endings which come a little too easily and frequently. These deficiencies mark a good deal of Mrs. Freeman's work, but in "A New England Nun" and in at least a few other stories she combined realism and artistry to a degree uncommon among the New England local colorists.

As in New England, a great number of local-color stories were published by southern writers during the last three decades of the nineteenth century. Many of these stories were printed in the *Atlantic* and other northern magazines, whose readers were thereby afforded fictional representations, sometimes realistic, sometimes highly romantic and idealized, of almost every part of the South and almost every aspect of its rich and varied culture, both past and present. The Creoles of Louisiana, the plantation aristocrats of Virginia and Kentucky, the mountaineers of Tennessee, the poor whites of Georgia—these and virtually every other distinctive type of southerner are portrayed in the stories and novels of George Washington Cable, Mary Noailles Murfree, Thomas Nelson Page, Joel Chandler Harris, James Lane Allen, and many lesser writers.

A profound interest in the history and traditions of Louisiana, particularly of the Creoles, laid the foundation for the writing career of George Washington Cable (1844–1925). While employed as a bookkeeper for a New Orleans cotton firm, he fashioned out of his reading and research the seven stories included in *Old Creole Days* (1879). Later editions contained also "Madame Delphine," first published in *Scribner's,* as were all but one of the other stories. A few other short stories and novelettes, a long novel of Creole life and character entitled *The Grandissimes* (1880), and several lesser novels make up Cable's fictional output. After 1879 he lived in the North; there, in addition to writing fiction, he appeared before lecture audiences in company with Mark Twain and published a number of magazine essays on southern history and social problems.

Cable's picturesque settings and characters are often vividly created, but his Creole stories are usually more romantic than realistic. Their situations are sometimes rather thin and often border on the implausible. Sentimental and melodramatic touches are frequent, and an element of mystery or a surprise ending may be introduced. In " 'Tite Poulette" a happy ending is achieved when a beautiful young girl, supposedly the daughter of a quadroon, turns out to be white and is thereby able to marry the young white man who has helped her and her mother out of difficulties. What appears to be deliberate mystification occurs in "Jean-ah-Pouquelin," which turns on an old man's devotion to his brother, who is revealed at the end of the story to be a leper. Sentiment in the manner of Bret Harte's "Tennessee's Partner" appears in "Belles Demoiselles Plantation," a romantic tale with legendary overtones of a Creole plantation owner whose plantation sinks into the Mississippi River taking with it to their deaths his seven beautiful daughters.

"Madame Delphine," which is of novelette proportions, well illustrates the characteristic aspects of Cable's fiction. Its plot is romantic, involving a pirate captain who reforms and becomes a respectable New Orleans banker after a beautiful young quadroon girl on a ship he has captured gives him a missal to read. Much mystery is attached to the actions of some of the characters, and the outcome of the story is similar to that of " 'Tite Poulette," with the revelation that the girl is really white and thus may marry the reformed pirate who has fallen in love with her. The story also has an interesting parenthetical comment by Cable on the Creole brand of English, which, as he represents it, is far from correct in grammar or pronunciation: "Both here and elsewhere let it be understood that where good English is given the words were spoken in good French."

IN *Balcony Stories* and in other works Grace Elizabeth

King (1851–1932), a native of New Orleans and herself part Creole, sought to correct what she regarded to be Cable's misleading representation of Creole character, manners, and traditions. While somewhat more realistic than those of Cable, her stories have far less action and dramatic quality, sometimes contain too much philosophizing and moralizing, and almost invariably depend for their effect on tragic or pathetic elements. Each of the tales in *Balcony Stories* (1893), as the author pointed out in her Preface to a new edition published in 1925, has "some unique and peculiar pathos to it." One story tells of an aristocratic old couple who are reduced to living on the charity of their landlord without being aware of it; in another a little girl is the school dunce; others are about the suffering of a crippled Negro woman, a convent girl who drowns, and a young bride whose husband is robbed and murdered.

ANOTHER WRITER of Creole stories popular in her day was Kate Chopin (1851–1904). Born in St. Louis of an Irish father and a Creole mother, she lived in New Orleans after her marriage and later in Cloutierville, a village in central Louisiana which provided the background for many of her stories. Among the forty-four pieces in her two collections, *Bayou Folk* (1894) and *A Night in Acadie* (1897), are some memorable portraits of the Creoles and Cajuns of the Louisiana bayous, for example, "A Gentleman of Bayou Teche" and "Neg Creol." "Desiree's Baby" and several other stories reveal an artistry which equals if it does not surpass that of Cable.

FLORIDA, THE DEEP SOUTH, and the mountain regions of Tennessee and North Carolina provide the varied settings for the stories in *Rodman the Keeper: Southern Sketches* (1880), by Constance Fenimore Woolson (1840–94), a local colorist who did not limit herself to one section of the country. Miss Woolson grew up in Cleveland and began her literary career by writing sketches and tales of the Great Lakes region. Most of these early pieces are highly romantic, and in several, such as "Misery Landing" and "The Lady of Little Fishing," the influence of Bret Harte is conspicuous. "Solomon," however, is a plain but moving tale of a coal miner who aspires to be a painter, and equally good is the characterization of the young minister who comes to a tragic end in "Peter and the Parson," a grimly realistic story whose locale is a lawless mining settlement on the shore of Lake Superior. Among her better local-color stories are "Up in the Blue Ridge" and a few other stories laid in the Tennessee Mountains.

Miss Woolson's best stories with southern settings are those in which she employed a realistic approach to show the effects of the Civil War and Reconstruction on the lives of individuals, as in "King David," in which a young New Englander goes down South after the war in a vain attempt to help educate the Negroes accept the responsibilities of freedom. These stories were followed by others laid in Italy, where Miss Woolson lived from 1880 until her death in 1894. In them considerable attention is given to setting, but there is also more concern with the psychological states of her characters than in her previous work. In this she may have been influenced by Henry James, who was a close friend during her years in Italy. Some of the situations in the Italian stories are also Jamesian, as, for example, that of the title story of *The Front Yard and Other Italian Stories* (1895), the tragedy of a New England woman who marries a worthless young Italian.

THE TENNESSEE mountains also provided the settings for

almost all the many short stories and novels of Mary Noailles Murfree (1850–92), whose pen name was Charles Egbert Craddock. Her first collection, *In the Tennessee Mountains* (1884), whose eight stories were first published in the *Atlantic*, is also her best. In this volume, as in much of her other work, are stories weakened by certain conspicuous faults: excessive description of the mountains and other scenery, a dialect so pronounced that the speech of the characters is often difficult to follow, a style which tends to be artificial and overly elevated, and a marked tendency to be moralistic and sententious. These qualities are especially evident in such stories as "The Star in the Valley" and "The Romance of Sunrise Rock," both of which are also melodramatic and overly sentimental. But in at least two or three other stories these qualities are either absent or appear in such slight degree that they do not appreciably lessen the artistic effect of the story. "Driftin' Down Lost Creek," which is considerably longer than the other stories, is the most outstanding of the latter. It is concerned with a mountain girl and her lover, who is convicted of a crime of which he is innocent, having pleaded guilty to protect his idiot brother. The heroine works untiringly against great odds to secure his pardon, but after he is released from prison, he remains in the valley and forgets her. There is nothing left for her to do but accept her fate. It is a story of pathos, but the pathos is legitimate, with an uncontrived, inevitable conclusion. Two other stories almost as noteworthy are "A-Playin' of Old Sledge at the Settlemint" and the long title story of *The Mystery of Witch-Face Mountain and Other Stories* (1895).

ALTHOUGH Joel Chandler Harris (1848–1908) did not specialize, as Miss Murfree did, in portraying southern mountaineers, he wrote two short stories laid in the Georgia mountains which invite comparison with her best work. As the creator of "Brer Rabbitt, Brer Fox, and the Tar Baby" and the other tales of Uncle Remus, Harris achieved lasting fame, but the short stories, most of which have their settings in middle Georgia, in *Mingo and Other Sketches in Black and White* (1884) and other volumes entitle him to a place in the first rank of the writers of southern local-color fiction.

"At Teague Poteet's: a Sketch of the Hog Mountain Range" is somewhat drawn out, and its love interest, which brings together, after much misunderstanding and complications, the beautiful daughter of a Georgia moonshiner and a young revenue agent, is conventionally treated, but the story is notable for its portrait of the girl's mountaineer father. In "Trouble on Lost Mountain" all the members of a mountain family are vividly portrayed. The reader sees Grandfather Hightower, his granddaughter Babe, and her father and mother mainly through the eyes of the northerner Chichester, an agent for a Boston syndicate interested in buying up mountain land for marble quarries. There is good description of the mountain scenery, but there is less of it and it is better integrated into the story than is usually the case in Miss Murfree's stories. Nor does Harris give undue emphasis, as Miss Murfree sometimes did, to the crudeness and ignorance of the mountain folk and to the drabness and sordidness of their lives. He also consciously endeavored to avoid sentimentality, as indicated by Chichester's observation that Babe's not having "gradually faded away, according to the approved rules of romance," when her jealous lover remained away from her "was entirely creditable to human nature on the mountain."

In a number of Harris' stories slaves and former slaves are the central characters. "Free Joe and the Rest of the World" describes the plight of a slave who is freed by his master in

1850 and who thereby becomes in effect a man without a country, looked on with suspicion by other Negroes and barely tolerated by the whites. His situation is full of pathos, but it is not sentimentalized. Also moving and sympathetically drawn, and only slightly sentimental and romantic, are the portraits of Negroes who are loyal and devoted to their masters in such stories as "Mingo," "Balaam and His Master," and "Ananias."

In several stories it was clearly Harris' intention to give the reader some idea of what the various classes of middle Georgia people were like and of the social conditions which prevailed a decade or so after the war. The best of them is "Azalia," in which a Boston girl, Helen Eustis, pays a visit to the piney-woods country for her health. Tolerant and not at all provincial in her outlook, she has ample opportunity to observe not only the plantation aristocracy as represented by General Peyton Garwood and his mother but also Mrs. Haley, who runs the inn where Helen and her middle-aged spinster aunt stay, the Negroes, and the Tackies, or poor whites. The story has a highly romantic outcome, but ordinarily Harris avoided contrived plots. His stories have an air of authenticity and a naturalness of tone not often found in the same degree in the other southern local colorists.

A QUITE DIFFERENT picture of southern plantation life from that depicted by Harris is found in the stories of Thomas Nelson Page (1853–1922). In "Marse Chan" an unidentified narrator comes upon an old Negro tending graves (the time is 1872), who describes the pre–Civil War days he knew in such speeches as the following:

> "Dem wuz good ole times, marster—de bes' Sam ever see! Dey wuz in fac'! Niggers didn' hed nothin' 't all to do—

> jes' hed to 'ten to de feedin' and' cleanin' de hosses, an' doin' what de marster tell 'em to do; an' when dey wuz sick, dey hed things sont 'em out de house, an' de same doctor come to see 'em whar 'ten to de white folks when dey wuz po'ly. Dyar worn' no trouble nor nothin'."

Of course, this idyllic existence was destroyed by the war. Sam's master, young Marse Chan, was killed in battle, and his sweetheart died of a fever. Now the two are buried side by side, and Sam is sure that they are married in heaven.

Two other stories in *In Ole Virginia* (1887), Page's first and best collection, employ much the same formula, except that they have happy endings. In "Meh Lady: A Story of the War" old Uncle Billy tells an unnamed listener how his young master, Marse Phil, was fatally wounded after fighting heroically and how Marse Phil's mother and sister endured much hardship during the rest of the war and later, the sister finally marrying a former Yankee officer, who, however, was "half Virginian." At the beginning and end of the story we see Uncle Billy preparing fishing poles for the couple's two boys, who are coming to the plantation for a visit. As in "Marse Chan," the burden of "Unc' Edinburg's Drowndin' " is that everything was much better in the old days before the war.

Less serious stories of plantation life are "Polly: A Christmas Recollection," also included in *In Ole Virginia*, and "George Washington's Last Duel," in *Elsket and Other Stories* (1891). Comic in intent, they show elderly planters and their Negro bodyservants on the most familiar terms, the latter helping themselves liberally to the whisky and other possessions of their masters, who in turn bluster and threaten dire punishment without really meaning it. There is love interest also, the girl being sweet and beautiful and her lover handsome and noble. Page could write stories which are interesting and have individuality, but his dependence on sentiment and romance make it difficult for the reader to take them very seriously.

EXCESSIVE SENTIMENT is also a prominent feature of much of the fiction of James Lane Allen (1849–1925). Such works as *Flute and Violin and Other Kentucky Tales and Romances* (1891), *A Kentucky Cardinal* and its sequel, *Aftermath*, both of novelette length, and a highly romantic novel, *The Choir Invisible* (1897), made him one of the most popular and widely read southern writers of his day. Critical of authors like Cable, Miss Murfree, and Harris, whom he regarded as too exclusively concerned with exploiting the peculiarities of their subjects, Allen's avowed aim was to give his stories of the bluegrass region of Kentucky not merely picturesque interest but significant meaning and literary quality as well.

Yet Allen was usually something less than successful in this endeavor, and the best story in the *Flute and Violin* volume, "Two Gentlemen of Kentucky," is also the simplest and least pretentious. In it, according to Allen's Preface, he endeavored to portray "a type of Kentucky gentleman farmer," who at the close of the Civil War abandoned the country for the towns, and led a rather idle, useless life. The story is a poignant one that idealizes somewhat, but avoids sentimentalizing, the characters of old Colonel Fields and Peter, the former slave who remains loyal to him. "King Solomon of Kentucky," the story of a good-for-nothing white man who redeems himself during a cholera epidemic in Lexington when he remains to dig graves and bury the dead, is also told with restraint, but in other stories the style is mannered and too deliberately poetic, the symbolism is too obvious, and Allen's favorite theme—that life is full of tragic pain—is so treated that overmuch pathos results. These qualities can be discerned in "Posthumous Fame; or, a Legend of the Beautiful," which is a Hawthorne-like parable; in "Flute and Violin," in which a minister tries unsuccessfully to win the love and trust of a fatherless crippled boy, who dies of remorse after he has committed a petty theft; and in "The White Cowl" and "Sister Dolorosa," which explore, respectively, the conflicts engendered in a man and in a woman who renounce the world for the religious life.

UNLIKE THE Far West, New England, and the South, the Middle West inspired relatively little local-color fiction. The frontier days of the region were depicted in the sketches of James Hall, Caroline Kirkland, and others, but after the Civil War, partly perhaps because the region was less picturesque than others and not as uniquely provincial, only a few writers were impelled to portray its farm and village life. Three who did, however, made important contributions to the regional fiction of the time: Edward Eggleston with *The Hoosier Schoolmaster* (1871); Edgar W. Howe also with a novel, *The Story of a Country Town* (1883); and Hamlin Garland with the short stories in *Main-Travelled Roads* (1891).

One of the most widely read American dialect novels, *The Hoosier Schoolmaster*, despite obvious artistic weaknesses, gives evidence throughout of being an honest attempt to give a comprehensive and authentic account of life in the Indiana backwoods. Howe's novel is more grimly realistic, and Garland's best stories are usually only a little less so. Garland acknowledged that he had been influenced by Eggleston, but, like Howe, he saw little that he could approve in the rural and village life he knew and stressed its inhibiting effects on the intellectual, emotional, and spiritual development of the individual.

THE SON OF a farmer who went west from Maine, Hamlin Garland (1860–1940) was born in Wisconsin and grew up on farms there and in Iowa and Dakota Territory. Ambitious to become a teacher and desiring more formal education than his studies at the Osage, Iowa, Seminary provided, he made his way to Boston in 1884. There he read voluminously in the public library and became a convert to certain radical doctrines of the time, soon to be reflected in his fiction. Among these theories were the views of Herbert Spencer and other evolutionists, Whitmanesque democracy as expressed in *Leaves of Grass*, and the arguments for the single-tax principle advocated by Henry George.

Eventually finding employment as a teacher and lecturer, Garland also tried his hand at writing and managed to sell a description of an Iowa cornhusking and several similar sketches to the *New American Magazine* and a long poem about the prairie to *Harper's Weekly*. It was a visit home in the summer of 1887, he said, which opened his eyes to the tragic lot of the prairie farmers and bred in him the resolve to protest against the social injustice of which they were the victims. As a result he produced between 1887 and 1891 most of the thirty-odd stories of prairie life included in *Main-Travelled Roads* (1891) and in two companion volumes, *Prairie Folks* (1893) and *Wayside Courtships* (1897).

Garland's best stories are found among the six he included in the first edition of *Main-Travelled Roads* (later editions contain three to six additional stories), and of these, partly because they embody less social protest than the others, "Mrs. Ripley's Trip" and "The Return of a Private" are the most artistically effective. The former, which has a simplicity and sympathy like that of Sarah Orne Jewett's best work, was Garland's first story and was suggested by an account his mother had given him during his 1887 visit of an elderly farm woman who had finally realized a long-cherished desire to make a trip back to her old "York State" home to visit relatives.

The situation of "The Return of a Private" came from one of Garland's most vivid memories of his early childhood, as the opening chapter of the autobiographical *A Son of the Middle Border* indicates. Although Garland made no mention there of his story, his account of the homecoming of his father, a volunteer soldier in the Civil War, corresponds in many of its details to the description of the return of Private Edward Smith. A significant difference, however, is that the social protest of the story is missing in the autobiographical version. Garland's father wanted to enlist at once but felt that he must wait until he could make the last payment on his mortgage. He came back apparently in good health, and the account of his return is concerned almost entirely with the excitement and joy which it occasioned. Smith, on the other hand, comes back sick and emaciated to a mortgaged farm, yet with the courage to resume "his daily running fight with nature and against the injustice of his fellow-men." Despite its note of protest, Garland did avoid making the story a tract, and he wisely omitted, after the first edition of *Main-Travelled Roads*, the Postscript he had affixed to the story: "He is a gray-haired man of sixty now, and on the brown hair of his wife the white is also showing. They are fighting a hopeless battle, and must fight till God gives them furlough." "The Return of a Private" shows Garland at his very best. There is a richness of detail, the characters and their actions always seem credible, their manner of speaking seems completely authentic, and there is an inevitable rightness in the dramatic ordering and development of the three main scenes: Smith's return with his fellow soldiers to LaCrosse; the terrible loneliness of the wife, relieved by the companionship and affection afforded to her and her children at Sunday dinner with the kind "Widder" Gray; and finally the reunion itself of the soldier with his family.

Although it, too, is remarkably effective in its narrative technique, "Under the Lion's Paw," whose theme is that of the prairie farmer at the mercy of the land speculator, does obviously have the tone of a tract. A compact little drama in four scenes, the story opens with the impoverished Haskins family, who have made their way to Iowa after having been driven off their Kansas farm by grasshoppers, being befriended by the benevolent farmer Stephen Council and his equally kind-hearted wife. With Council's assistance Haskins establishes himself on a rented farm and by dint of the hardest kind of labor makes many improvements on it, only to discover when he believes himself in a position to buy it that its land-speculator owner has doubled the original sale price. The impact of the story is powerful, but Garland insists too much on telling the reader how hard Haskins and his family work; the conclusion smacks of melodrama; Butler, the land speculator, is too villainous; and the Councils are too good to be entirely credible characters.

If "Under the Lion's Paw" has an air of contrivance, that is not true of "Up the Cooly," which also focuses on the hopeless struggle of the average middle western farmer. It is also the most personal of Garland's stories, and may have been written because of the need to purge himself of a sense of guilt caused by the feeling that in leaving his family, and particularly his mother, he had deserted them. Although the Howard McLane of the story, who revisits his Wisconsin farm home after ten years, is an actor and has achieved considerable material success, essentially he is his creator thinly disguised. He reacts in the same way that Garland did during the summer visit of 1887—as described in *A Son of the Middle Border*—to the beauty of the rural landscape, the squalid aspect of the town where he gets off the train, and the sordidness and endless drudgery of farm life. Like Garland, he resolves to make amends for neglecting the mother who is proud of her successful son and to whom it does not even occur to reproach him.

But merely to model his situation on his own visit would not give Garland the story he wanted, and so he invented Grant McLane, the younger brother who has stayed behind to work the farm. He and most of his neighbors have no prospect of anything but eking out the barest kind of living from their toil. Howard can buy back the family farm on which the mortgage had been foreclosed and give further financial assistance to Grant, but that will not solve Grant's problem. It is not charity he wants but the opportunity, of which he is deprived by economic injustice, of earning his own living. "A man like me is helpless," says Grant. "Just like a fly in a pan of molasses. There ain't any escape for him. The more he tears around the more liable he is to rip his legs off."

Among other things Howard discovers during his visit is that most of the younger men of the region, to avoid becoming Grant McLanes, have gone west. One of them might well have been Rob Rodemaker, the likable hero of "Among the Corn-Rows." Because there was no "chance" for him there, he had left Wisconsin to stake out a claim on the Dakota prairie. "Land that was good was so blamed high," he says, "you couldn't touch it with a ten-foot pole from a balloon. Rent was high, if you wanted t' rent an' so a feller like me had t' get out, an' now I'm out here, I'm goin' t' make the most of it." He works hard, has good wheat crops, and finds satisfaction in being his own master, but after he becomes established, he realizes that he is lonely and needs a wife, and he goes back to Wisconsin, where "girls are thick as huckleberries," to find one. How he carries off Julia Peterson, whose Norwegian father has made her drudge in the house and in the cornfields, makes for a love story of great charm and poignancy. Perhaps we should assume from Garland's other stories that Julia will inevitably know drudgery and loneliness as a prairie wife, but that would be another story which only by a great stretch of the imagination can be inferred from this one.

"A Branch-Road" is also a love story, its situation being the familiar misunderstanding between lovers. Yet its treatment is far from commonplace. The opening section is especially notable, depicting a day's activity on the farm during the threshing season, one of the finest scenes Garland ever wrote. At the same time it portrays with remarkable psychological insight the perversity that causes young Will Hannan to work himself up into a jealous rage against Agnes Dingman, the girl he loves. But the promise of this opening is not quite fulfilled by the remainder of the story. Ashamed of his conduct, the repentant Will intends to make amends when, several days later, he takes Agnes to the county fair. There follows an incident that has a little too much the appearance of being necessitated by Garland's plot. An accident to Will's buggy delays his arrival by two hours. Agnes has gone with his rival, and overcome with rage, shame, and the desire to make her suffer, Will writes her a cruel letter and goes off to the Southwest. Seven years later he returns, successful and prosperous, to find her a worn and wasted farm wife, abused by her husband and nagged at by his old parents. Filled with compassion and considering himself responsible for her plight,

Will persuades the hesitant Agnes that she has a right to the new life he is ready to give her, and, taking her child with her, she abandons her husband for an illicit but better future with her old lover. Garland's story thus turns into a kind of little problem play with a regional flavor. To find Garland resolving his plot conflict by having his characters flout the moral code is not a little surprising, since there was a strong strain of the moralist and puritan in him. Garland no doubt was sincere in advocating his radical solution for Agnes' problem, but it is presented in such a way that it does not seem entirely convincing or strongly felt. One has the feeling that Garland's youthful radicalism compelled him to follow the lead of Ibsen or Thomas Hardy but that he felt a little uncomfortable in doing so.

In addition to the short stories picturing prairie life, Garland showed the middle western farmer fighting a losing battle against the forces of nature and the land speculator in the novel *Jason Edwards* (1892), and he endeavored to advance the Populist cause in *A Spoil of Office* (1892); but by the mid-nineties his reform fervor had noticeably begun to wane, and the uncompromising realism (Garland called it "veritism") to which he had seemingly been dedicated had all but disappeared from his writing. From that time on uninspired, conventional romances flowed from his pen. He was attracted especially by the picturesque settings which Colorado and the Southwest afforded, and utilized them in such novels as *The Captain of the Gray Horse Troop* (1902) and *Hesper* (1902) and in *They of the High Trails* (1902), a volume of short stories which portray the cow boss, forest ranger, prospector, and other mountaineer types. Later he devoted himself to the writing of a long series of volumes of autobiography and literary reminiscences. One of these, *A Son of the Middle Border*, complements and deserves to rank very close to

*Main-Travelled Roads*, in that it describes with great effectiveness the frontier experience which gave Garland his materials, and the mood of youthful rebellion which gave him the impetus to fashion them into stories with a distinctive regional flavor, and, in at least a few instances, both social and literary significance.

# 6
# The Rise of the Journalistic Short Story
## O. Henry and His Predecessors

ALTHOUGH LOCAL COLOR and regionalism were predominant in American short fiction from the Civil War to almost the end of the century, the same period also saw the parallel development of a very different kind of story flourishing on a smaller scale. Its antecedents can be at least vaguely discerned in Irving and Poe, but in certain respects it was a new kind of story. Ingeniously and carefully plotted, often culminating in a surprise ending, it was written more to entertain than to be taken seriously. Wit and humor, vivacity and lightness of touch, and an urbane manner and cultivated style were its other principal characteristics.

The first to write in this vein was Edward Everett Hale (1822–1909), a Unitarian clergyman and a prolific author who achieved fame with his "The Man Without a Country," a tremendously popular fictionalized tract on the necessity of northern citizens to be loyal to their government during the Civil War. Hale's best humorous story, "My Double and How He Undid Me," published in the *Atlantic* in 1859, is an extravagant but diverting account of a clergyman who hits on the idea of using a double to represent him at the innumerable public meetings and social functions he is expected to attend. The double is an Irishman with little imagiiation or intelligence, but with the aid of a few phrases taught him by the clergyman he manages to take the minister's place without giving himself away. On one occasion, however, he is called on for a speech and gets into an argument with his audience, whom he reviles in a rich Irish brogue. The minister is of course undone and leaves town posthaste. From the wilds of Maine he writes his story to dispel the impression "that the Rev. Frederic Ingham had lost all command of himself in some of those haunts of intoxication which for fifteen years I have been laboring to destroy."

HALE'S STORIES are less witty and urbane than those of Thomas Bailey Aldrich (1836–1907), who besides writing stories and novels was a poet and holder of various editorial positions, including the editorship of the *Atlantic Monthly* from 1881 to 1890. "Marjorie Daw," his most artfully contrived story, appeared in the *Atlantic* in 1873 and was one of the most admired pieces of short fiction of its day. Written in the form of letters between two young bachelors, Edward Delaney and John Flemming, it deals with Delaney's efforts to relieve the boredom and irritation of his friend, who is confined to his New York townhouse because of a broken leg. From his New Hampshire summer cottage Delaney laments that he has nothing entertaining to write about. If he were only a novelist, he would write his friend a story. This, in effect, is what he does by creating the fictitious Marjorie Daw and describing her so convincingly that Flemming falls in love with her. Carried away by his imagination, Delaney writes that she has also fallen in love with Flemming, but his innocent hoax backfires when Flemming recovers sufficiently to travel and telegraphs his intention of coming to see the wondrous Marjorie. Delaney's emphatic protests are of no avail, and when Flemming arrives, the abashed and contrite Delaney has fled, leaving a final note: "Oh, dear Jack, there isn't any colonial mansion on the other side of the road, there isn't any piazza, there isn't any hammock,—there isn't any Marjorie Daw."

Besides lesser stories on the order of "Marjorie Daw," Aldrich wrote several highly sentimental tales like "The Little Violinist," in which a boy violinist dies from exhaustion brought about by being compelled to overwork as an entertainer, while other boys, like the son of the narrator, are cherished and well cared for. More realistic, however, and somewhat reminiscent of the manner of Sarah Orne Jewett, are "Quite So," a tale of a Civil War soldier from Maine, and "A Rivermouth Romance," the story of a middle-aged Irishwoman who is deserted by her shiftless husband. But for the most part Aldrich was scarcely one of those writers "who are honest dealers in real life," as he claimed to be in an aside in the last story, and the interest his stories have depends more on their manner than their matter.

FRANK R. STOCKTON (1834–1902) also achieved a contempory fame with a short story that is a whimsical *jeu d'esprit*. Readers found highly intriguing his "The Lady, or the Tiger?" published in *Century* in 1862, with its mock-serious description of a "semi-barbaric" king's method of determining the guilt or innocence of a subject charged with committing a serious crime. The accused was put in the public arena and given the choice of opening one of two doors. Behind one is a tiger who will tear him to pieces as punishment for his guilt. Behind the other is "a lady, the most suitable to his years and station that his majesty could select among his fair subjects," and to her he will be married immediately as a reward for his innocence. The method is applied to a handsome young courtier of low station who is charged with being the lover of the king's daughter. The princess has learned the secret of the doors, and her lover senses this when he enters the arena and meets her glance. He unhesitatingly advances to the door which she indicates by a gesture perceptible only to him and opens it. "Now the point of the story," says Stockton, "is this: Did the tiger come out of that door, or did the lady?" Taking into consideration both the princess' great love for the courtier and her jealous, half-savage nature, did she save her lover or let him die rather than have him marry another? Since the question, involving as it does a study of the human heart, is too difficult for Stockton, he will leave the answer up to the reader. Pressed by many readers to give them a conclusion to his story, Stockton instead wrote "The Discourager of Hesitancy," a pallid imitation that poses a similar dilemma. Stockton wrote a few serious stories, the best of them being "The Griffin and the Minor Canon," an indictment of man's selfishness and cowardice, but he was mainly a facile journalist and entertainer in his short fiction and in longer works like *Rudder Grange* (1879), sketches of a family who live on a houseboat.

A SIMILAR FACILITY is seen in the work of Henry Cuyler Bunner (1855–96), who contributed stories to the comic weekly *Puck*, which he also edited, and to *Harper's* and other magazines. With Brander Mathews, who was the first to give currency to the term "short story" and who promulgated a set of principles for the short story derived from Poe, Bunner collaborated in producing *In Partnership: Studies in Story Telling* (1884). In the stories in *Short Sixes* (1891) and other volumes Bunner achieved the dexterity he admired in Aldrich, and he wrote skillful imitations of Guy de Maupassant in *Made in France: French Tales Told with a United States Twist* (1893), although his tone was lighter and less sardonic. In a number of stories Bunner wrote about life in New York City in much the same way that O. Henry was to do a few years later, and he also anticipated that writer in the artful mingling of pathos, irony, and humor.

An artifice which sometimes approaches being art distinguishes the short stories of Ambrose Bierce (1842–1914?), although in their tone they are a far cry from the stories of Aldrich, Stockton, and Bunner. Bierce's predilection for the grotesque and the macabre and his frequent depiction of the effects of fear and terror put him in the tradition of Poe and Fitz James O'Brien, an Irishman who came to America in 1852 and settled in New York City. O'Brien achieved a considerable reputation as a playwright, journalist, and story writer before his death of a Civil War wound ten years later. Although less Gothic than Poe, as was Bierce, O'Brien exploited with no little skill Poe's techniques in stories like "The Diamond Lens," whose mad scientist murders to gain a diamond out of which he can construct a microscope of great power and covers up his crime cunningly. There is also a Poe-like effect of verisimilitude in "What Was It?" which in its account of an invisible presence with a horrible yet human shape anticipates Bierce's "The Damned Thing" and also Maupassant's "Le Horla." These two stories of O'Brien and Bierce aim only at striking terror in the reader, whereas Maupassant's story is a powerful study of incipient madness, the narrator fearing that under the domination of his invisible alter ego he will lose his will and reason.

Born on a farm in Ohio, Bierce served in the Civil War and began a long career as a journalist in San Francisco. In time he became a kind of literary dictator of the Pacific Coast, noted for his cynicism, misanthropy, and the cruel wit and bitter invective of his writing. Toward the end of his life he was a Washington correspondent for a New York newspaper and a magazine contributor, and in 1913 he vanished under mysterious circumstances in Mexico, probably meeting his end at the hands of one of the factions engaged there in civil war.

Bierce's atmospheric effects in "The Death of Halpin Frayser" and in some of the other stories of the supernatural in *Can Such Things Be?* (1893) are worthy of Poe, and he tellingly exploited the gruesome and the macabre in such stories as "The Man and the Snake" and "The Boarded Window" in *In the Midst of Life*, first issued in 1891 as *Tales of Soldiers and Civilians* and retitled a year later. Fifteen of the twenty-six stories in this volume are of war, and, to judge from the frequency with which it has been anthologized, the best known and most admired of these war stories is "An Occurrence at Owl Creek Bridge," a tour de force of undoubted technical brilliance, which surprises and shocks with its revelation in the final sentence that Peyton Farquhar, the Confederate plantation owner and would-be saboteur, has not, as the reader has been led to believe, escaped hanging by his Union captors. Compelling the reader to participate emotionally in Farquhar's "escape" through the suspense and vividness with which he renders it, Bierce gives to its details an artful mixture of the real and the unreal which hints at and yet conceals that it is taking place only in Farquhar's mind as he drops to the end of the rope.

Bierce's most gruesome episode of war is "Chickamauga," in which a six-year-old boy wanders away from his plantation home while playing soldier with a wooden sword and falls asleep in a nearby forest. A terrible battle rages around him, but, being a deaf mute, he is not disturbed. When he wakes, he sees a group of maimed, bleeding survivors crawling toward a brook to slake their thirst. He plays that he is their leader, placing himself at their head and brandishing his sword. "To him it was a merry spectacle. He had seen his father's Negroes creep upon their hands and knees for his amusement—had ridden them so, 'making believe' they were his horses. He now approached one of these crawling figures from behind and with an agile movement mounted it astride." Flinging the child off, the man turns upon him a face from which the jaw has been shot away. Other ghastly details follow, culminating in the child making his way home, only to discover the building ablaze and his mother killed and horribly mutilated by a cannon shell.

"One of the Missing" portrays the demoralizing effects of fear on Jerome Searing, an intrepid Federal scout who is trapped when the building in which he conceals himself to observe the movements of the enemy is struck by a shell and he is imprisoned in the debris. Only slightly injured, he views his situation calmly until he observes the barrel of his rifle protruding from the rubbish covering his body and pointed directly at the center of his forehead. If he struggles to free himself, the rifle, which he had cocked just before the explosion, may be discharged. The menacing stare of the gun eventually reduces this brave man to a creature insane with fear. He cannot push the rifle aside with the board he grasps in his one free hand, but there is a last resort. He thrusts the board against the trigger. "There was no explosion; the rifle had been discharged as it dropped from his hand when the building fell. But it did its work." This ironic outcome is reinforced by a sequel. Searing's brother, an officer, passes the building with his troops some twenty minutes after the shell had struck it. Observing the corpse half buried in the debris, he does not recognize its distorted face and comments, "Dead a week."

Except for two or three stories like "One Officer, One Man," a fine study of a captain who cannot stand the strain of waiting for an attack that never comes and kills himself with his sword, Bierce's ironic episodes of war are too contrived and theatrical, too dependent on coincidence, on sudden revelations intended to shock and surprise, and on other tricks. But in their unity and concision, their striking imagery, the authentic ring of their circumstantial details concerning men at war, and the remarkable clarity and precision of their style, they are the work of a master technician. They have, too, the virtue of never making war seem glamorous, picturesque, or romantic.

By the end of the nineteenth century the carefully made, ingeniously plotted story had become a well-established tradition, but it was during the first decade of the twentieth century that the type was carried to its ultimate lengths in the stories of O. Henry. None of his predecessors exploited the contrived story with quite such deliberate calculation or with more facility, and none achieved anything like the phenomenal popularity of O. Henry, who produced his stories for mass-circulation magazines and newspapers with the intent, as he put it, of pleasing "Mr. Everybody."

O. Henry (1862–1910), whose real name was William Sydney Porter, grew up in Greensboro, North Carolina. In Texas, where he went in 1882 for reasons of health and to seek his fortune, he lived for a time on a ranch, was a bank teller in Austin, edited a short-lived humorous weekly called the *Rolling Stone*, and wrote a daily column, filled mostly with humorous anecdotes, for a Houston newspaper. Indicted for the alleged embezzlement of funds from the bank that had employed him, he maintained his innocence but fled to Honduras instead of standing trial. On his return he was convicted, and during his three-year imprisonment in the federal penitentiary in Columbus, Ohio, he began to write and sell stories to the magazines. Soon after his release he settled in New York City, where he wrote prolifically, the rapidity with which his work was turned out being illustrated by his feat of producing a story at the rate of one a week over a period of thirty months for the Sunday edition of the *New York World*. *Cabbages and Kings* (1904), a collection of loosely integrated stories about adventure and revolution in Latin America, was his first book. Twelve more volumes, several of which were published posthumously, were required to collect the remainder of his short stories.

O. Henry's stories have a variety of settings, but most of them are laid in either New York City or Texas. His characters include shopgirls and millionaires, policemen and burglars, cowboys and tramps, confidence men and southern gentlemen, and assorted other types. His manner is usually that of the garrulous taleteller, and his style is almost invariably breezy, flippant, and slangy, with puns, malapropisms, and big words being used for humorous effect. His stories are liberally sprinkled with asides in which he addresses the reader in a familiar and chatty tone. Literary allusions, often made facetiously, are common, and there are many references to other writers. Kipling, whom O. Henry greatly admired, is either mentioned or quoted frequently. In "A Municipal Report," for example, besides a quotation from Kipling, both Frank Norris and Tennyson are cited, the latter being referred to as "My old friend, A. Tennyson," and there are in addition allusions to characters in two of Dickens' novels. Fond of referring also to *The Arabian Nights,* O. Henry often called New York City "Bagdad-on-the-Subway" and likened the wealthy New Yorkers in his stories to caliphs.

Although he usually used stock story formulas, O. Henry had an undoubted gift for devising ingenious variations on them. Coincidence figures largely in his stories, and they often have a surprise twist, or "snapper," as O. Henry called it. Unabashed sentiment and the broadest kind of comedy and burlesque are other conspicuous ingredients. In addition, O. Henry usually made his contrived stories illustrate some more or less serious theme. Most of his many stories of New York City, found mainly in *The Four Million* (1906), *The Trimmed Lamp* (1907), and *The Voice of the City* (1908), make the point that the humble, insignificant little people of New York are just as admirable and their lives as worthy of attention and interest as the members of the Four Hundred. Typical is "The Gift of the Magi," O. Henry's famous story of the young married couple, each of whom sells a treasured possession to obtain money to buy a Christmas present for the other. Della sells her beautiful long hair to buy a platinum chain for Jim's watch, only to discover that he has sold it to buy jeweled tortoise-shell combs for her hair. O. Henry builds up to his surprise twist very artfully, and with deft touches he elicits the reader's admiration and sympathy for his young couple, whose love for each other more than compensates for Jim's meager salary, their shabbily furnished apartment, Della's old brown hat and jacket, and the fact that Jim needs a new overcoat and has no gloves. Artfully, too, O. Henry does not end on the note of irony and surprise but gives to what he calls his "uneventful chronicle of two foolish children" the appearance of a little parable with a significant meaning. The magi, he reminds the reader, were the wise men who brought gifts to the Christ child, and thus invented the giving of Chrismas presents. As for Jim and Della, "in a last word to the wise of these days let it be said that of all who give gifts these two were the wisest. Of all who give and receive gifts, such as they are wisest. . . . They are the magi."

"A Municipal Report," another of O. Henry's best-known stories, provides an especially good illustration of virtually all his mannerisms and devices. The story takes its cue from a statement of the novelist Frank Norris, quoted at the beginning, to the effect that there are only three big cities in the United States that are "story cities," New York, New Orleans, and San Francisco. "Fancy," Norris had said, "a novel about Chicago or Buffalo, let us say, or Nashville, Tennessee!" This, suggests O. Henry, is a rash statement, and he proceeds to tell a tale refuting it. It is part of O. Henry's irony that he leads the reader to believe in the first part of the story that Nashville is a humdrum place, this being the initial impression of the first-person narrator, who gets off the train in Nashville one evening and after settling himself in his hotel can find nothing of interest to observe or do. But then comes a striking contrast when O. Henry manufactures a plot utilizing coincidence and surprise, which indicates that there can be excitement and romance aplenty in this apparently dull town. And, says O. Henry at the end, "I wonder's what's doing in Buffalo!"

O. Henry's stories of Texas and of Central and South America often have much vivid descriptive detail, and their backgrounds seem authentic enough, but they are like his other stories in that they have little realism in their characters and actions. In "A Double-Dyed Deceiver," for example, a desperado known as the Llano Kid kills a man in Texas and flees to South America. There an unscrupulous American consul persuades him to pose as the lost son of a wealthy couple, with the idea that the consul and the Kid will rob them. But the Kid has a heart of gold under his hard exterior, and is so moved by the joy of the woman who believes him her son that he refuses to go through with the plot. Furthermore, it turns out that it was the lost son whom the Kid had killed back in Texas, and therefore the Kid, to make restitution, will take his place. None of O. Henry's grafters, burglars, and robbers are really bad men either. O. Henry is said to have got the ideas for some of his stories of these characters from his prison experience, but if he achieved any insight there into criminal mentality and psychology, he made no attempt to portray it in his fiction. Instead, he wrote stories like "Babes in the Woods," in which a confidence man comes from the West to New York confident that he can find all kind of dupes on whom to practice his trade, but who is himself taken in. Some of these stories, however, like "The Man Higher Up," which is one of several about a grafter named Jeff Peters, are among O. Henry's most cleverly done pieces. An especially artful con-man yarn, and quite possibly O. Henry's funniest story, is "The Ransom of Red Chief," in which a kidnaping plot ludicrously boomerangs on its perpetrators." And there is the gentleman burglar who goes straight in "A Retrieved Reformation," whose name became a household word when O. Henry's story was made into the highly popular play *Alias Jimmy Valentine*.

Besides his great popularity with readers, O. Henry also received much adulation from contemporary critics. He was often spoken of a "a Yankee Maupassant" and praised for his literary artistry and broad understanding of humanity. Although no one today would attribute these qualities to his stories, they have a special verve, freshness, and good humor which make for their continued readability, as witnessed by the fact that they are still frequently anthologized and even occasionally issued in new editions.

# 7
# The Short Story as Fine Art

## Henry James

To write a series of good little tales I deem ample work for a lifetime," said Henry James in 1870, near the beginning of his long career. He was to realize this ambition and a great deal more. The work of his lifetime consisted of more than one hundred short stories and longer tales, some twenty novels, and a great quantity of nonfiction, including literary criticism, travel sketches, plays, biographical, and autobiographical works. James wrote many "good" tales, but relatively few of them are "little." Of his eighty or so stories under twenty thousand words which may be classified as short, many run to at least fourteen or fifteen thousand words. The longer stories, not a few of which are from thirty to forty thousand words, are by modern standards short novels, but James thought of them as examples of what he called "the beautiful and blest *nouvelle*," a tale of no definitely prescribed length, which had for him "the value above all of the idea happily developed."

Henry James (1843–1916) was born in New York City, the son of a well-to-do father with an intellectual bent, who devoted himself to the study of philosophy and theology. Along with his older brother William, who was to become the famous philosopher and psychologist, he was educated by tutors and in private schools in America and abroad. After studying painting briefly and then law at Harvard, he published his first tale anonymously in an obscure New York periodical in 1864. His second appeared the next year under his own name in the *Atlantic Monthly* and was followed by a dozen more before 1870, most of them in the same magazine. By 1876, the year he settled permanently in England, he had increased his output of stories to almost thirty, along with publishing much dramatic and art criticism, many book reviews and travel essays, and two novels, *Watch and Ward* (1871) and *Roderick Hudson* (1875).

The early stories were the product of much experimentation, James striving to emulate such diverse writers as Balzac, George Eliot, and Hawthorne and trying his hand at developing a variety of themes and situations. "The Story of a Year," James's first *Atlantic* tale, is a lengthy study of a shallow girl whose lover dies of a Civil War wound. The war also takes the life of the officer whom the wealthy young heroine loves in "Poor Richard," although the story focuses mainly on the young farmer, Richard, one of her rejected suitors. In "A Most Extraordinary Case" a convalescing Union officer loses his will to live when the girl he loves marries another man. These stories are rather wooden and lacking in any real sense of tragedy, but it is to James's credit that he did not make them unduly sentimental. Of two tales told in diary form, "A Landscape Painter" depends on a twist ending, but in "A Light Man" the device is used more effectively to reveal the opportunistic, dishonest character of the diarist, who vies with his friend for the inheritance of their elderly patron. There is also good character portrayal in "The Story of a

Masterpiece," in which a man learns that appearance and reality are not the same and which is interesting also because its situation somewhat resembles that of one of James's best mature stories, "The Liar."

Other early stories narrate odd or unusual happenings, usually somewhat less than convincing and seldom of particular significance. "Osborne's Revenge" is a long-drawn-out, inconclusive story of a man who intends to take revenge on a woman, because of whom his friend has killed himself, but who falls in love with her instead. "A Problem" is a contrived tale of a young married couple who are worried by the prophecy of a fortuneteller that their child will die and that each of them will marry twice, forecasts which come true when the child dies after the parents separate and when the parents later remarry. In the melodramatic "Master Eustace" a boy grows up to discover that he is the illegitimate son of his mother's second husband, and in "De Grey: A Romance" young lovers are the victims of an ancestral curse. In "A Romance of Certain Old Clothes" a wife returns after death to kill her successor, while in "The Ghostly Rental" an apparently supernatural occurrence turns out to have a natural explanation, a daughter posing as a ghost to take revenge on her father for his cruel treatment.

James made two trips abroad between 1869 and 1874, travels which resulted in the use of European settings in most of the stories written during these years. The narrators of "Travelling Companions" and "At Isella" are traveling in Italy for the first time (as James himself did in 1870); they describe at length the scenes they visit and record, often rapturously, their sensations and impressions. The narrator of "The Sweetheart of M. Briseaux" is a traveler in France who loves "to ruminate the picturesque." These stories have slight, romantic plots, as do "The Last of the Valerii" and

"Adina," both of which again have American narrators and are laid in Italy. In "A Passionate Pilgrim," however, the extensive appreciation of English sights and scenes, although it delays James in getting into his narrative and makes it overly long, is in the main an integral part of the story, since the pathos, as well as the irony, of the situation of the American Clement Searle results from his regarding England as his spiritual home and from his obsessive desire to feed his imagination on the richness of its historical and romantic associations.

Like Searle, the painter Theobald in "The Madonna of the Future" deplores his American origin. Domiciled for twenty years in Florence, he contends that the American soil is devoid of everything that nourishes and inspires the artist and that in order to excel he "has ten times as much to learn as a European." Only when it is too late are his eyes opened to the fact that, as the narrator puts it: "Nothing is so idle as to talk about our want of a nutritive soil, of opportunity, of inspiration, and all the rest of it. The worthy part is to do something fine. There's no law in our glorious Constitution against that. Invent, create, achieve!" Theobold leaves behind at his death only a bare canvas, cracked and discolored by time, instead of the magnificent madonna he had deluded himself into believing he was preparing to create. Theobald's unrealized painting was suggested by a tale of Balzac's, which James alludes to in the story, but the implications of Theobald's experience were not. They were of special importance for James as an aspiring literary artist, and they are made very meaningful in this strongly felt and finely executed parable.

The hero of "Eugene Pickering" and the heroine of "Madame de Mauves," like Searle and Theobald, are the victims of romantic illusions. They are also the first of a long line of Jamesian Americans in Europe whose fate is often, in James's words, to be "insidiously beguiled and betrayed" and to be "cruelly wronged" by those "pretending to represent the highest possible civilization and to be of an order in every way superior to his own." This character is only very faintly drawn in "Eugene Pickering," in which a naïve young man is thrown over by a European woman ten years older than he. A much better illustration of the beguiled American is Euphemia Cleve in "Madame de Mauves." Educated in a Parisian convent, she dreamed of marrying a title "because she had a romantic belief that the best birth is the guaranty of an ideal delicacy of feeling." In her innocence and inexperience she cannot foresee what lies in store for her when she marries the Baron de Mauves, even though she is warned by his grandmother. He has married her for her money and continues having affairs with other women after marriage. Yet he is disturbed by the stoicism and reserve with which she masks her disillusionment, and he and his family endeavor to encourage the friendship between her and the American Longmore in the cynical hope that she too will prove unfaithful. But, demonstrating an integrity and sense of duty characteristic of many a later Jamesian heroine, she sends Longmore away, winning a victory which James unnecessarily and inartistically underscored in the story's sequel, in which the baron repents and kills himself when she will not restore him to her favor. Nevertheless, "Madame de Mauves" is one of the best stories of James's early period, for in it we can see him formulating the general situation, creating the character types, and perfecting the style and technique which he later elaborated and refined to a high degree in the novels *The American* (1877) and *The Portrait of a Lady* (1881).

James's output of stories during the middle and late 1870's includes some more or less mediocre ones, among them "Crawford's Consistency," "Benvolio," "Longstaff's Marriage," and

"The Diary of a Man of Fifty," but they are more than offset by a number of variations on his "international subject" published during the same period and during the early 1880's. James's story of Miss Caroline Spencer, the little New England spinster of "Four Meetings," is one of his most moving and ironic accounts of Americans who harbor ultraromantic conceptions of Europe. "Daisy Miller" is the first of several notable studies of American girls, who, although they vary in their traits from story to story, are almost invariably young women of a free and independent spirit. This *nouvelle*—for James it was "a signal example in fact of that type"—achieved a wider notice than anything he had previously written, partly because it was regarded by some critics and readers as a slander on American girlhood. There is little, if anything, in the portrait of Daisy, as was pointed out by James's more discerning contemporaries, to justify this view, although there is an uncertainty about her true character that exists until almost the very end in the mind of Winterbourne, from whose point of view the story is told. An American in his late twenties who has lived for a number of years in Europe, he becomes acquainted with Daisy in Vevey without benefit of a formal introduction. He is amused, perplexed, and above all charmed by this pretty girl from Schenectady on her first trip to Europe. He does not agree with his aunt, Mrs. Costello, that Daisy, her mother, and her little brother, Randolph, are "very common," "the sort of Americans that one does one's duty by not accepting."

Later, however, in Rome where Daisy consorts with the fortune hunter Giovanelli and as a result is snubbed by her American compatriots, Winterbourne finds it increasingly difficult to hold to his belief that she is really only very ignorant and very innocent, and he feels that he must discard this opinion altogether when late one night he encounters Daisy alone with the Italian in the Colosseum. This indiscretion is her last, since she contracts the Roman fever and dies a few days later. Had the realization that Winterbourne no longer held her in esteem deprived her of the desire to live? Too late he perceives that she had both wanted and merited his respect, that she was not only pretty and charming but, as Giovanelli testified after her death, "also—naturally!—the most innocent."

Giovanelli's "naturally!" which James added when he revised the story, points up Daisy's "naturalness," which is the key to her character and actions. Because she is a work of nature rather than art, as one character describes her in the revised version, Daisy seems at times capricious, perverse, and deliberately defiant of convention, and yet there is logic and consistency in her behavior. It is natural for her to want to enjoy herself and do what she likes. It is not natural to expect her not to appear in public when her mother is unable to accompany her. And it would have been unkind—that is to say, unnatural—to abandon Giovanelli during their walk in the Pincian Gardens by allowing herself to be driven off in the scandalized Mrs. Walker's carriage. Daisy's naturalness is also evidenced in her belief that the people who criticize her only pretend to be shocked. This does not mean that she is naïve and stupid—there is ample evidence that she is actually quite shrewd and perceptive—but that she is ignorant of, or at best unconcerned about, the premium society puts upon appearances. Daisy's moral sense is certainly more admirable than that of her critics, and she is an appealing example of the nonconformist free spirit in opposition to convention, tradition, and decorum. Yet there is also the implication in her character and career that to have no sense of their necessity and inevitability in society is a deficiency. In Winterbourne's eyes, Daisy conforms to certain generalizations, not all of them complimentary, concerning American girls, and he is puzzled exactly "how far her eccentricities were generic, national, and how far they were personal." The reader, too, may find it difficult, if not impossible, to separate these qualities. Both Daisy's distinctive American traits and the "poetical terms," to use her creator's phrase, through which her personal quality shines forth make her a character of great vitality and charm.

Although Bessie Alden, the heroine of "An International Episode," is more cultivated and sophisticated than Daisy Miller, she is almost as ingenuous and in her way just as independent. We like her because, as her English admirer, Lord Lambeth, puts it, "She is not afraid, and she says things out, and she thinks herself as good as anyone." But she is also a somewhat too serious, if not humorless, young woman, the reason being, it is suggested, that she has lived a good deal in Boston. Her story gives us, mainly in terms of social comedy, an especially acute and detailed comparative study of American and English manners, but she does not win our sympathy to the extent that Daisy does, nor is her experience as moving. We are left instead with only a mild feeling of regret that two such fundamentally nice and well-meaning young people as Bessie and Lambeth should have had their relationship broken off.

Both American and foreign manners and attitudes are examined extensively in the witty and satirical "A Bundle of Letters," which portrays a group of persons of various nationalities living in a pension in Paris, and in its even more penetrating and amusing companion piece, "The Point of View," which is also in part a sequel to still another story on the same order, "The Pension Beaurepas." In "The Point of View," written mainly to express his varied reactions to America gained from a return visit in 1881–82 after a six-year absence, James even made himself one of the objects of his satire in having a French literary critic say: " 'They've a novelist with pretensions to literature who writes about the chase for the husband and the adventure of the rich Americans in our corrupt old Europe, where their primeval candour puts the Europeans to shame. *C'est proprement ecrit* but it's terribly pale.' "

James might humorously deprecate his international subject, but he continued to write about it frequently during the 1880's, varying his approach from story to story with considerable imagination and ingenuity. Mrs. Headway, in "The Siege of London," might almost be a Daisy Miller twenty years older. Being "a genuine product of the far West—a flower of the Pacific slope; ignorant, audacious, crude but full of pluck and spirit, of natural intelligence, and of a certain intermittent, haphazard good taste," she is able to overcome the handicap of a past which includes divorcing several husbands, thereby becoming the favorite of English high society and the wife of a nobleman. Although James said there were no actual cases of the sort to guide him at the time he wrote "Lady Barbarina," he made the story answer affirmatively the question, Would a young Englishwoman of noble birth who married an American, even though he was wealthy and socially acceptable, be unhappy in her husband's country and want to return to her own? In "The Impressions of a Cousin" and "A New England Winter," Europeanized Americans view their native country unfavorably, and in "Pandora" there is a picture of Washington society as seen through the eyes of a young German diplomat. Of particular interest to him is Pandora Day, who, he is told, is representative of a new American type—the self-made girl, in this case a once unprepossessing young woman from Utica who had so impressed the President of the United States that at her request he appointed her fiancé minister to Holland. Also of the sisterhood of Pandora, Daisy Miller, and Bessie Alden is Francie Dosson, the gentle and charming heroine of *The Reverberator*. Like Bessie, Francie is regarded as socially inferior if not unacceptable by her lover's American expatriate family and French relatives, but happily, despite her lowly estate and her innocent *faux pas*, which results in the publication of certain of their family secrets in the scandal-mongering press, her suitor is able to put Francie before loyalty to family.

A very different kind of heroine is the sensitive, tormented Laura Wing of "A London Life," a story which in its moral and social complexities and obliquities is characteristic of much of James's later fiction. As James pointed out, there was no good reason for making the story international since all his characters could have been English. Laura is like two of his English heroines, Fleda Vetch of *The Spoils of Poynton* and Rose Tramore of "The Chaperon," all three being endowed with "acuteness and intensity, reflexion and passion." The anecdote told to James by Paul Bourget, which served as the germ for "A London Life," ended with the girl committing suicide. Perhaps one of the reasons why James rejected this means of resolving Laura's situation was that he had already employed just such a denouement in "The Modern Warning." There is no question about the international character of this story, with its two diametrically opposed characters, the intensely nationalistic Macarthy Grice, who holds Englishmen in contempt, and Lord Rufus Chasemore, the author of a book that stringently criticizes American manners and institutions and is intended to warn his fellow Englishmen of the contaminating influence of democracy. Between these two is James's heroine, the sister of the former and wife of the latter. Only the very credulous reader, however, can believe she would be so torn between patriotism and affection for her brother and loyalty and love for her husband that she would commit suicide.

In two of James's best later stories with international ramifications, mothers exert a baneful influence on their children. The somewhat sententious declaration with which "Louisa Pallant" begins—"Never say you know the last word about any human heart!"—may seem trite, but the narrative illustrating it is far from being so. " 'Europe' " has a brevity which is rare in James's later stories; it was kept to seven thousand words by "innumerable chemical reductions and condensations." It is also, like the earlier and similar "Four Meetings," one of his most moving studies of privation and the pathos attendant on it; both stories employ with great skill a favorite Jamesian device, the sympathetic observer-narrator who fully engages the emotions of the reader.

Although in the 1880's and later James devoted himself increasingly to writing other kinds of stories, he never abandoned altogether his international theme. As late as 1900 he portrayed once again the independent American girl with a mind of her own in "Miss Gunton of Poughkeepsie," in which the heroine turns down an Italian prince for reasons not too different from those that cause Bessie Alden to reject Lord Lambeth in "An International Episode." "Fordham Castle," published four years later, was the result, not altogether satisfying to James, of an effort to offset his creation "of too unbroken an eternity of mere international young ladies." James's portrayal, however, of the self-effacing Americans Abel Taker and Mrs. Magaw, both of whom consent to being exiled in a Swiss pension—Taker so that his wife, masquerading as a widow, can establish herself in English society, and Mrs. Magaw so that her daughter can achieve a match with an English lord—is neither one of his more inspired treatments of the American in Europe nor one of his more successful later stories dealing with the theme of the person who lives a kind of death in life. During this same period, on the other hand, James produced his three later major novels—*The Wings of the Dove* (1902), *The Ambassadors* (1903), and *The Golden Bowl* (1904)—and in them, with consummate narrative artistry and profound moral insight, he gave rich new dimensions to his international subject.

Of the many stories other than international ones that James wrote during the 1880's and 1890's, a considerable number are about writers and artists. They constitute a particularly interesting and important, as well as eminently readable, group of James's fictions, portraying, frequently in comic and ironic terms and in a form akin to fable or parable, the aesthetic and moral problems and dilemmas the writer or artist must almost inevitably face, both in the pursuit of his vocation and in his relation to society. These stories furnish also occasional illuminating insights into the nature of art and the mysteries of the creative process.

Several of them are about novelists who bear at least a general resemblance to James himself. One is Mark Ambient, in "The Author of Beltraffio," a serious literary artist who knows "as much torment as joy" in the act of literary creation and who is not a popular success. Another is Dencombe in "The Middle Years," which reflects perhaps more than any of James's other stories his most deeply felt attitudes and emotions as a writer. As Dencombe reads his newly published novel, completed before the serious illness from which he is convalescing, he regards it as James might will have contemplated a work of his own over which he too had suffered and struggled: "His difficulties were still there, but what was also there, to his perception, though probably, alas! to nobody's else, was the art that in most cases had surmounted them." Dencombe's novel has been critically acclaimed, and he has the satisfaction of knowing that his work has cast a spell over the young doctor, a lover of literature, who nurses him when his illness returns and who even throws over material interest for his sake. But what concerns Dencombe is the thought that he could write a much finer work if he only had the opportunity which he knows is going to be denied him. Before he dies, the revelation comes to him that the artist labors under a delusion in believing that he can ever be satisfied. He can never have, Dencombe realizes, a second chance! "There never was to be but one. We work in the dark—we do what we can—we give what we have. Our doubt is our passion and our passion is our task. The rest is the madness of art."

The reader may well wonder why young Doctor Hugh, with all his intelligence, does not discern on a first reading of Dencombe's novel what the author had "tried for." Too infrequently, James felt, did readers do an author justice in reading him closely enough to perceive fully his intention and meaning. In "The Figure in the Carpet," which James rightly called an "ironic and fantastic fable," the characters who relentlessly and singlemindedly seek to discover the meaning of the novelist Hugh Vereker are at the opposite remove from the generality of readers and critics, who were disinclined to read with "anything like close or analytic appreciation." Another unhappy aspect of the writer's relation to the public which James's experience had forcefully brought home to him was that even though the writer might be known to the public, and might even be a celebrity, that did not mean that there would be more than a few people who would care about his work or even be acquainted with it. This is the theme of the highly comic and ironic "The Death of the Lion," in which the obscure middle-aged novelist Neil Paraday is suddenly catapulted into prominence when he is "discovered" by a great newspaper desperately in need at the moment of some personality to play up in its columns.

Comedy, irony, satire, and complication of plot mark several other stories in which James called attention to the great popularity achieved by hack novelists and, conversely, the lack of notice which seemed inevitably to be the fate of the serious writer. Mrs. Stormer, the best-selling novelist of "Greville Fane," has no concern for style and form and the attempt to relate one's writing directly to life, but she has a sure-fire formula—"Passion in high life"—and the facility of invention and industry to grind out story after story embodying it. In "The Next Time," another popular lady novelist is contrasted with her brother-in-law, who wants to write novels which will sell, but no matter how much he tries to debase his talent the result always has too much literary quality to be popular. In "The Lesson of the Master" the wife of the novelist Henry St. George is the making of him, in a material sense, in persuading him to be prolific and superficial instead of continuing to aim at perfection as he had done in his early work.

"John Delavoy" expresses a view of the writer which is reflected frequently, if less pointedly, in James's other fiction. The sister of a recently deceased novelist refuses to write the biographical and anecdotal account of her brother that a magazine editor wants to give his readers. To do so would be an injustice, she maintains, since the artist can really be known only in his work and in fact, *is* his work. This view is also expressed in "The Real Right Thing," in which a biographer abandons his work on a lately deceased novelist because he believes he can sense the presence of his subject and that it disapproves of the project. "The artist," he recalls the novelist saying, "was what he *did*—he was nothing else."

Although James put no writer in it, he made "The Story

in It" a little parable amounting to an oblique statement of his belief that it is felt experience which evokes art and that only art can give the experience meaning and significance. Mrs. Dyott, who is having a secret affair with Colonel Voyt, tells him that her friend Maud Blessingbourne is in love with him but would never dream of admitting it. In a talk with the two ladies the colonel had maintained that the French novels Maud reads and finds wearisome—because they always show men and women in the same relationship—follow an inevitable law: if the stories are to have adventure, romance, and drama, they cannot portray virtuous women. Maud is not convinced, however; she believes that it depends on what one regards as adventure, romance, and drama. Later Voyt admits to Mrs. Dyott that Maud's consciousness of her love does in a way constitute "a kind of shy romance," but what a pale thing it is compared to a romance like their own, and who, he asks, "would see the shadow of a 'story' in it?" It was to such inner dramas as Maud's that James's art gave life, interest, and importance.

Another important article of James's aesthetic faith is expressed in "The Real Thing," generally acknowledged to be one of his finest stories. Its narrator is an illustrator of books and magazines, and it has to do with the difficulties he encounters in using a middle-aged gentleman and his wife as models. No matter how the artist arranges Mrs. Monarch, she always comes out the same way in his drawing, which has the appearance of a photograph or a copy of one. "She was always a lady certainly," the artist says of her, "and into the bargain was always the same lady. She was the real thing, but always the same thing." He has the same trouble with Major Monarch, who, like his wife, is endowed with good looks, an impressive manner, and perfect gentility. Because they lack the sense of variety and talent for imitation possessed by his professional models, the cockney Miss Churm and the Italian boy Oronte, the artist concludes that he cannot go on using them to illustrate a novel, even though it deals with the very social class to which they belong. The decision is not an easy one, since the Monarchs are in straitened circumstances and have come to depend on him, but he cannot ignore the warning of his friend Hawley, whose judgment he trusts, that his amateur models will not do; and the publisher's dissatisfaction with his illustrations makes clear that if he persists in using the Monarchs he will not be commissioned to do the remaining volumes in the edition of the novelist. The artist can admire the perfection of the Monarchs which makes them the real thing and feel compassion for their plight, but he can also be bored by them and irritated by their inability to understand why they will not do as models. Ultimately, however, they acknowledge their failure by voluntarily taking over the household duties performed by the professional models. "They had," says the artist, "bowed their heads in bewilderment to the perverse and cruel law in virtue of which the real thing could be so much less precious than the unreal." Unable to keep them indefinitely in such a capacity, he gives them what money he can spare when he sends them away, and if his experience has done him harm as an artist, he is "content to have paid the price—for the memory."

In most of James's other stories about the experiences of artists, problems of the artist as artist, if they are raised at all, are subordinate to the moral and social implications of the narrative. A question which becomes of obsessive interest to Oliver Lyon in "The Liar," another of James's best stories, is how the wife of a compulsive liar can either endure or condone what her husband does. It is a natural enough interest for a portrait painter, heightened by the fact that Lyon is still in love with the wife, who turned him down some years before when he was a struggling artist because she felt obliged to marry well for the sake of her family. How can Mrs. Capadose, whom Lyon had known as the most candid and honest of women, profess to love and admire her husband so much? Is it because of her pride, or has her character altered? Hoping to find out, and even more to elicit some admission from her that she has cause to regret having married the colonel instead of him, Lyon arranges to do the colonel's portrait and thereby expose the true nature of the man. The colonel brings his wife to Lyon's studio during the artist's absence for her first view of the almost completed portrait. By coincidence, Lyon arrives just in time to witness unobserved that she clearly recognizes what he has done, and to watch the colonel, when he sees his wife is horrified, although he does not know why, destroy the portrait. Lyon does not reveal himself and makes no effort to save the picture, since it has served its purpose in proving that Mrs. Capadose is ashamed of her husband. Ironically, however, Lyon discovers that he has failed in his ultimate aim. When he speaks to the colonel later about the mysterious destruction of the portrait, the latter does more than merely maintain that he has no knowledge about it. He puts the blame on a supposedly disgruntled artist's model, and his wife backs him completely. It is a bitter pill for Lyon to swallow when he comes to realize fully the power of her love. That his success should turn to ashes in his mouth is surely a just retribution for his machinations. Colonel and Mrs. Capadose are of course culpable, but Lyon's obsession makes him at least equally so. For Mrs. Capadose, the artist's portrait is not only a hideous and cruel exposure of the colonel's failing, leaving out everything else that makes up the man, but also a revelation of the duplicity and deceit that the artist has been guilty of in his relationship with her and her husband.

A comparison may be drawn between "The Liar" and "The Tree of Knowledge," in which, with fewer moral complexities and a lighter irony, James portrayed another wife who knows her husband for what he is, a sculptor devoid of talent, but who sacrifices honesty in loyally sustaining him in his delusion that he is a genius whose work is so fine that the public as a matter of course cannot appreciate it and therefore will not buy it.

The plots of "The Tone of Time," "The Special Type," and "The Beldonald Holbein" have perhaps a little too much of the tour de force about them, but their artist-narrators are men of "imagination and observation," as one of them describes himself, and it is because of their sensibility and awareness that the events they narrate take on interest and significance. This is particularly true of the last-named story, of which the "Holbein" is Mrs. Brash, brought over to London from America by Lady Beldonald, whose practice is to engage the plainest companions she can find as foils for her beauty. The exquisite quality of this little old lady, with her wrinkled face framed by the black velvet bonnet she wears, makes her, as the narrator had predicted to Lady Beldonald, a social celebrity, especially in the art world. To him this is "an example of poetic, of absolutely retributive justice," and he asks himself what had been responsible for making Mrs. Brash what she was:

> The only thing to be said was that time and life were artists who beat us all, working with recipes and secrets that we could never find out. I really ought to have, like a lecturer or a showman, a chart or a blackboard to present properly the relation, in the wonderful old tender, battered, blanched face, between the original elements and the exquisite final "style."

If he is unable to do justice to Mrs. Brash in words, the artist is certain he can on canvas, but he never gets to paint her, Lady Beldonald shipping her back to America to die, having found a satisfactory successor in the person of a pretty woman whom no one notices. But since Lady Beldonald has commissioned her own portrait, the artist will at least be able to do justice to her by showing her as she really is, by bringing out "the real thing."

During the early 1890's, James began producing a number of stories which were, to use his description, "tales of the quasi-supernatural or 'gruesome' order," which might for the sake of convenience be called ghost stories, although this was "roughly so to term them." James had an extensive acquaintance with this kind of story as written by Wilkie Collins, Dickens, Le Fanu, and others, and he took considerable interest in the investigations carried out by the Society for Psychical Research concerning ghosts, haunted houses, clairvoyance, telepathy, and other such phenomena. Given sufficent ingenuity, he felt there was a possibility for one to refine upon the gross and obvious characteristics commonly associated with the supernatural tale and thereby to achieve perhaps something in the way of a renewal of the form.

James's "ghostly" tales show that their author exercised much ingenuity, but with a few notable exceptions not a great deal more than that. Both "Sir Edmund Orme" and "Owen Wingrave" are melodramatic, and their apparitions put a considerable strain on the reader's credulity despite strenuous efforts on James's part to make them convincing. There is the implication in "Nona Vincent" that clairvoyance is at work, but the story is of interest chiefly for its picture of the trials and tribulations of a playwright during the production of his first play, many of the details of which, as Leon Edel has shown, appear to have been suggested by James's own experience in the theater in 1891, when his dramatic version of *The American* was produced. "Sir Dominick Ferrand" is a highly contrived tale, in which the heroine's sixth sense enables her to divine the existence of certain letters which, if made public, would destroy her late father's reputation. "The Private Life" resulted from James's interest in the personalities of two famous men of his time, the poet Robert Browning and the painter Lord Frederick Leighton. A little fantasy on the alter-ego theme it is, as James felt it should be, "very brief —very light—very vivid."

"The Friends of the Friends" is a more conventional supernatural story, employing a situation common in ghost stories and psychical case studies, that of a person, as in Defoe's "The Apparition of Mrs. Veal," who has just died but who appears as still alive to another person who is unaware that the death has occurred. Set down in the form of a diary extract, the story is more than a psychical case study, being primarily a self-portrait that reveals that the writer had allowed herself to become the victim of jealous fears and thereby lost the man she loved. Her account also raises some doubt about its reliability, since there appears to be the possibility that the diarist's jealousy and obsession may have prevented her from always putting the proper construction on events and appearances.

A similar uncertainty is engendered by "The Turn of the Screw," James's finest and certainly his most provocative ghost story. Its genesis was a story told James by the Archbishop of Canterbury concerning the corruption of young children brought about by wicked servants in whose care they had been entrusted. The servants had died, but their ghosts had returned in an effort to continue their evil hold on the children and lead them to destruction. Whether they had succeeded the story apparently did not make clear, and it was because of the vagueness of his source, said James, that he could give his imagination free play in making the most of "the haunted children and the prowling servile spirits." What he had done was to write a "fairy-tale," which was "a piece of ingenuity pure and simple, of cold artistic calculation, an *amusette* to catch those not easily caught," the tone of which was that "of suspected and felt trouble, of an inordinate and incalculable sort—the tone of tragic, yet of exquisite mystification." As for his narrator, the governess, who tells of her desperate struggle to protect little Miles and Flora from the ghosts of Peter Quint and Miss Jessel, James considered it essential that her record of the story's anomalies and obscurities be kept clear, but her explanation of them, he seems also to say, is something else again. As for his ghosts, they differ from the negative and undramatic apparitions of psychical case records because it was necessary that they be agents in the action: "there would be laid on them the dire duty of causing the situation to reek with the air of Evil." For this "Evil" to have a maximum effect, the general vision of it must be made intense, but it must remain unspecified, the reader thus being stimulated to imagine its particulars. "This ingenuity," James said, "I took pains—as indeed great pains were required—to apply; and with a success apparently beyond my liveliest hope."

How are we to regard the governess and her remarkable narrative? Most critics have not accepted Edmund Wilson's theory that the governess is "a neurotic case of sex repression, and that the ghosts are not real ghosts but hallucinations of the governess," since in certain respects his Freudian interpretation seems clearly to run counter both to what James said about the story and to the story itself. But not all the critics who accept the ghosts as real, or at least as something more than mere hallucinations, have been content with reading "The Turn of the Screw" as simply a tale of the supernatural. It has been variously interpreted as a kind of morality play, the theme of which is the struggle of evil to bring about the damnation of the human soul; as an allegory in which the governess personifies ignorant, incompetent authority, destroying instead of saving the children by attempting to deny them knowledge; as a parable on the sin of pride, with the overprotective governess so hounding and harrying the children in her determination to save them unaided that she is more injurious to them than the ghosts; and even as an unconscious expression by James of the doubts and fears of his own "haunted mind."

Clearly no one view of the governess or of her experience will ever satisfy all readers. Most would probably agree that she has at least some admirable qualities and should be regarded as providing a generally reliable account. But at the same time they are likely to be disturbed by her tendency to employ hyperbole, to express opinions as if they were facts, to dramatize herself overmuch, and to give way to fanciful imaginings—all of which makes for a lingering doubt about the possibility of saying positively that her dilemmas and predicaments, her hopes and fears are all fact or all fancy or some compound of the two. What is important, finally, is that James did achieve his desired effect. The terror and horror are there for the governess and through her for the reader.

James's final ghost story, "The Jolly Corner," is one of his very best tales. It tells of Spencer Brydon, who after living abroad for many years returns to New York. Though it may be a morbid obsession, as he admits to Alice Staverton, his old friend and confidante, he wants to know more than anything else what he would have been like had he never gone away. The house on the "jolly corner" where he was born and grew up stands vacant, and he takes to roaming it at night in an effort to waylay the alter ego he becomes convinced inhabits it and seeks to evade him. Brydon's actions, as he is aware, would seem ridiculous in the light of day, and yet they seem altogether fitting in the atmosphere created within the house, and the scene in which the confrontation between Brydon and his alter ego occurs is as suspenseful and dramatic as anything else in James's fiction. At the same time that Brydon sees the awful figure, "spectral yet human," Alice Staverton sees it in a dream. Through her and her love Brydon comes to understand that the figure is not dreadful, but pitiful, for Brydon himself might have suffered, as the figure's ruined sight and mutilated hand seem to signify, from the pressures and materialism of American life.

"Crapy Cornelia" is in the nature of a companion piece to "The Jolly Corner," it too reflecting certain feelings evoked in James by his visit to America in 1904–1905. White-Mason, a middle-aged bachelor, chooses the society of his old friend Cornelia, who recalls cherished memories for him, instead of marriage with the wealthy Mrs. Worthington, who represents the modern American life he dislikes. There are also unfavorable reflections on the new America in "A Round of Visits," an ironic and somewhat melodramatic story of a man who, because he has been wronged, feels a compulsion to seek sympathy from others but finds they are preoccupied with their own troubles and usually looking for sympathy themselves. These stories in which the hero returns to America are complementary to *The American Scene* (1907), a work which examines in a more impersonal and objectively analytical manner an American society which James found had radically changed.

Besides White-Mason and Brydon there are other "poor

sensitive gentlemen," as James called them, preoccupied by the past and haunted by its ghosts. George Stransom's consuming interest, in the rather morbid "The Altar of the Dead," is to perform his ritual of burning candles on a church altar in memory of his dead friends. Marmaduke, in "Maud-Evelyn," comes to believe that he has been the lover and later the husband of a dead girl whom he has never seen. A more significant experience, and in some instances a tragic one, for certain of these gentlemen is to become aware that they have not really lived. James's finest creation of this kind is Lambert Strether in *The Ambassadors*. A lesser Strether is Sidney Traffle, in "Mora Montravers," who sees that his niece, in running off to live with a painter, has demonstrated that she possesses a "sense of life," a quality which Traffle realizes to his regret he has never had. Strether and Traffle are to some extent able to benefit from their knowledge, but not so John Marcher in "The Beast in the Jungle," a penetrating and subtly ironic story which ends with Marcher's horrified realization that the something rare and strange he had always thought would happen to him has happened without his being aware of it. His egotism and self-centeredness blind him to May Bartram's love and his need for her until it is too late.

Other stories besides international and "ghostly" tales and stories of writers and artists came from James's pen during the 1890's and later. As Morton Dauwen Zabel has observed, many of them picture the comedy and tragedy of society. In "The Chaperon," for example, it was James's intention to portray "the observed London world—'society' *telle que je l'ai vue*." Rose Tramore, with her integrity, intelligence, and determination, is one of James's more admirable heroines, but what gives the story its chief interest is its social satire. Rose's efforts to restore her ostracized mother to the good graces of society seem doomed to failure until a hostess has a happy inspiration. "With a first rate managerial eye she perceived that people would flock into any room—and all the more into hers—to see Rose bring in her dreadful mother." Another such hostess appears in "Mrs. Medwin." The feature whom Lady Wantridge wants for the amusement of her guests at her country place (fittingly named Catchmore) is Scott Homer, an impecunious American with a dubious past and an ingratiating manner. As a result, he becomes an asset instead of a liability to his half-sister, Mamie Cutter, who supports herself by introducing people into English society, enabling her to achieve the apparently hopeless task of getting a Mrs. Medwin accepted by Lady Wantridge and her friends and to collect a double fee from a grateful client in the bargain.

Those stories in which James combined criticism of society with study of character tend toward tragedy rather than comedy. In "The Marriages," Adela Chart has such a fanatical reverence for the memory of her dead mother that she cannot tolerate the idea of her father marrying Mrs. Churchley, whom she regards as the antithesis of her beloved mother. She tells Mrs. Churchley that Colonel Chart had made her mother's life wretched by his cruel treatment, and this lie, which at first appears to achieve the result Adela desires, in the end makes for unhappiness for all concerned. Although the manner in which retribution is brought down upon Adela's head is none too convincing, the story is one of James's best studies of obsession. Its point of view is confined to that of Adela, but at the same time the reader is made fully aware of how distorted it is. "Brooksmith," on the other hand, is a first-person narrative on the order of "Four Meetings" and " 'Europe' " and an equally moving study of privation. A remarkable feature of the story is that James was able to make its unusual situation entirely convincing, namely, that it was the artistry of the butler Brooksmith which was responsible for the perfection of his master's salon, a place in which the social amenities held sway and the finest talk flourished, and that this unobtrusive member of the salon, without ever mingling in the conversation, received as much as he gave by breathing in its atmosphere to such an extent that it became a necessity for him.

Of the many representations in James's fiction of moral deceit and confusion in society and their effects, perhaps none is more penetrating and profoundly affecting than "The Pupil," in which we see, as James pointed out, the precocious young Morgan's parents and their lack of moral sense through "Morgan's troubled vision of them as reflected in the vision, also troubled enough, of his devoted friend," the tutor, Pemberton. In the short novel *What Maisie Knew*, on the other hand, there is only Maisie's troubled view of her fashionable but disreputable and corrupt elders. Along with the little girl, who is aided by her developing moral sense, the reader comes to perceive the truth about them. In "In the Cage" the sordid affairs, the extravagances, and the selfish materialism of the same kinds of people as Maisie's parents and step-parents are seen through the eyes of a young woman who works in a Mayfair telegraph office. The governess in "Paste," through her association with a society matron, Mrs. Guy, and with her cousin, Arthur Prime, has a vivid revelation of the duplicity human nature is capable of. James here set himself an exercise in reversing the situation of Maupassant's "The Necklace," a string of pearls thought to be paste turning out to be real, and he made the story a convincing illustration of the unhappy truth that to be scrupulously honest is likely to result in inviting others less honest to take advantage of one.

Certain of his society stories show James at something less

than his best. History repeats itself in a most unlikely fashion in "The Wheel of Time," and a none-too-plausible story situation is employed in "Lord Beaupre" to illustrate James's observation that the eligible young English bachelor was in great danger from designing mothers with daughters to marry off. Even if the present-day reader were not accustomed to the universal wearing of glasses by women, he would probably still not find very affecting the plight of Flora Saunt in "Glasses." This beautiful but shallow and self-centered girl loses a wealthy nobleman's affections when he discovers that she has deceived him about the glasses her failing eyesight requires, and the conventional ending, in which Flora finds happiness with the faithful suitor who has remained patiently in the background, does nothing to redeem the story.

In "The Great Condition" the contrast between the two suitors of the same woman seems exaggerated. One has complete faith in her; the other becomes obsessed with the idea that there is something objectionable in her past. The story is perhaps the weakest of James's many studies of obsession. "The Two Faces" is one of James's more vivid pictures of the workings of London society, but it suffers from a certain triteness in the mode of revenge a woman takes on a man who has jilted her and from the fact that she is able to achieve it much too easily.

Although it is convenient to place James's stories in one category or another when describing them, they often do not, as F. W. Dupee has observed, adhere strictly to type, and there is in most of them "a kind of cross-reference." Some defy classification except that they deal, as so many of James's stories do, with the moral sense. One such is "The Aspern Papers," a *nouvelle* suggested by an account told James of the efforts of an American admirer of Shelley to obtain some letters of Shelley and Byron in the possession of Byron's mistress, Jane Clairmont, who at an advanced age was living in seclusion in Florence with her middle-aged niece. He had persuaded the old woman to take him in as a lodger and after her death not long afterward had approached the niece, who told him that she would give him the letters only if he would marry her. In addition to the scheming lodger who gave him his plot, James saw in the anecdote an opportunity to project something of the Byronic past into the present. He masked his source by placing his story in Venice and by making his long-dead poet, whom he called Jeffrey Aspern, and his two ladies —the Misses Bordereau—Americans.

If James went too far in asking the reader to believe that early-nineteenth-century America had produced a poet who ranked with Byron and Shelley, this flaw is more than compensated for by the other elements in the story: the description of the dilapidated old Venetian palazzo that harbors the two women; the gruesome relic of a distant romantic past that is the older Miss Bordereau, who keeps the eyes which had once inspired a poet's immortal verses covered with "a horrible green shade" and counters the "publishing scoundrel" with a cupidity and cunning equal to his own; the younger Miss Bordereau, who is almost unbelievably helpless and obtuse and yet is sympathetic as a character and convincing as the agent responsible for the denouement; and the narrator's avid desire for the papers and his schemes for obtaining them, with their attendant duplicities, the rationalizations employed to justify them, and the predicaments in which they result. It is possible to see various symbolical meanings in "The Aspern Papers," but perhaps the reader should be content with the explicit comment the narrator makes regarding the "extravagant curiosity" of his friend Cumnor and him concerning Aspern's relics: "We had more than enough material without them and my predicament was the just punishment of the most fatal of human follies, our not having known when to stop."

There is "cross-reference" aplenty in "The Birthplace." Its hero, Morris Gedge, has some of the characteristics of the "poor sensitive gentleman," and his predicament involves a question of right conduct. There is the theme, often expressed in the stories of writers and artists, that what is important is the author's work and not the facts of his life or his legend; and, as in such stories as *The Reverberator*, "The Papers," and "Flickerbridge," Henry James condemned the "deadly" publicity of his day for pandering to the low estate of public opinion and taste. The story's source was an account James had heard of a married couple in charge of the Shakespeare house, who had given up their position because they were disgusted by the humbug about the poet they were expected to give the public. Gedge is similarly affected, but when the directors warn him that he is giving away "the show" and "that it simply won't do," he resolves for his wife's sake to stifle his critical sense and amazingly becomes an accomplished dramatic actor whose dissimulation holds his audience in thrall. This, according to Quentin Anderson, makes the story a parable of the mystery of artistic creation, since we cannot say how Gedge became an artist, just as we cannot answer that question for Shakespeare. It seems unlikely, however, that James would have employed the tone of a *jeu d'esprit* if that had been his intention. At the heart of this wonderfully comic and ironic story is the idea that humbug purveyed with Gedge-like artistry, or even something less, does bring in the receipts, and what is stressed in the closing pages of the story is that no matter how far Gedge goes he cannot overdo his act to an extent which will give him away. Mrs. Gedge's fear that this may happen makes for suspense about the outcome, but when it is emphatically dispelled by the action of the directors in doubling Gedge's salary out of gratitude to him for increasing the receipts, Gedge has the last word: "And there *you* are!"

"The Aspern Papers," "The Birthplace," "The Pupil," and the many other stories which show James at his best, or very nearly so, testify both to his great technical virtuosity and to his penetrating psychological and moral insight. James, in addition to other alleged shortcomings, has been charged with being overly preoccupied with narrative techniques and devices, with having cultivated too mannered and abstract a style in his later fiction, and with being too narrow in the range of experience he portrayed. Yet it is nearly always conceded that these strictures do not apply to his best stories. They give us interpretations of human nature and experience which have life and truth, and they are the products of high art, so much so that, as the British critic Cyril Connolly has said, James was "a man, who, if he had never written a novel, would be considered the first of short story writers."

Although James spoke of a story as a "situation revealed," he often relied heavily on plot complication, and thus he did not contribute as much to modifying radically the short-story form as writers like Chekhov, Joyce, and Katherine Mansfield did. Nevertheless, his technical innovations, especially his practice of limiting the point of view instead of employing omniscient narration—and, even more, his conception of fiction as a fine art—have exerted a profound influence on twentieth-century writers. The best work of those who have emulated him, like his own, corroborates James's assertion that "the figures in any picture, the agents in any drama, are interesting only in proportion as they feel their respective situations. . . . Their being finely aware—as Hamlet and Lear, say, are finely aware—makes absolutely the intensity of their adventure, gives the maximum of sense to what befalls them."

# 8 The Short Story in Transition

## Stephen Crane
## Jack London
## Edith Wharton
## Willa Cather
## and
## Theodore Dreiser

THE PERIOD FROM 1890 to 1920, although overshadowed by the work of Anderson, Hemingway, Faulkner, and other writers of the 1920's, for which it did much to prepare the way, is in its own right of considerable significance in the history of the American short story. Henry James reached the height of his powers as a master of the short story of character, manners, and moral and ethical issues. Hamlin Garland added a new dimension to the regional story by making it express social protest without appreciably sacrificing narrative art. Stephen Crane's brief but brilliant career during the 1890's resulted in short stories which make him one of the few great masters of the form and which in their tone and technique look forward to Hemingway and other moderns. Jack London is more nearly akin to Harte and Kipling than to Crane, but in at least a few of his adventure stories of the Far North written around the turn of the century, he portrayed with power and originality men in conflict with nature and with each other. Edith Wharton and Willa Cather, both of whom had established themselves as distinguished short-story writers and novelists before James's death in 1916, carried on his tradition of the realistic story. Theodore Dreiser's unrelenting emphasis on biological and socioeconomic determinism in his novels and stories, beginning with *Sister Carrie* (1900) and culminating in *An American Tragedy* (1925), makes him the foremost exponent of American literary naturalism, a movement which had its beginnings with Stephen Crane and Frank Norris in the 1890's and whose tendencies, although they have seldom been as fully exhibited in the work of writers other than Dreiser, continue to be one of the most pervasive influences in American fiction.

THE SON OF a Methodist clergyman, Stephen Crane (1871–1900) was born in Newark, New Jersey. After attending briefly first Lafayette College and then Syracuse University, he became a cub reporter for a New York newspaper but soon turned to free-lance writing, finding subjects for articles and descriptive sketches in the slums of New York's Lower East Side, where he lived for a time close to poverty and where, he said, he began his "artistic education." Renouncing what he called "the clever school in literature" and believing that a writer should strive to show life as it really is, including the ugly and unpleasant, he wrote such pieces as "The Men in the Storm," which pictures a crowd of down-and-out men waiting long hours in a blizzard to be admitted into "a charitable house where for five cents the homeless of the city could get a bed at night and in the morning coffee and bread." Of "An Experiment in Misery" he said: "I tried to make plain that the root of Bowery life is a sort of cowardice. Perhaps I mean a lack of ambition or to willingly be knocked flat and accept the licking."

These sketches, which exhibit a remarkable talent for etching a scene in sharp, vivid detail, are companion pieces to the short novel *Maggie: A Girl of the Streets* (1893), written, said Crane, "to show that environment is a tremendous thing in the world and frequently shapes lives regardless." So strongly naturalistic was his story of a girl, who after being seduced and abandoned by a bartender becomes a prostitute and eventually a suicide, that Crane could not find a publisher and had to pay a printer to bring it out. Whether Crane knew Zola's *L'Assommoir*, which in some respects *Maggie* resembles, is not certain, but at any rate the novel, along with Frank Norris's *McTeague* (1899), marks the beginnings of literary naturalism in America. *Maggie* sometimes verges on melodrama for it is too consciously impressionistic and too heavily ironic. But it has dramatic power and vitality, and it is superior to Crane's other short novel of the Bowery, *George's Mother* (1896), which portrays the relationship between a pious and overdevoted mother and her son, a young man for whom she has great hopes but who finally ends up a member of the neighborhood gang and frequenter of the corner saloon.

Crane's masterpiece is, of course, the short novel *The Red Badge of Courage* (1895), an account of a young Civil War soldier's thoughts and actions before, during, and after his first battle. Crane's best short stories of war, some of which, like the longer work, were written out of his reading and imagination before he saw battle as a war correspondent, resemble the novel in their objectivity and irony, their acute penetration into the psychology of fear and other emotions of men at war, their vivid impressionistic description, and their brilliant use of symbolism and imagery. In "A Mystery of Heroism" the soldier Collins, who is constantly complaining of thirst, is taunted by his comrades with being afraid to walk through the line of enemy fire to a well to get water, and responds to their dare. He thinks at first that he must be a hero because he feels no fear and that therefore "heroes were not much," and yet he cannot really believe himself one since "heroes had no shames in their lives, and, as for him, he remembered borrowing fifteen dollars from a friend and promising to pay it back the next day, and then avoiding that friend for ten months." Then, while getting the water, he is suddenly filled with fear, but running back awkwardly carrying a full bucket, he pauses, although still terror-stricken, to give a wounded officer a drink. When he returns safely to his fellows, the bucket is upset by two of the soldiers playfully struggling for it, and the water is lost. Like *The Red Badge of Courage* the story makes effective use of irony and paradox to suggest how much there is of the equivocal in the phenomenon called heroism.

The brief "An Episode of War," deservedly called by Richard Chase "as perfect a thing as Crane wrote," turns on the ironic situation of a Civil War lieutenant who is wounded in the arm by a stray bullet while he is dividing up his company's supply of coffee beans with his sword. The lieutenant is cast under a kind of spell. His sword, which he must take in his left hand, now seems a strange thing, and he is unsuccessful in his awkward attempts to replace it in its scabbard. His wound also sets him apart from his comrades, who, although they sympathetically proffer assistance, stand in awe of him. As he makes his way to the rear, he is able "to see many things which as a participant in the fight were unknown to him." And in the presence of the doctor at the field hospital it is as if his wound put the lieutenant "on a very low social plane." Certainly Crane's concluding sentence, "And this is the story of how the lieutenant lost his arm," must be regarded as highly disingenuous. What we are vividly shown is how a wound can alter everything for a man, along with all that is paradoxical, ambiguous, and unreal in the experience.

Crane's other notable war stories include "The Price of the

Harness," a moving tribute to the regular army soldier in the Spanish-American War; "Death and the Child," an account of a Greek war correspondent who is filled with an excess of patriotism and tries ineffectually to take part in a battle in the Greco-Turkish War; and "The Upturned Face," perhaps the sparest, most restrained, and yet one of the most impressive of all Crane's stories, in which two officers bury a fellow officer while under enemy fire.

Although not about war, "The Open Boat" was a by-product of Crane's war-correspondent activities. In early January, 1897, Crane sailed in search of story material from Jacksonville, Florida, on a ship carrying a cargo of arms and munitions for revolutionaries in Cuba. A few hours out of port the ship began leaking badly in heavy seas and eventually sank. Some of the crew members were lost, and Crane spent thirty hours in a ten-foot dinghy, with the captain, an oiler, and the cook. Crane first recounted the experience in a cabled dispatch to a New York City newspaper, entitled "Stephen Crane's Own Story." A vivid piece of journalistic writing, it deals almost entirely with what happened before the ship was abandoned, with only a brief reference in the last two paragraphs to the ordeal endured by Crane and his companions in the dinghy and their ultimate rescue. In his story Crane rendered this last part of the experience with high art. To some extent it is told from the point of view of the correspondent, since from time to time we have glimpses into his mind, but essentially it is objective and collective, all four men being involved equally in the struggle for survival. "None of them," reads the opening sentence, "knew the color of the sky," and the final statement of the story, after the rescue is accomplished, reads, "When it came night, the white waves paced to and fro in the moonlight, and the wind brought the sound of the great sea's voice to the men on the shore, and they felt that they could then be interpreters." What they have learned is that, in man's struggle with the forces of nature, nature remains unconcerned and indifferent to the outcome. The tower on the distant shore, which the correspondent wonders if anyone ever ascends to look seaward, symbolizes this theme. It "was a giant, standing with its back to the plight of ants." But the men have also gained a sense of brotherhood, a "comradeship, that the correspondent, for instance, who had been taught to be cynical of men, knew even at the time was the best experience of his life." Even so, their victory is not complete, the price of it being the life of the heroic and gallant oiler, who is drowned when the dinghy is swamped in the surf. The story has many memorable aspects—its effective use of understatement, its symbolism and imagery, its irony, its individualized and differentiated portraits of the four men, and its vivid description of the omnipresent, ever-menacing sea. It is a virtually perfect story in every detail and cannot fail to strike us as it did Conrad, who said, "The deep and simple humanity of its presentation seems somehow to illustrate the essentials of life itself."

"The Blue Hotel," Crane's other masterpiece of short fiction, again deals with the relationships of a group of men, this time in a Nebraska prairie town. Again the force and violence of nature, in the form of a blizzard, is present, although it is with violence in human actions and its consequences that the story is primarily concerned. The men involved are Pat Scully, proprietor of the hotel; his son, Johnnie; and three travelers, who are induced by Scully to interrupt their railroad journey westward and stop off temporarily at his hotel. The travelers are a Swede, who has worked for ten years as a tailor in New York, a cowboy on his way to Dakota, and a "little silent man from the East."

During a game of cards it is revealed that the Swede thinks men have been killed in the hotel and that his life is in danger from his companions. Determined to leave, he goes upstairs to get his bag, followed by the protesting Scully, and during their absence the easterner tells the others that the Swede is probably frightened because he "has been reading dime novels, and he thinks he's right out in the middle of it—the shootin' and stabbin' and all." Upstairs Scully gives the Swede whisky to placate him and persuade him to stay.

When the Swede comes back downstairs, he is drunk, and now, instead of being frightened, he is arrogant, belligerent, and profane. Another card game begins and the Swede suddenly accuses Johnnie of cheating. Johnnie hotly denies the charge, and the two go outside, followed by the others, to settle the issue by fighting. The Swede is the victor. He leaves the hotel and makes his way through the blizzard until he finds a saloon. There he drinks heavily and boasts of his victory. Enraged because none of the four men in the place, all sitting together at a table, will drink with him, he seizes one of them, a professional gambler, by the throat. "There was a great tumult and then was seen a long blade in the hand of the gambler. It shot forward, and a human body, this citadel of virtue, wisdom, power was pierced as easily as if it had been a melon. The Swede fell with a cry of supreme astonishment."

Some months later the easterner and the cowboy meet. They have learned that the gambler was sentenced to three years in prison for the death of the Swede. The cowboy declares that it was all the Swede's fault for accusing Johnnie of cheating and acting like such a fool. Angrily the easterner calls the cowboy a fool and says:

> "Johnnie was cheating. I saw him. I know it. I saw him. And I refused to stand up and be a man. I let the Swede fight

> it out alone. And you—you were simply puffing around the place and wanting to fight. And then old Scully himself: We are all in it! . . . Every sin is the result of a collaboration. We, five of us, have collaborated in the murder of this Swede, . . . and that fool of an unfortunate gambler came merely as a culmination, the apex of human movement, and gets all the punishment."

To which the cowboy replies, mystified, his feelings wounded, "Well, I didn't do anythin', did I?" Objection has sometime been raised to this conclusion as too deliberately pointing a moral, but Crane's story is nevertheless one of great intensity and power, and the figure of the Swede, first craven and fearful and then infected by hubris, is unforgettable.

Crane wrote several other western stories, including "Horses—One Dash!", "A Man and Some Others," "Five White Mice," and "Twelve O'Clock." They are all somewhat weak and have little that is distinctive in them. But "The Bride Comes to Yellow Sky" fully merits the high praise it has almost invariably been accorded by critics, some of whom have even placed it above "The Blue Hotel." Although the setting and characters are of the kind that Bret Harte might have treated, Crane's method is the antithesis of Harte's. Nearly everything that Harte would have played up is played down. The characters are not glamourized, nor are they made to seem odd or unusual. Jack Potter's bride "was not pretty, nor was she very young." At the end of the story Potter is revealed to be a man of great courage, but until then he seems ordinary and not particularly heroic. On their train trip from San Antonio to Yellow Sky, the newly married Potters feel the mingled happiness and self-consciousness natural to their state. They are the objects of amused glances from the other passengers and of patronizing but kindly attention from the porter and dining-car waiter. But Potter's nervousness is occasioned by something more: "He, the town marshall of Yellow Sky, a man known, liked, and feared in his corner, a prominent person, had gone to San Antonio to meet a girl he believed he loved, and there, after the usual prayers, had actually induced her to marry him, without consulting Yellow Sky for any part of the transaction."

Full of guilt feelings for not having performed a duty he obviously owed his friends but unable to muster the courage to face the tumultuous reception and enthusiastic congratulations they will give him, he can only hope that he and his bride can make their way unobserved to the security of his house. Ironically, he is accorded a very different kind of reception. The streets of Yellow Sky are deserted because one Scratchy Wilson is on the warpath. Scratchy is kind and gentle when sober, "the nicest fellow in town," as the bartender in the saloon, whom Crane employs to provide much of the necessary exposition, tells a visiting drummer. From time to time he gets drunk and tries to shoot up the town, and on these occasions Jack Potter has had the task of subduing him. When the drunken Scratchy finds the saloon shut against him, he takes it into his head to go to his enemy's house, and thus it is that when the newlyweds arrive they are confronted by Scratchy.

The tension and suspense are relieved when Scratchy is put to rout upon learning that Potter has just been married and is not carrying a gun. "He was not a student of chivalry; it was merely that in the presence of this foreign condition he was a simple child of the earlier plains. He picked up his starboard revolver, and placing both weapons in their holsters, he went away. His feet made funnel-shaped tracks in the heavy sand." And so concludes a story which we feel ought to turn into one in the tradition of Harte or Mark Twain, and which seems sometimes on the verge of doing so but never does, its humor, irony, understatement, and imagery all differing strikingly from theirs.

Crane wrote no other short fiction which comes up to "The Open Boat," "The Blue Hotel," "The Bride Comes to Yellow Sky," and his better war stories. He thought highly of *Whilomville Stories*, his last work, but it is hard to understand why, since these thirteen more or less related stories—mostly about the antics and escapades of small boys—although they make mildly amusing reading, are conventional in their situations and in their child psychology. Young Jimmie Trescott is the chief actor in several of them and appears in others. In one story he and his friends go lynx hunting and by accident shoot a cow instead, and in another Jimmie suffers acute stagefright when called upon to recite "The Charge of the Light Brigade." Some of the other stories, notably "The Knife" and "His New Mittens," are better, but they are not quite in the same class with *Tom Sawyer* or *Penrod* or even the Basil Duke Lee stories of F. Scott Fitzgerald.

Although not published with them, "The Monster" also belongs to the group of Whilomville stories. In this long story the Trescott house catches fire, and Henry Johnson, the Negro hostler of the Trescotts, is trapped, along with Jimmie's father, Dr. Trescott, when he tries to rescue the sleeping Jimmie. All three are rescued by the firemen and recover, but Henry's face is burned away and his mind crazed. The townspeople who lauded Henry for his bravery now visit their opprobrium on Dr. Trescott for refusing to send this awful-appearing, though actually harmless, creature away. The story thus has more serious and significant implications than the other Whilomville stories, but it has obvious artistic weaknesses. It is too long and diffuse. For example, at one of the most dramatic moments—when Jimmie and Henry Johnson are in the burning house—Crane pauses to devote two or three pages to describing the reactions of the small boys of Whilomville to fires. The story calls to mind Mark Twain's "The Man That Corrupted Hadleyburg," but it is a less powerful and convincing indictment of the popular mind and morality.

Crane's best work does not bulk large, but it makes an impression out of all proportion to its quantity. Crane is one of the most original and distinctive of American writers. In his style and narrative method he seems to owe almost nothing to those who preceded him. As John Berryman has pointed out, plot manipulation, contrivance, romantic love, the usual conflicts between characters, characters as we expect them to be in fiction—these and other elements that we are conditioned to look for in stories are not in Crane. What he has given us is a style of great purity, clarity, and intensity and a profound insight into both the irony and the pathos of the human condition.

IN THE LIFE of Jack London (1876–1916) there was as much—if not more—adventure and variety of experience as in his volumes of fiction. Born in San Francisco, during his youth he worked long hours for low wages at various jobs, was a sailor and a hobo, prospected for gold in the Klondike, lectured on behalf of socialism, voyaged to the South Seas in his own sailing ship, and was a war correspondent. In addition, his boundless energy and industry enabled him to produce during a sixteen-year writing career some twenty novels, numerous collections of short stories, and a considerable amount of nonfiction.

London found no gold when he took part in the Klondike gold rush in 1897, but he did find the materials for the short stories and novels of the Far North which established his reputation and for which he is best remembered. Between 1898 and 1900 the *Overland Monthly* and the *Atlantic* published his first Klondike stories, which were collected in *The Son of the Wolf* (1900). Reviewers praised the volume and likened London to Kipling, but in most of the stories the characters, situations, sentiment, and style are more reminiscent of Bret Harte. The Malemute Kid, who figures more or less prominently in several of the tales, although a colorful and even memorable character, is essentially a Harte-like stereotype. Rough and strong, as a man must be to survive in the Klondike, the Kid also has sympathy for his fellow man, as is shown in "The Man on the Trail." There he aids an unfortunate prospector, who has stolen money under extenuating (for the Kid, at least) circumstances, to elude the Northwest Mounted Police. In "The White Silence" the Kid is called upon to abandon his fatally injured partner on the trail if he is to save himself and his partner's wife. The internal conflict in the Kid is overexaggerated, the emphasis on the dying partner's love for his wife and the unborn child he will never see makes for too much sentiment, and the personification of the "white silence"—the cold and isolation of the frozen North—is overdone, but London's depiction of the basic conflict of man versus nature gives the story considerable power. The same quality does much to redeem "An Odyssey of the North," a romantic tale somewhat reminiscent of Kipling's "The Man Who Would Be King," of an Indian who loses his betrothed to a white man and pursues them over land and sea before he finally achieves revenge.

Besides *The Son of the Wolf*, London published half a dozen more collections of short stories and five novels with Yukon settings, but only the short novel *The Call of the Wild* and a few of the stories have real distinction. Two of the latter are "Love of Life" and "To Build a Fire." In them London eschewed the sentiment, stock characters, contrived plots, trite humor, and overliterary language that weaken many of his stories. Instead he depicted vividly and dramatically man struggling against nature for survival. In the former an injured man is deserted by his partner in the wilderness, but, clinging tenaciously to life, he overcomes hunger and exhaustion to make his way to succor. The miner in the second story, London's masterpiece, is not as fortunate. Foolhardy in believing he can disregard with impunity the law that no man must travel the trail alone in the Klondike when the temperature is more than fifty degrees below zero, he is unsuccessful in his desperate efforts to build a fire to ward off the extreme cold and freezes to death.

The sled dogs of the Yukon appear in many of London's stories, and in some they are even the chief actors. "Bâtard" (originally entitled "Diable—A Dog"), despite its somewhat contrived twist ending, is a gripping and suspenseful tale of a fierce and diabolically shrewd dog—half-wolf, half-husky—and of his brutal, sadistic master who is determined to break the dog's spirit. *The Call of the Wild,* London's first best-seller (said to have sold over six million copies since its publication in 1903), is the history of the dog Buck who is stolen from a Santa Clara Valley ranch and taken to the Klondike, where he learns to survive in a hostile environment by following "the law of club and fang" and wins fame for his feats of strength and endurance. Finally, after his master dies, the primitive instincts of his remote ancestors having strongly made themselves felt, he becomes the leader of a wolf pack. *The Call of the Wild* is not a great short novel, but it is a very good one. Its adventure is genuinely dramatic, it has some of London's finest descriptions of the beauty and grandeur of the Far North, and its theme of ruthless individualism is artistically developed rather than philosophically labored, as it is somewhat unduly in London's otherwise very fine semi-autobiographical novel *Martin Eden* (1909), and to almost a ridiculous degree in the adventure novel *The Sea Wolf* (1904).

London also wrote about thirty adventure tales laid in the South Seas which he collected in *South Sea Tales* (1911) and other volumes. The twenty-five-month South Pacific voyage (1907–1909) that provided the background for these stories is described by London in *The Cruise of the Snark*, a volume which makes for more interesting reading than most of the tales. As in his Far North stories, London is usually at his best when he shows his characters in conflict with the hostile forces of nature, as in "The House of Mapuhi," a vivid description of a hurricane and the struggle of an old native woman to survive the death and destruction it brings to her atoll.

In writing some of his South Sea stories, London may well have been influenced by both Conrad and Robert Louis Stevenson, but he made little attempt to explore moral ambiguities as did Conrad, and Stevenson also on occasion, limiting himself rather to adventurous incident and action. Although a very readable story, there is none of the richness of implication, for instance, of Conrad's "Youth" in London's "The Seed of McCoy," which, like Conrad's story, uses the situation of a ship with its cargo on fire attempting to reach the safety of port. London's South Sea writings rival Stevenson's in their picturesque local color, but they are not primarily concerned, as is the case with Stevenson's "The Beach at Falesá," with the moral problems involved in white-native relations. Several of London's tales, however, do graphically portray the white man's exploitation of the natives. White traders pillage and burn native villages and massacre their inhabitants, and of them London makes a South Sea islander say: "White men are hell. I have watched them much, and I am an old man now, and I understand at last why the white men have taken to themselves all the islands in the sea. It is because they are hell." There are also ironic references to "the inevitable white man," with his racial egotism, cruelty, stupidity, and greed, but London's attitude in the main seems to be that, since he is an accomplished fact, there is not much point in reprehending him.

In addition to *The Iron Heel* (1908), a novel which envisions the ultimate triumph of socialist democracy after a long period during which the workers are suppressed and enslaved by the capitalist oligarchy, London wrote a number of proletarian short stories. None of them rises much above the level of propaganda except for "The Apostate," a powerful indictment of the evils of child labor, expressed with an artistry and restraint worthy of Stephen Crane. Especially in his later stories, London often wrote, as he frankly admitted, merely to make the money he needed to build his expensive boat, the *Snark*, a costly ranch house, and so on—and the result inevitably was a great deal of unabashed hackwork. London's great energy and considerable narrative talent made him one of the best-paid and best-known writers of his day, and had he not succumbed to the temptation to write so much and so fast he might well have produced a body of work of more enduring merit.

A PROLIFIC AUTHOR of both short fiction and novels, Edith Wharton (1862–1937) published eleven volumes of short stories between 1899 and 1936. A number of her stories, especially the earlier ones, reflect the influence of Henry James. In "The Pelican," for example, a characteristically Jamesian blend of irony and pathos, like that in "Four Meetings" and

" 'Europe,' " is evoked when a superficial but highly successful lady lecturer's style of lecture ("It was her art of transposing second-hand ideas into first-hand emotions that so endeared her to her feminine listeners") goes out of fashion, and she loses her audiences. The situation of "The Descent of Man" is also of the wryly ironic sort that James was fond of developing. A biology professor writes a book parodying popular books on science, but his publisher does not realize that it is satirical until he is told and then maintains that it will be a great success which readers will take seriously. All of this happens when the book is published, and the professor, beset by money problems, succumbs to the temptation of writing another book of the same kind instead of the serious scientific book he wants to do.

Like James, Mrs. Wharton also wrote a number of stories about artists. They are usually concerned primarily with moral and ethical problems and conflicts, but several deal with the problems of the artist as artist. In "The Recovery" a painter realizes that he must transcend his provincial American background if he is to do significant work. The moral of "The Verdict" is that the true artist must possess more than skillful technique. "The Pot-Boiler" treats the familiar Jamesian theme of the general public's lack of appreciation for the work of the serious artist. Other stories by Mrs. Wharton that are strongly reminiscent of James's style include "The Quicksand"; "The Dilettante"; "In Trust"; *Madame de Treymes*, a novelette whose situation recalls James's "Madame de Mauves" and *The American*; and "The Last Asset," which also treats the international theme of America and Europe. Yet merely to point out parallels between James and Mrs. Wharton is to do her an injustice. Her novel *The Reef* (1912), although markedly influenced by James, and her better Jamesian short stories have their own style and demonstrate a storytelling gift and a psychological and moral insight which fall only a little short of his.

Mrs. Wharton was both an acute and a none-too-sympathetic observer of the New York polite society into which she was born, and in several of her best novels, such as *The House of Mirth* (1905), she exposed with telling irony its narrowness, lack of cultivation, and false values. In these works and in the novelette *Ethan Frome* (1911), perhaps her finest piece of fiction, the dramatic conflict comes about because seemingly the only way for a character to find happiness is to defy the accepted mores, usually the conventions concerning marriage. The same kind of conflict figures in a number of her short stories. In an early story, "Souls Belated," a woman leaves her husband for a lover and maintains that, since her act has demonstrated that she and her lover care nothing for the opinions of others, they should continue to live together without marrying after her husband divorces her. Her experience teaches her that it is better to bow to the conventions than to flout them, a lesson also learned by a number of heroines in Mrs. Wharton's later stories and novels. Indeed, Mrs. Wharton was particularly concerned with the consequences arising from unhappy marriages, love outside of marriage, and divorce; among her better stories testifying to this preoccupation are "The Reckoning," "Autres Temps," "Bunner Sisters," "The Other Two," "The Long Run," and "Atrophy."

From time to time Mrs. Wharton wrote romantic historical tales somewhat in the manner of Stevenson, but she did nothing particularly worthy of note in this vein, and with the exception of "The Eyes," a powerful study of egotism, she was no more than moderately effective in the supernatural stories which she essayed on a number of occasions. Furthermore, from the early 1920's to the close of her career most of her fiction was written for mass-circulation magazines, and the result was usually a slick commercial approach. But despite her decline as a serious writer, her best work has considerable distinction in both style and subject, and she remains one of the best American writers of the short story of manners and moral conduct.

DURING HER LONG career Willa Cather (1876–1947) published three volumes of short stories. These, plus a fourth, posthumously published, volume, *The Old Beauty and Others* (1948), contain seventeen stories, about half of her total output of short fiction. During her student days at the University of Nebraska she wrote sketches of the immigrant farmers whom she had observed during the years she was growing up in Red Cloud, Nebraska. Shortly after her graduation she went to Pittsburgh, where she did newspaper work, taught high-school English, and contributed poems and stories to *Cosmopolitan, McClure's Magazine*, and other periodicals. For several years, both in Pittsburgh and in New York, where she was for a time editor of *McClure's*, she wrote stories usually centering on artists or persons of artistic temperament, some of which, as well as her first novel, *Alexander's Bridge* (1912), show the influence of Henry James.

The most obviously Jamesian of the stories in Miss Cather's first collection, *The Troll Garden* (1905), is "Flavia and Her Artists." A wealthy matron named Flavia Hamilton gives one of the house parties at which it is her practice to assemble celebrities, in this instance an actress, a painter, a musician, a singer, a novelist, and a scientist. An additional guest is Imogen Willard, a young woman who has known Flavia and her husband for some years and who has recently returned to America after a period of study in France. Imogen feels that Arthur, Flavia's husband, is surely too intelligent not to perceive that his wife neither understands nor is interested in the work of these guests and that as a result they regard her with something akin to contempt. The novelist, a Frenchman, leaves the party several days before the others, and an article by him, ostensibly on "The Advanced American Woman" but actually a satirical characterization of Flavia and her Philistinism, appears in the newspaper. Although everyone else sees the article, Arthur manages to keep it from his wife's knowledge, and at dinner he denounces the novelist and his work. This confirms for the obtuse Flavia, she confides to Imogen, her belief that Arthur is utterly prejudiced and lacking in any aesthetic sense. Imogen attempts to defend him, but she cannot of course reveal that he has sacrificed himself to spare his wife's vanity. The story is expertly handled but leaves the feeling that James might have done more with its ironies and made the reader feel more strongly the profound effect the incident has had on Imogen.

The non-Jamesian stories in *The Troll Garden,* "The Sculptor's Funeral" and "A Wagner Matinée," are reminiscent of Hamlin Garland in their depiction of middle western farm and village life as ugly, mean, and provincial. There is powerful feeling behind the first story, but its thesis might have been made more convincing and its small-town characters more believable if Miss Cather had been more restrained in her indictment. The other story is more artistically effective in its portrayal of a woman who for thirty years has been starved for art and beauty by her bleak, grim life on the prairie.

Likewise denied by his environment the aesthetic experiences he craves—or rather the glamor and romance he associates with them—is the boy in "Paul's Case," probably Miss Cather's best-known story. A maladjusted youth of high-school age, Paul is unhappy both at home and in school, and through his friendship with an actor in the local stock company in Pittsburgh and through his job as an usher at Carnegie Hall he finds in the theater and in music some of the romance he seeks:

> Perhaps it was because in Paul's world, the natural nearly always wore the guise of ugliness that a certain artificiality seemed to him necessary in beauty. Perhaps it was because his experience of life elsewhere was so full of Sabbath-School picnics, petty economies, wholesome advice as to how to succeed in life, and the unescapable odors of cooking, that he found this existence so alluring."

When Paul's teachers finally decide that his case is hopeless, his father removes him from this existence, and he is put to work as an office boy. He absconds with money entrusted to him to deposit in the bank and goes to New York, where he finds for a few days the environment he craves, living in a luxurious hotel and wearing expensive clothes before committing suicide in preference to returning to his dreary former life. "Paul's Case" is the only instance in which Willa Cather wrote in a naturalistic vein, and it constitutes an impressive example of this kind of story.

Miss Cather reprinted four of the stories in *The Troll Garden* ("Paul's Case," "A Wagner Matinée," "The Sculptor's Funeral," and " 'A Death in the Desert' ") and added to them four new stories in *Youth and the Bright Medusa* (1920). The latter, all of which deal with either opera singers or painters, show no advance over what she had already done in the short story and give no indication that she had developed the individual style and high degree of artistry manifested in her two early novels of Nebraska farm life: *O Pioneers* (1913) and *My Ántonia* (1918). The three stories in *Obscure Des-*

*tinies* (1932), however, compare favorably with
and are similar to them, being based on Miss C
ollections of childhood and youth and concerne
with character portrayal and depicting a way
theme of "Neighbor Rosicky," the story of a C
in Nebraska, is that, despite its difficulties and hardships, rural life provides a better opportunity than industrial, urban living for achieving security, human dignity, and family solidarity. If the story exhibits some tendency to picture country life more favorably than may be warranted, it nevertheless avoids the sentimentality of stories with a "life can be beautiful" theme, and its warmly sympathetic portrait of Rosicky makes him a memorable character. A similar talent for bringing characters to life is displayed in the portraits of the heroic, self-sacrificing grandmother in "Old Mrs. Harris," a study of a Tennessee family transplanted to a small town in Colorado, and of the two Kansans in "Two Friends," a banker and a cattleman, who had represented for the author in her youth "success and power" and whose friendship came to "a stupid, senseless, commonplace end" in a disagreement over William Jennings Bryan and his political views.

Two of the three pieces brought out after Miss Cather's death in *The Old Beauty and Others* did little to add to her reputation as a short-story writer. The title story appears to have been one of her favorites, but its history of a once-famous beauty, now an old woman living in retirement at Aix-les-Bains, lacks the interest and poignancy one expects of such a story. Nor is the reader likely to be particularly moved by the situation of the prosperous but emotionally disturbed businessman in "Before Breakfast," who is finally able to resolve his inner conflicts and believe once again that living can be a challenge and delight. "The Best Years," on the other hand, not only has much in common with the stories in *Obscure Destinies* but also compares favorably with them, particularly in the vividness and charm of its portrait of the fourteen-year-old girl Lesley Ferguesson, who teaches in a Nebraska country school.

Miss Cather's shorter fiction is overshadowed by her novels, but her achievement as a short-story writer, despite her limited and somewhat uneven output, was more than a minor one. She wrote with honesty and insight and usually with great technical skill. Like Sarah Orne Jewett, who encouraged her to write stories like "Neighbor Rosicky" and "Old Mrs. Harris" and from whom she learned much, she is one of our most distinguished writers of regional fiction.

BORN in Terre Haute, Indiana, of German stock, Theodore Dreiser (1871–1946) was a journalist in his late twenties, who had worked in St. Louis, Chicago, and Pittsburgh when he began to write fiction. He was unfavorably regarded for a number of years by many readers and critics because of the uncompromising naturalism and alleged immorality of *Sister Carrie* (1900), suppressed after publication by the publisher and not reissued until 1907, *Jennie Gerhardt* (1911), and later novels. It was not until the 1920's, by which time he had published his best-known novel, *An American Tragedy* (1925), that he attained a substantial measure of popularity and prominence.

Dreiser's earliest short stories, written at about the same time as *Sister Carrie*, are notable for their variety but are a good deal less impressive than that novel. "When the Old Century Was New" is a somewhat stilted historical narrative laid in New York City at the beginning of the nineteenth century. "McEwen of Shining Slave Makers" is an allegory with a deterministic theme, in which a man falls asleep on a park bench and dreams he is an ant fighting with his tribe against another tribe of ants. The conflict, characterization, and setting of the much more realistic "Old Rogaum and His Theresa" are handled convincingly, although Dreiser perhaps overemphasizes his point that if the strict old German father had not relented after locking his rebellious teen-age daughter out of the house one evening when she does not come in as soon as he calls, she would have been compromised by the young tough she has been meeting on the street and would have suffered the fate of other young girls for whom such a situation had been the first step in becoming a prostitute. "Nigger Jeff," based on a lynching which occurred during Dreiser's early newspaper days in St. Louis, has less impact than later lynching stories by Erskine Caldwell and William Faulkner but achieves considerable force by focusing on the reactions of a somewhat naïve young reporter whose belief that justice prevails is destroyed by the event.

*Free and Other Stories* (1918) contains these early stories and seven others. Two or three of the latter are little more than journalistic pieces. Dreiser employed the manner of O. Henry in "A Story of Stories," an entertaining account of the rivalry between two newspaper reporters. Very different is the poetic tone and poignant situation of "The Lost Phoebe," in which an old man suffers from a hallucination that his dead wife is still alive. Also moving without being sentimentalized is the plight, in "The Second Choice," of the working girl who is thrown over by the man she loves. There is likewise little hope for happiness for the young husband in "Married," a story which appears to reflect Dreiser's own unhappy first marriage. A musician married to a farm girl, whose background prevents her from sharing his aesthetic interests, the husband increasingly feels their incompatibility, but because of her devotion and her obvious fear that she will lose his love, he cannot bring himself to leave her. In "Free," also, a sense of duty has long tied a much older man to a woman whose attitudes and values he could never agree with, even though he deferred to them. Should he not, he reflects, have ignored convention and left her? Yet his wife had tried to do her best according to her lights, and he reproaches himself for hoping that she will die now that she is seriously ill. She does die, and he is free, but, ironically, free only to die also. "Now the innate cruelty of life, its blazing ironic indifference to him and so many grew rapidly upon him."

The retrospective method of narration in "Free" has certain drawbacks. When a character is made to review past actions in his mind, the story is likely to seem tedious and lacking in dramatic quality, and if the actions are a cause for regret there is a danger that the character will be made to seem too sorry for himself. Yet Dreiser manages to a considerable extent to make a virtue of the method. The very weight of the slow-paced, sometimes repetitious, relentless accumulation of details bearing on the situation of Rufus Haymaker gives power to his story.

Dreiser was so preoccupied with the theme of unhappiness in marriage that he also treated it in seven of the fifteen stories collected in *Chains, Lesser Novels and Stories* (1927). Like "Free," three of them—"Chains," "The Old Neighborhood," and "Fulfilment"—are retrospective in their telling, repeat its tone of irony and futility, and emphasize that life traps and deludes us. More like some of Sherwood Anderson's stories are "Convention" and "Marriage—For One." In the first story, of which Dreiser said, "I set it down as something in the nature of an American social document," a man is unfaithful to his dull, drab wife. When the wife sends herself a box of poisoned candy, attempting to make it appear that it came from the other woman, the affair is exposed and brought out in the newspapers. What concerns the narrator is the effect this incident has on the husband, who is so bound by convention that he can easily abandon the mistress whom he had loved to go back to his wife. He is a psychological mystery to the narrator, who is left feeling "cold and sad." In the second story the narrator is profoundly moved by "the despair, the passion, the rage, the hopelessness, the love," of a man whose wife has left him. The other stories in *Chains* either are journalistic pieces written to entertain or treat themes which Dreiser had developed to better advantage in his novels. "The Victor" is a miniature companion piece to *The Financier* (1912) and *The Titan* (1914), Dreiser's two lengthy novels of unscrupulous financial dealings, being the history of a shrewd and ruthless financier who becomes a multimillionaire oil king. "Typhoon" and "Sanctuary," stories of girls betrayed by faithless lovers, are reminiscent in some respects of *Jennie Gerhardt* and *An American Tragedy*.

*Twelve Men* (1919) and *A Gallery of Women* (1929) are other collections of Dreiser's shorter pieces, but they are not properly short stories. The former is made up of sketches of actual persons who were Dreiser's friends and acquaintances, while the latter contains descriptive portraits—partly factual, partly fictional—of the personalities of various kinds of women. Dreiser once said, in explaining why he did not write more short stories, "I need a large canvas." His novels and stories seem to confirm that this statement was usually true, though certainly not in every instance. The comment of one critic that Dreiser was never at home in the short story does not give us a fair picture of his shorter work. It has obvious limitations—an almost relentless and sometimes tiresome imposing of his deterministic philosophy on the reader, a lack of psychological penetration into his characters, and stylistic lapses—yet Dreiser succeeds nevertheless in leaving an impression on us, and few other short-story writers have written more powerfully and movingly on the theme of entrapment.

# 9
# The Liberation of the Short Story

## Sherwood Anderson

EVEN MORE THAN Theodore Dreiser, whom he greatly admired, Sherwood Anderson (1876–1941) revolted against the stereotyped and conventional fiction of his time. Maintaining that "there were no plot short stories ever lived in any life I had known about," Anderson endeavored in *Winesburg, Ohio* (1919), a collection of related stories about maladjusted people in a small middle western town, to employ a form and a narrative technique which grew freely and naturally out of his materials and the reaction of the narrator to them. The individuality, freshness, and lyric quality of the best stories in this volume and in his other short-story collections and the example and inspiration they provided for writers who came after him give Anderson a prominent place in the history of the short story.

Anderson was born in Camden, Ohio, the son of a harness-maker and house and sign painter. The father, an easygoing man with a reputation as a storyteller, moved his family from one small Ohio community to another during Anderson's early boyhood, finally settling in Clyde. Anderson's schooling was irregular and limited, and he worked at various jobs while growing up. He came to the writing of fiction relatively late, making his first efforts in Elyria, Ohio, where he was manager of a paint factory. Leaving his position—though less dramatically than the legend would have it, which pictures him as the artist in rebellion walking out abruptly on business and family—in 1913 he went back to writing advertising copy in Chicago for a firm which had employed him some years earlier. Soon he began to publish articles and stories, including some of his *Winesburg* pieces, in the *Little Review* and in other small literary magazines.

Two novels written in Elyria, the semiautobiographical *Windy McPherson's Son* (1916) and a labor story, *Marching Men* (1917), as well as a small volume of free verse, *Mid-American Chants* (1918), preceded the publication of *Winesburg, Ohio*, but they give little indication of what Anderson was to achieve in that work. In Chicago his association with artists and writers was a powerful stimulus in his development as a literary artist, and his reading of Mark Twain, Ivan Turgenev, and Gertrude Stein taught him that a writer must develop his own style. The general idea for the *Winesburg* stories may have been suggested by Edgar Lee Masters' *Spoon River Anthology* (1915), with its epitaphs in verse describing the inner lives of persons buried in a cemetery in a small middle western town. It is tempting to see in Anderson's treatment of his characters the influence of Freudian psychology, with which he seems to have had at least a general acquaintance.

The *Winesburg* stories present a procession of characters who are, to use Anderson's term, "grotesques." In his introductory sketch, "The Book of the Grotesque," Anderson portrays an old man who regards all the people he has ever known as grotesques and who writes an unpublished book developing his theory about what made them so. There were, he believed, many truths in the world, but "the moment one of the people took one of the truths to himself, called it his truth and tried to live his life by it, he became a grotesque and the truth he embraced became a falsehood." But this explanation does not really apply to most of the *Winesburg* grotesques. Instead, their warped personalities and abnormal actions are the result of their inability to satisfy certain powerfully felt but vaguely understood needs in their natures. They are inhibited, frustrated, isolated creatures, to whom is denied, more than anything else, the need to be understood and to love and be loved.

Most of the *Winesburg* stories follow the pattern of "Hands," which appears first in the volume, although not all of them are as artistically effective. This story may seem discursive and loosely organized, but it has a definite form and is by no means artless in its selection and arrangement of details. The opening paragraph is a vivid little scene which suggests at the outset the isolation of the fat little man, Wing Biddlebaum. As he walks up and down on the veranda of his small house outside Winesburg, nervously stroking his bald head, a wagonload of shouting, laughing boys and girls returning from berrypicking comes along the highway, and one of the girls jibes at him, "Oh, you Wing Biddlebaum, comb your hair, it's falling into your eyes."

Then come several paragraphs of exposition in which we learn that Wing has lived in Winesburg for twenty years without becoming a part of the life of the community. Full of self-doubts and timid in the presence of others, he can express himself only to George Willard, the young reporter on the Winesburg newspaper, who sometimes comes to take a walk or spend an evening with him. Wing's story lies in his hands, which "made more grotesque an already grotesque and elusive individuality." They are constantly active, and without them Wing cannot talk, but their movements alarm him, and he strives constantly but unsuccessfully to keep them concealed. Wing's peculiarity is brought out dramatically in a scene in which he admonishes George almost fanatically not to be like the other people in the town but instead to follow his inclination to be alone and to dream dreams. Forgetting his hands, Wing lays them on George's shoulders. Then, realizing what he has done, he is horrified and abruptly leaves the boy. "His hands," George concludes, "have something to do with his fear of me and of everyone." Why they do is then explained in a passage which tells of Wing's history before he came to Winesburg—how as a young schoolteacher in Pennsylvania, in expressing his ideas and dreams to the boys who were his pupils, he had touched them as he had George. This had been misunderstood, and he had been driven out of town, narrowly escaping being lynched. Finally, in the last long paragraph, we are brought back to the story's beginning, when we see Wing once more pacing his veranda before his simple evening meal and again before he goes to bed, isolated and denied his need to express his love of mankind.

George Willard is employed as a confidant in several of the other *Winesburg* stories, although it is not always made clear why the grotesque should be especially drawn to him. The grotesque one, in attempting to express himself or to explain his past to George, is likely to become excited, almost frenzied, and to speak incoherently and vaguely. In "The Philosopher," Dr. Parcival wants George to look on others with hatred and contempt because it will make him their superior. We are told, "He seemed intent upon convincing the boy of the advisability of adopting a line of conduct that he was himself unable to define." And there is Enoch Robinson, in "Loneliness," who having found that he could not express his thoughts and feelings to real people filled his lonely room with creatures of his imagination. In trying to make George see how happy and important this solution made him feel, he says: "You'll understand if you try hard enough . . . It isn't hard. All you have to do is to believe what I say, just listen and believe what I say, just listen and believe, that's all there is to it."

Others who confide in George Willard, or with whom he has encounters, are Wash Williams, in "Respectability," whose wife's infidelity had made him a fanatical woman hater and who wants his story to be a warning to George; Seth Richmond, the young man in "The Thinker," who, in contrast to George, feels that he is an alien in Winesburg; and Elmer Cowley, in " 'Queer,' " who is tormented by his feeling that the town regards him and his family as abnormal and by his inability to express himself to anyone except a half-witted farmhand. After making an ineffectual attempt to confide in George Willard, Elmer resolves to escape to Cleveland, where he can be like other men and find friends. At the railroad depot where he has asked George to meet him, he is again unable to declare his determination not to be queer. Overcome by his emotions, he strikes George violently and repeatedly, jumps aboard a passing freight train, and cries, "I guess I showed him I ain't so queer."

For Elmer Cowley, George Willard represents the town, and to strike him is to hit back at the town itself for having condemned the Cowleys as queer. "Elmer Cowley could not have believed that George Willard had also his days of unhappiness, that vague hungers and secret unnamable desires visited also his mind." George Willard is not a grotesque, but that he also needs to fulfill himself is evidenced by four of the *Winesburg* episodes in which he is the principal character. The last of these stories "Departure," shows him leaving

Winesburg to seek his fortune. "Nobody Knows" portrays his nervousness and trepidation before working up the courage to have an affair with a girl of loose morals who has offered herself to him, and after the event his mingled feelings of masculine vanity and fear that the girl may expose him. In "An Awakening," excited by his belief that he is able to think far more profoundly about life than formerly, George insists to Belle Carpenter, the older woman he takes walking one evening, that he is different, and declares, "You've got to take me for a man or let me alone." Having taken the woman to a lonely place, George, filled with a sense of masculine power, believes that she will give herself to him, but at that moment her sweetheart, the bartender Ed Handby, appears, thrusts George rudely aside, and takes her away, leaving George filled with hatred and humiliation.

George Willard's growing sense of maturity is treated again in "Sophistication," one of the most perceptive and poetic of all the *Winesburg* pieces. The sense in which Anderson uses the term "sophistication" is indicated in a passage near the beginning of the story, in which he says in part:

> There is a time in the life of every boy when he for the first time takes the backward view of life. Perhaps that is the moment when he crosses the line into manhood. The boy is walking through the street of his town. He is thinking of the future and of the figure he will cut in the world. Ambitions and regrets awake within him. Suddenly something happens; he stops under a tree and waits as for a voice calling his name. Ghosts of old things creep into his consciousness; the voices outside of himself whisper a message concerning the limitations of life. From being quite sure of himself and his future he becomes not at all sure. If he be an imaginative boy a door is torn open and for the first time he looks out upon the world seeing as though

> they marched in procession before him, the countless figures of men who before his time have come out of nothingness into the world, lived their lives and again disappeared into nothingness. The sadness of sophistication has come to the boy.

When this moment comes for George, he seeks out Helen White, the banker's daughter, and she, because she is experiencing a similar change, is instinctively drawn to him. The final scene, in which Anderson shows them together in the dark, deserted Winesburg fairgrounds—awed by the solemnity and significance of their thoughts and emotions, and then giving vent to their animal spirits in laughing and running about before being recalled to their new-found maturity —is beautifully handled throughout.

"Mother" and "Death" are devoted to the history of George Willard's mother. Old and neurotic at forty-one, she had sought blindly and unsuccessfully as a young woman "some hidden wonder in life." Her marriage to Tom Willard, who allows the hotel she inherited from her father to deteriorate while he dreams vainly of becoming a congressman or governor, has not brought her happiness. She hates the boy's father for counseling him to be ambitious for material success, and she prays that the defeat she has known will not be the fate of her son and that he will be able "to express something for us both." She is too inhibited to be able to express these feelings to him, but is made happy by George's obvious desire to get away from Winesburg. For a period before her death she finds a confidant in the old physician Dr. Reefy (whose habit of writing his thoughts on pieces of paper and stuffing them in his pockets, where they become hard balls and are eventually thrown away, is described in "Paper Pills"), and it is almost as if he becomes for her the lover she had never found. Anderson's description at the end of "Death" of the effect of her passing on her son is one of the finer passages in *Winesburg.*

There are other women in Anderson's gallery of grotesques. Alice Hindman, in "Adventure," believes for many years that the man who had left town after she had given herself to him as a young girl will return and marry her. When she finally realizes that she has deceived herself, she feels that she will never know happiness. She tries to overcome her sense of isolation by making friends, but she needs something more. She is doomed, says Anderson, like many others, to live and die alone.

The desire of Kate Smith in "The Teacher" to be loved by a man also goes unfulfilled. Unknown to her, the Reverend Curtis Hartman in another story, "The Strength of God," feels an overwhelming passion for her. He can see into her room from his study window in the church tower, and he prays to God to help him resist temptation. After a prolonged conflict he believes himself saved—that God has manifested himself in Kate, when he sees her throw herself naked on her bed and appear to pray.

Hartman has strong religious feelings—he doubts that the spirit of God is really in him, and he longs for the day when he will be infused with power to spread God's word—but he is not a grotesque, as is the prosperous farmer and religious fanatic Jesse Bentley, in "Godliness." This long, four-part story is one of the inferior pieces in *Winesburg*—others are "Tandy" and "A Man of Ideas"—but "The Untold Lie" is a poignant and very good story of a middle-aged farmhand with a wife and a half-dozen children who feels trapped by life. He wants to advise a younger man not to marry but can't bring himself to do so and decides that it is just as well, saying, "Whatever I told him would have been a lie."

In one of his comments on the *Winesburg* episodes Anderson said, "I felt that taken together they made something like a novel, a complete story," giving "the feeling of the life of a boy growing into manhood in a town." But *Winesburg* is hardly a novel, nor is it primarily concerned with George Willard and his development. Though there are occasional glimpses of what village and rural life was like in the Middle West around the turn of the century, it was obviously Anderson's intention not to make a sociological study of a small town but rather to penetrate beneath the exteriors of its more maladjusted townspeople, to do presumably what Kate Smith advises George Willard to do when he tells her he wants to be a writer: "You must not become a mere peddler of words. The thing to learn is to know what people are thinking about, not what they say." Perhaps, however, Anderson was too preoccupied with looking for the grotesque in human nature to see its reality fully; but although, as Lionel Trilling has said, the portraits of his grotesques may pain and puzzle us, *Winesburg* also, as the same critic has admitted, "has its touch of greatness."

Besides *Winesburg, Ohio*, Anderson published three other collections of stories: *The Triumph of the Egg* (1921), *Horses and Men* (1923), and *Death in the Woods* (1933). These stories, along with some additional stories which were never collected, vary greatly in quality, but at least a dozen of them are noteworthy, and several are superior to even the best of the *Winesburg* narratives. Those which resemble the *Winesburg* sketches are usually longer and more detailed in their portrayals of the repressions and maladjustments of their principal characters. "Unlighted Lamps," for example, is a lengthy study of the relationship between eighteen-year-old Mary Cochran, who lives in a small Illinois town neaı Chicago, and her cold, inarticulate father, who is unable to give his lonely, motherless daughter the love she needs. "Out of

Nowhere into Nothing" also deals with a young woman who finds the influence of her small-town upbringing oppressive. Still another case of repressed desires and emotions is that of the Vermont spinster of thirty-five, in "The New Englander," who goes with her elderly parents to live near a brother who has settled on a middle western farm. In this almost too elaborately symbolical story, the vast Iowa cornfields, of which she is afraid, stand for the kind of life she has desired but never experienced.

Anderson's fullest and frankest treatment of the effects of unsatisfied sexual desire in a woman is found in "Seeds." Suggested, according to Anderson, by a conversation he had with the psychoanalyst Dr. Trigant Burrow, the story begins with a prefatory discussion between the narrator and a psychoanalyst, who declares that he is weary and sick of going "beneath the surface of the lives of men and women" and wants to be "free" for a while. When the narrator maintains that the doctor reacts thus because no one can "venture far along the road of lives" and that the "universal illness" cannot be cured, the doctor retorts that the narrator has missed the point. "The lives of people," he says, "are like young trees in a forest. They are being choked by climbing vines. The vines are old thoughts and beliefs planted by dead men. I am myself covered by crawling creeping vines that choke me." The story proper, which the narrator has been told by a painter named LeRoy, reiterates the views of both the narrator and the doctor. In the Chicago boardinghouse where he lived, LeRoy met a young woman with a deformed foot who had come from an Iowa town, ostensibly to study methods of teaching music. She feared men but wanted to be loved by one. Sex, explains LeRoy, permeated her being, and when the narrator suggests that, since LeRoy had been kind to her and had won her confidence, he might have been her lover, the painter points out that this would not have been enough:

> "She needed to be loved, to be long and quietly and patiently loved. To be sure she is a grotesque, but then all the people in the world are grotesques. We all need to be loved. What would cure her would cure the rest of us also. The disease she had is, you see, universal. We all want to be loved and the world has no plan for creating our lovers."

"Seeds" is perhaps less interesting as a story than it is as a commentary on the themes of *Winesburg, Ohio*.

In two of his finest stories Anderson eschewed almost entirely his *Winesburg* manner. In "The Egg" the first-person narrator describes the unsuccessful attempts of his parents to get ahead in the world. Reasonably contented with his lot as a farmhand, the father had become ambitious after his marriage, mostly at the instigation of his wife, and for ten years they had struggled unsuccessfully to make a chicken farm pay. Then, the mother having become ambitious for her son, they had moved to town and gone into the restaurant business. There were few customers, and it finally occurred to the father that if he was more cheerful and provided his patrons with entertainment there would be more business. One night he had tried to put his idea into practice for the benefit of a customer by first attempting to make an egg stand on end and then heating an egg and forcing it into the neck of a bottle. Neither attempt was successful, and the father was frustrated. The narrator concludes:

> I wondered why eggs had to be and why from the egg came the hen who again laid the egg. The question got into my blood. It has stayed there, I imagine, because I am the son of my father. At any rate the problem remains unsolved in my mind. And that, I conclude, is but another evidence

> of the complete and final triumph of the egg—at least as far as my family is concerned."

Although the experiences of his parents have had a lasting effect on the narrator, he can take a detached view of them. The father's well-meaning but absurd attempts to express himself are both comic and pathetic, as are so frequently the efforts of persons who want desperately and try hard to succeed, but are doomed to fail. The triumph of the egg seems to symbolize this state of affairs.

"I Want to Know Why" is one of several stories dealing with the experiences of adolescent boys and young men. In this story a fifteen-year-old boy extends his love of thoroughbred racehorses to the man who is their trainer, only to be disillusioned when he sees the man consorting with prostitutes. Everything was changed for him:

> At the tracks the air don't taste as good or smell as good. It's because a man like Jerry Tillford, who knows what he does, could see a horse like Sunstreak run, and kiss a woman like that the same day. I can't make it out. Darn him, what did he want to do like that for? I keep thinking about it and it spoils looking at horses and smelling things and hearing niggers laugh and everything. Sometimes I'm so mad about it I want to fight someone. It gives me the fantods. What did he do it for? I want to know why.

As an example of the kind of twentieth-century story that has come to be known as the story of initiation, "I Want to Know Why" has no peer, with the possible exception of Hemingway's "My Old Man," which in certain respects it closely resembles.

Not quite as affecting and meaningful, although one of Anderson's most popular stories, is "I'm a Fool." It is another oral narrative whose language and speech rhythms closely approximate those of *Huckleberry Finn*. The nineteen-year-old narrator recounts a foolish act he has committed and wishes with all his heart that he could undo. He tells a girl who is "nice" and above him in social station, whom he meets at the races and with whom he falls in love, that he is the son of a wealthy racehorse owner. He realizes after spending the evening with her and with her brother and another girl that she likes him for himself and not because of the lies he has told to impress her, but he cannot bring himself to confess that he works in a livery stable. After she boards the train for her home in a nearby town, having told him that she knows he will write and that she will write him, he reproaches himself bitterly. His foolish act and his resulting state of mind are entirely understandable, and, though we may feel that his mood will certainly pass in time, we sympathize with him fully.

The conflicts attendant on a boy growing into manhood are also portrayed in "The Man Who Became a Woman" and "The Ohio Pagan," both of which are concerned mainly with the growth of sexual desire and its effects, and in "The Sad Horn Blowers," in which a number of the details of the family life of Will Appleton parallel Anderson's own as a youth. The conflict in this story arises from Will's leaving home to work in a factory and the unsettling effect on him of the realization that he is no longer a boy but not yet quite a man, a feeling finally alleviated for him when he becomes aware that something of the child remains in all men.

"Death in the Woods" is called by its narrator a simple tale, but its richness of meaning and artistry of composition make it one of Anderson's finest stories. It is about an old farm woman whom the narrator, when he was a boy, had seen on her trips to town to trade the eggs from her few hens for a little meat, sugar, and flour. A bound girl in the employ of a

German farmer, she had later married a brutal and shiftless farmer. Her son had grown up to be like his father, and the two spent much of their time away from home "trading horses or drinking or hunting or stealing," leaving her with the problem of feeding the livestock, the farm dogs, and the two men themselves when they are home. "She had to scheme all her life about getting things fed."

One bitterly cold day she makes a trip to town followed by the four gaunt farm dogs. A kindly butcher gives her some liver and dog meat, which, with the provisions for which she has exchanged her eggs, she puts in a grain sack which she ties to her back. She must get home to feed the stock and perhaps her husband and son, who may come home drunk. Tired and unwell, she makes her way painfully through the deep snow, and when she comes to a clearing sits down to rest against a tree and falls asleep. The dogs go off to chase rabbits, and when they return to the clearing it is night. Playfully they run around the tree in "a kind of death ceremony," and after the old woman dies, they drag her body out into the clearing so that they can get at the meat in the sack. They tear her worn dress from her shoulders but leave her body untouched, and when it is found later, frozen stiff, it appears "like the body of some charming young girl." With his brother, the narrator had gone to the woods with the group of men and boys led by the hunter who had discovered the body, and the scene had affected him strangely, as it had the others. It became, he says, "the foundation for the real story long afterward." Somehow the story of the old woman, destined to feed animal life even after her death, had, despite its elements of ugliness and sordidness, a completeness that gave it beauty. The need to try to understand is what impels the narrator to tell the story. He blends together what he had heard about the old woman, what he has observed, and, most important of all, his imaginative identification with the old woman. "Death in the Woods" was Anderson's favorite story, and he rewrote it several times (another version, told in the third person, appears in the semiautobiographical *Tar, A Midwest Childhood*). It is, of all his stories, the most brilliant application of the way in which Anderson believed that the storyteller's imagination should work on personal experience, a method which he endeavored to explain in *A Story-Teller's Story* and elsewhere, but with far less success than he was able to illustrate in this story.

"Brother Death," the other outstanding story in *Death in the Woods,* is concerned with the inexplicable character of family relationships and conflicts. The girl Mary comes to realize that there can be more than one kind of death—the death-in-life her older brother suffers when he is compelled to bow before the will of a stern father in order to be sure of his inheritance, and the actual death of her younger brother, who suffers from heart disease but who "would never have to face the more subtle and terrible death that had come to his older brother." A simpler but very poignant story is the uncollected "The Corn-Planting," which tells of the grief of an old farm couple whose only son goes to Chicago to study art and is killed in an automobile accident. After receiving the news, the old people do something that the narrator finds hard to understand. They go out at night and plant corn, apparently trying to overcome death by making life grow in the field.

"A Meeting South," first published in *Sherwood Anderson's Notebook* (1926) and later reprinted in *Death in the Woods,* is one of several pieces in the latter volume which are more nearly impressionistic autobiographical sketches than stories, and it is the only successful one. It evokes vividly the atmosphere of New Orleans, as well as the characters of Aunt Sally, an erstwhile proprietor of a gambling and bawdyhouse, and of the young southerner David (presumably modeled on William Faulkner, with whom Anderson became acquainted in New Orleans), who wants to write like Shelley and who cannot sleep unless he is a little drunk because of the pain of his war wounds.

Anderson's reputation was very great during the early 1920's. He appeared then to be very much of an innovator, as indeed he was. During the next decade his reputation declined, partly because he wrote little which matched his earlier work, and partly because there was less interest during the socially conscious thirties in the Anderson kind of story, a fact which Anderson referred to sardonically in "A Walk in the Moonlight," in which he made a doctor who aspires to be another Chekhov say: "Nowadays, it seems that there is not much interest in human relations. Human relations are out of style. You must write now of the capitalists and of the proletariat. You must give things an economic slant. Hurrah for economics! Economics forever!" At this distance we are perhaps in a better position to assess Anderson's achievement. He had his limitations, and he is not a major figure like Hemingway. His range was limited, his mysticism and questioning sometimes grow tiresome, and he does not tell us as much about life as we might expect from a truly great writer. Yet he gave a new impetus and direction to the short story, and he produced short fiction which is unique in American writing in both its matter and its manner.

# 10
# Short-Story Writers of the 1920's

## Wilbur Daniel Steele
## Ring Lardner
## F. Scott Fitzgerald
## Conrad Aiken
## and
## Stephen Vincent Benét

THE DECADE OF THE 1920's is one of the great periods in the development of the American short story. It saw the publication of much of Sherwood Anderson's best work and of two volumes of short fiction by Ernest Hemingway, and before its end William Faulkner and Katherine Anne Porter had begun to publish short stories. Several other writers of only a little less distinction came to the fore during the decade. Wilbur Daniel Steele was a virtuoso of the plotted story and wrote with dramatic power and psychological insight. Ring Lardner brilliantly employed the short story as a vehicle for humor and satire. F. Scott Fitzgerald indelibly recorded in many of his short stories the manners and morals of the Jazz Age. Conrad Aiken found in Freudian psychology, as did Stephen Vincent Benét in American history and folklore, the materials for notable short fiction.

WILBUR DANIEL STEELE (1886–1970) published his first story in 1912 in the *Atlantic Monthly* and within a decade had became one of America's most widely known and highly regarded short-story writers. When Edward J. O'Brien began the publication of his annual *Best Short Stories* volumes in 1915, he included Steele's "The Yellow Cat," and thereafter for almost twenty years Steele was represented frequently in O'Brien's selections as well as in the annual *O. Henry Memorial Award Prize Stories.* In 1919, the first year of the O. Henry competition, Steele won second prize with "For They Know Not What They Do." In 1921 he received a special O. Henry award for maintaining the highest level of merit for three years among American short-story writers, and he later won O. Henry prizes for the best story of the year: "The Man Who Saw Through Heaven" (1925), "Bubbles" (1926), and " 'Can't Cross Jordan by Myself' " (1931). Typical of the many critical tributes paid his work during the 1920's was that of Katherine Fullerton Gerould, herself a well-known and respected writer of fiction, who described Steele in 1924 in a lengthy *Yale Review essay* on the short story as "at present our American best."

During the late 1930's, believing that his kind of short fiction had gone out of fashion, Steele virtually gave up writing stories and turned to playwriting for some years, an endeavor which was not very successful, however, and which Steele called a waste of time. The kind of story Steele specialized in was, as Mrs. Gerould pointed out, well made and had those elements deemed necessary by most readers of the day for an effective story, namely, situation, suspense, and climax. Given Steele's inventiveness and his great technical skill, it seems reasonable to believe that he could have continued indefinitely to write stories which would have been publishable, if not in *Harper's*, *Atlantic*, and other such magazines which had printed much of his work, at least in the mass-circulation magazines. Why he did not do so is perhaps explained by certain remarks attributed to him in the Introduction to the twenty-fifth anniversary edition of the *O. Henry Memorial Award Prize Stories of 1943*, for which he acted as one of the judges. He wrote that, although the "older" plotted story, related as it was to drama and even melodrama, had been appropriate to its day, it had been superseded by a "newer" story which was akin to poetry, truer, more perceptive, and more mature. Presumably because he felt incapable of writing this more meaningful and significant kind of story, Steele preferred not to write stories at all.

Although Steele was undoubtedly overrated during the 1920's, he is deserving of something better than the disparaging criticism his work usually receives nowadays, when it is noticed at all. Besides the stories already mentioned there are several others capable of making a strong impression on the reader. A late story, "How Beautiful with Shoes," published in 1933, is perhaps his finest and also reveals most fully his gifts as a storyteller. As with a number of his other stories, it deals with madness, but, as is usually the case with Steele, the situation and its outcome are of primary importance. Told from the point of view of the slow-minded farm girl Amarantha, who is lifted out of herself to a vision of love and beauty by her encounter with the crazed Humble Jewett, a murderer escaped from an asylum, who alternately addresses her in the love poetry of Lovelace and the Song of Solomon and seems on the verge of killing her, the story is full of suspense, vivid description of setting, and emotional intensity which combine to give it a powerful effect despite its somewhat melodramatic undertones.

Steele's virtuosity is impressive, even though he relied too much on contrivance and especially on surprise endings and other twists. His stories have varied settings—New England, the South, the Middle West, the South Pacific, and Africa—and although he is certainly not of the same stature as Conrad, some of his sea stories are very nearly as good as those of the latter. Steele can perhaps be best described as a kind of American Somerset Maugham, a highly gifted teller of very readable if not profound tales. (It is interesting to compare "The Man Who Saw Through Heaven" with Maugham's famous "Rain.")

A JOURNALIST TURNED short-story writer, who began his careers as a sportswriter and newspaper columnist in Chicago, Ring Lardner (1885–1933) first achieved popularity with a series of humorous letters purportedly written by one Jack Keefe, a bush-league pitcher who is signed by a big-league team in Chicago and who reports on his experiences to a friend back in his Michigan home town. Published in the *Saturday Evening Post*, beginning in 1914, they were collected in *You Know Me Al* (1916) and later volumes. Most of Lardner's short stories about baseball players follow the pattern of the *You Know Me Al* letters, having the same kinds of characters, situations, and complications. The bush-league pitcher in "Hurry Kane," for instance, is almost an identical twin of Jack Keefe. Although we may feel a little more kindly toward him than toward his prototype, he is essentially the same kind of boob and braggart. As is the case with Jack, his stinginess, vanity, and gullibility make him a natural target for practical jokes, and the plot complication parallels one which Lardner developed in a certain sequence of the Jack Keefe letters. Both are the pitching mainstays of their respective clubs, both become dissatisfied and threaten to quit, and both are dissuaded by similar arguments and ruses devised by their teammates.

None too admirable baseball players are also the principal characters in "Alibi Ike," "Women," and "My Roomy"; but in "Harmony" and "Horseshoes," Lardner varied his formula somewhat. Both stories are perhaps a little too drawn out, but the first is an amusing portrait of a baseball player who would rather sing in a barbershop quartet than eat or play ball, and the second shows Lardner for once at least in a baseball story, instead of focusing on satirical portraiture, telling a good story which has considerable suspense and human interest.

An even larger number of Lardner's stories than those about baseball players are about married couples and their relations with each other or with other married couples. These stories fall roughly into two groups, the first stemming from the five closely related stories in *Gullible's Travels, Etc.* (1917). The first-person narrator in these stories may be lacking in formal education—his grammar, for instance, leaves much to be desired—but he is far from being a boob, and he has some sense of the decencies and amenities which should prevail in social relationships and would like to see others observe them, even though that is more than one can reasonably expect. For the most part in these stories it is Lardner the humorist rather than Lardner the short-story writer at work. "Carmen," in which the narrator and his wife and another married couple attend the opera for the first time, relies for its effect mostly on a burlesque paraphrase of the operatic work, but Lardner's humor is put to more effective use in the two stories in which the social ambitions of the narrator and his wife backfire on them, "Three Without, Doubled" and the title story. In the latter they go high-hat and take a Florida vacation, feeling themselves too good to associate with their Chicago friends. After spending much money and not making friends with any people of wealth and social position, as they had hoped to do, they come back to Chicago humbler than when they left.

Lardner's later stories on the *Gullible's Travels* pattern, which include "Mr. and Mrs. Fixit," "Liberty Hall," "Reunion," and "Contract," have more finesse and restraint. In some of them Lardner also eschewed his wisecracking, slangy, ungrammatical monologuist, using instead a straight third-person style. In the main, the irony in these stories is lighter and the satire less biting than in the husband-wife stories, in which marriage is shown to be an unhappy affair or at best a pretty drab business. Sometimes Lardner treated this theme in too contrived a fashion, as in "Now and Then," or too much in the *Gullible's Travels* manner, as in "Who Dealt?" but he makes us feel pity for the wife in "Anniversary" who has a bore and miser for a husband, and feel sympathy for the parents in "Old Folks' Christmas," whose children are thoughtless and self-centered.

Two of Lardner's finest stories belong in this group. "The Love Nest" is similar to "Anniversary," but the plight of the unhappy wife is represented with more powerful and trenchant irony. "The Golden Honeymoon" is the story of an elderly married couple's month's vacation in Florida, told by the garrulous husband. The old man's account of the railroad trip and of how the two of them occupy themselves in St. Petersburg is sprinkled with much minute and trivial detail —such matters as the times of arrival and departure of trains, the prices of cafeteria meals, and the fact that one of the sons of a couple with whom they become acquainted "lives in Providence and is way up in the Elks as well as a Rotarian" —all clearly of great importance to the narrator. He further relates how they spend the days with the other oldsters in the park, where such diversions as checkers, horseshoes, and band concerts are available, and how they spend their evenings playing cards or attending meetings of either their own state society or those of their friends, at which they can hear their contemporaries give talks on business conditions and on such subjects as "Rainbow Chasing," as well as sing grand-opera selections and imitate bird calls. The enjoyment the old couple at first derive from these activities is subsequently considerably lessened by their associations with another couple, the husband turning out to be, to the surprise of all concerned when they first meet, "the man who was engaged to Mother till I stepped in and cut him out, fifty-two years ago." The relationship with the other couple results in a quarrel between "Mother" and the narrator, but they have the good sense to make up and not spoil their golden honeymoon. It is the old couple's affection for and understanding of each other that belies the critical opinion sometimes advanced that a great contempt for them underlies the story. Lardner can satirize the banality of their lives, but underlying his satire is something akin to tenderness.

Most of Lardner's best stories are monologues or employ a similar form of narration, such as the story told in the form of diary entries or in the form of letters. "Haircut," probably Lardner's best-known story, is the monologue of a barber who tells an out-of-town customer about the late Jim Kendall, a one-time traveling salesman whom the barber admiringly refers to as "a card," "a character," and "a caution." Besides his wisecracks, which the barber had considered very funny indeed, Jim's forte had been practical jokes, the barber himself having been the victim on one occasion when Jim had fooled him into going out in the country on a cold winter day to shave a dead man who turned out to be very much alive. More typical of Jim's jokes, however, were those he played on a town boy not quite right in the head, and one he played out of spite on his wife and children, telling them that he would meet them with tickets at the circus and then deliberately not showing up. This vicious, vengeful character finally plays one practical joke too many; it boomerangs, and he is killed. More significant than this irony, however, is the irony that, though Jim may have been "kind of rough," he remains for the barber and the hangers-on at the barbershop, for whom he provided so much sport and diversion, "a good fella at heart."

In "A Caddy's Diary," on the other hand, the narrator does not remain oblivious to the significance of the actions of the people he writes about. He is an ambitious and in some respects shrewd sixteen-year-old who hopes not only to become a golf professional but also to write newspaper articles on how to play golf. To prepare himself, he follows the suggestion of a club member that he keep a diary, writing down whom he caddies for and what they do. Most of the golfers he caddies for cheat on their scores, and if he wants to be sure of a tip he must be an accessory. The club champion and the professional are exceptions, mainly because, as Joe, another caddy, points out to the narrator, they play "too good" to be able to get away with cheating: "To hear Joe tell it pretty near everybody are born crooks, well maybe he is right." Joe also points out something else to the narrator that hits closer to home. When the latter says that he cannot understand why the club members should be so indignant and self-righteous in condemning the club champion, who has absconded with eight thousand dollars from the bank that employed him, when they have sold their souls merely to win an insignificant prize or bet at golf, Joe reminds the caddy that he has been just as culpable.

The caddy's manner of expression—his misspellings, bad grammar, misused words, and the like—is used mainly for humorous effect, as, for instance, when he expresses the hope that he will get on to the "nag" of "writeing" by means of the practice he will gain in keeping a diary. In the letters which make up "Some Like Them Cold," on the other hand, the effect of the similar language of the girl, Mabelle Gillespie, and of the man, Charles F. Lewis, on whom she tries so hard to make a favorable impression, is both humorous and something more. Not only by what they say but also by how they say it they unwittingly lay bare their real selves—her coyness, affectation, and false modesty, his conceit, crudeness, and vulgarity.

Lardner's work in the short story has certain limitations. Neither in Lardner's people nor in the situations in which he puts them do we find as much variety of human nature and experience as we might wish for, and in some instances the lesser stories duplicate the character types or story situations of the better ones. "Mr. Frisbie," for example, is similar though inferior to "A Caddy's Diary," and the girl in "Travelogue," who talks all the time about herself and the places she has visited, and the shallow, vain nurse in "Zone of Quiet" are pallid imitations of Mabelle Gillespie.

Lardner shows his people from the outside rather than getting inside them, one reason for this being, perhaps, that for the most part the kinds of people he wrote about have very little inside them to bring out. Nonetheless, as Edmund Wilson has pointed out, it is to Lardner's credit that, when he comes close to Sinclair Lewis in his satirical portraiture, he is less likely to falsify and caricature, being more concerned with studying a type than in making an indictment. As for Lardner's alleged misanthropy, it is true that he portrayed a good many boobs, fools, egoists, and even on occasion vicious or reprehensible characters, such as Jim Kendall, the boxer Midge Kelly in "Champion," and the theatrical producer in "A Day with Conrad Green." But in "Haircut," Jim Kendall is offset by the kindly and compassionate Doc Stair, and at the same time that Lardner holds Mabelle Gillespie up to ridicule, he makes her seem pathetic. Lardner did not hate people as such, but he hated the pretentiousness, cruelty, stupidity, fraud, and humbug that people can be capable of, and it is these qualities in human nature that he indicted in his best work with a genius and artistry only a little less than that of Swift or Mark Twain.

F. Scott Fitzgerald (1896–1940) published his first novel, *This Side of Paradise*, in 1920 and in the same year made an impressive debut as a writer of short stories, publishing no less than sixteen in such magazines as the *Saturday Evening Post, Scribner's* and *Smart Set*. Some of these stories are reminiscent of O. Henry, an example being the amusing and very readable "The Off-Shore Pirate." Others are more serious in tone but rely too much on contrivance, an instance being "The Cut-Glass Bowl," in which a huge bowl received as a wedding present is made too obviously and gratuitously the symbol of fate. However, in two stories, "May Day" and "The Jelly Bean," with its fine portraits of a rich wild flapper and the poor boy who loves her from afar, Fitzgerald demonstrated he could be more than a facile storyteller.

Into "May Day" Fitzgerald projected a good deal of his experience and many of his attitudes, something he did frequently in his stories and to a greater extent than most other writers. As his biographer Arthur Mizener has pointed out, Fitzgerald seems to have felt a strong sense of self-condemnation for being so much like the hero of this story—the weak-willed, self-pitying Gordon Sterrett, who after he comes back from the war loses his job because he goes to too many parties, drinks too much, gets involved with a woman he does not want to marry, tries unsuccessfully to borrow money from a wealthy Yale classmate, and in the end shoots himself in the head. Sterrett's actions are only a part of the futile, senseless activity that the story suggests was characteristic of the times, as indicated by such episodes as that of the two returning soldiers who join a mob attacking a socialist newspaper, one of the soldiers being accidentally killed, and that of the two Yale men who go on an all-night drunk after a fraternity party at Delmonico's and tear up a restaurant in the process. "May Day" is not a perfect story from the standpoint of narrative technique, but it is a vivid and powerful one.

During the early 1920's, Fitzgerald continued to publish much short fiction. He was paid exceedingly well for his stories but usually spent money as fast as he made it, and he turned out much hackwork. However, his second novel, *The Beautiful and the Damned* (1922), was an advance beyond *This Side of Paradise*, and occasionally, too, in a short story, there was seriousness of tone and strength of feeling. One of these is "Absolution," a story of an eleven-year-old boy and a priest who is close to madness. The boy tells the priest a lie in confession and believes that he has thereby committed a mortal sin. The action takes place in the Dakota wheatlands, and there are psychological undertones reminiscent of Sherwood Anderson. "The Baby Party" is also very different from the general run of Fitzgerald's stories, there being a Crane-like irony and understatement in its portrayal of an altercation between two families, culminating in the two fathers feeling foolish and embarrassed for having senselessly fought each other after their wives quarrel at a children's party when their children fight over possession of a teddy bear.

More in Fitzgerald's usual vein is "The Sensible Thing," which reflects the author's own despairing actions and sense of hopelessness at a time when he was engaged to Zelda Sayre, the girl who was to become Mrs. Fitzgerald, but had neither enough money nor the prospects of making it to get married. Ultimately the hero gets a good job in South America, and the broken engagement is renewed, but though there is still love, the freshness of the first love is gone forever. In "Winter Dreams," Dexter Green also experiences a sense of loss. As a youth he dreams of having the things the rich have, and after college he prospers in business and gains what he wants, except for the girl Judy Jones, who twice throws him over after leading him to believe that she might marry him. Seven years later he learns that she is unhappily married and has lost her beauty. "He had thought that having nothing else to lose he was invulnerable at last—but he knew that he had just lost something more, as surely as if he had married Judy Jones and seen her fade away before his eyes." His dream of her as she had been is gone, and though he wants to care he cannot. There is poignancy in Dexter Green's realization that he cannot feel a sense of loss, but the resulting grief which overwhelms him, as Mizener has remarked, does not seem altogether warranted under the circumstances.

Although Fitzgerald never quite succeeded in a short story in doing full justice to the theme that the past can never be recaptured, he did so brilliantly in *The Great Gatsby*. This short novel, his masterwork, was published in 1925 and was followed the next year by a story which has certain affinities with it and is also one of his finest, "The Rich Boy." "Let me tell you about the very rich," says Fitzgerald in his Prologue. "They are different from you and me." But besides making this statement, which has achieved a certain fame through frequent quotation, Fitzgerald emphasizes that his story is about an individual. The rich boy, Anson Hunter, early develops a sense of his difference, "and a sort of impatience with all groups of which he was not the centre—in money, in position, in authority—remained with him for the rest of his life."

His sense of superiority and a perverse kind of pride make him feel that everything should be given to him freely and on his own terms. As the story richly shows, he is "a mixture of solidity and self-indulgence, of sentiment and cynicism." He has his faults—his heavy drinking is a factor in his loss of Paula Legendre, the girl he keeps believing he will some day marry—but he has his good points also. He is intelligent, well liked by most people, and a hard-working, successful stockbroker. Yet there is something about his nature, about his attitude toward women, which seems to destine him never to marry, never to know the kind of happy married life which Paula achieves briefly before her death in childbirth. All Anson will ever have is someone like the girl he picks up on shipboard at the end of the story, someone who will nourish his sense of superiority.

After "The Rich Boy" several years passed before Fitzgerald wrote another story with as much substance and artistry. In 1928 and 1929 he published eight of a group of nine stories about a boy named Basil Duke Lee, which reflect experiences of his boyhood and young manhood in St. Paul, at an eastern preparatory school, and at Princeton. They are readable and often very amusing, but even the best of them, "The Freshest Boy" and "The Captured Shadow," are too much like the Penrod stories of Booth Tarkington to be really individual or distinctive.

A year later Fitzgerald published five more stories of the same sort about a girl named Josephine, but only one, "A Woman with a Past," in which the heroine, a boarding-school girl, goes to her first college dance, merits comparison with the better Basil stories. During this time Fitzgerald was also still writing an occasional variation on his favorite theme of regret for the loss of past happiness, but he now treated it with more restraint and more subtle irony than he had usually done earlier, as is apparent in such stories as "The Last of the Belles," whose narrator can recount with a certain measure of detachment his unsuccessful attempt to recapture the past, and in "The Bridal Party," whose hero, though he loses the girl he loves because he is poor, is able to get over it.

Again, in "Babylon Revisited," published in 1931, the past plays a part very different from that in the earlier stories. Here, as had happened to Fitzgerald himself, the past with its irresponsible actions and unfulfilled opportunities comes back to haunt and spoil the present. Three years before the story opens Charlie Wales had embarked on a career of heavy drinking and squandering money in Paris which had ended with his wife dead of a heart attack, himself in a sanitarium, and his young daughter in the legal custody of his sister-in-law. But now for a year and a half Charlie has been on the wagon and is doing well in business in Prague. He returns to Paris, hoping to persuade the sister-in-law to let him have the child back, and, although she is still strongly prejudiced against him she consents, on the condition that she retain the guardianship for another year. When two drunken friends from Charlie's past appear in search of him at her apartment, her suspicions that he has not really reformed are confirmed, and Charlie must go back to Prague without the child. "Babylon Revisited" is a beautifully executed story without a single false note, and its artistry and depth of feeling make it one of the great modern American short stories.

Fitzgerald wrote two other notable stories about the same time. In "Family in the Wind" he put a good deal of his own feelings into his hero, a doctor who had been a celebrated surgeon before committing professional suicide by taking to cynicism and drink, but who at the end of the story is on the way to starting life over. "Crazy Sunday" penetrates deeply into the complex psychological conflicts involving a Hollywood screenwriter, a movie director, and the director's actress wife.

For various reasons Fitzgerald was not as prolific a short-story writer during the 1930's as he had been during the previous decade. He was occupied with writing his most ambitious novel, *Tender Is the Night* (1934), his wife was mentally ill, he was plagued by recurring illnesses and financial worries, and during much of the time from 1937 to his death in 1940 he worked as a movie scenario writer. Nevertheless, during the middle and late thirties he managed to publish a good many articles, sketches, and stories, most of them in *Esquire*, although none of these pieces come up to the best of his earlier shorter work. Several of the series of seventeen Pat Hobby stories about a broken-down screenwriter who lives by his wits are clever commercial pieces, and there are a few serious stories worthy of mention, such as "An Alcoholic Case," "The Long Way Out," and "The Lost Decade." In them Fitzgerald adopted a new manner of telling a story—one which is somewhat reminiscent of Hemingway—and they make worthy companion pieces to the revealing and profoundly felt autobiographical articles Fitzgerald wrote during his last years, "The Crack-up" and "Afternoon of an Author."

For some years after his death Fitzgerald was a much underrated and almost forgotten writer, but more recently Edmund Wilson, Malcolm Cowley, Arthur Mizener, and others have labored to correct what was undoubtedly a false image of Fitzgerald—that of a man who was merely a superficial historian of the Jazz Age, who wrote mostly poor and mediocre stories for commercial magazines. As a result he has been the beneficiary of a revival of interest in his life and work resulting in a popularity equal to and perhaps even greater than that he enjoyed in the 1920's. The inevitable effect has been that he is overrated in some quarters; it has even been claimed that as a short-story writer he is of the same rank as Hemingway and Faulkner. Fitzgerald did serious and important work in the short story, but his commercial fiction bulks very large, and there are only a few superb stories like "The Rich Boy," "Babylon Revisited," and *The Great Gatsby* to set against it.

ALTHOUGH SOME of the stories of Wilbur Daniel Steele, mentioned earlier in this chapter, deal with madness and with the effects of the unconscious, and with the dream life of a person, they do not give the impression of being influenced by psychoanalytical theory to any very great extent, as do certain of the stories of Conrad Aiken (1899– ). By the time Aiken graduated from Harvard, he had become acquainted with Freudian doctrines and was convinced that they were of profound importance in understanding man and his consciousness. Aiken's continuing preoccupation with psychoanalysis is reflected in the long poem "The Jig of Forslin" (1916) and later poetry, in much of the short fiction he published during the 1920's and 1930's, and in such novels as *Blue Voyage* (1927) and *Great Circle* (1933).

Aiken's most artistically effective developments of psychoanalytic theory are to be found in "Mr. Arcularis" and "Silent Snow, Secret Snow," both published in 1932. The first of these frequently anthologized stories depicts a dream within a dream which the main character has under anesthesia during an unsuccessful operation to save his life. The story bears a superficial resemblance to Bierce's "An Occurrence at Owl Creek Bridge" in the revelation at the end that Mr. Arcularis has died on the operating table and has not gone on a sea voyage at all. As in Bierce's story, there are many details which prepare for the surprise ending—the cold wind of which

Mr. Arcularis is almost always conscious despite the fact that it is June; the pain, vagueness, and dizziness he feels intermittently; his sleepwalking and strange dreams; and other unusual and eerie aspects of the voyage. But Aiken's story is more complex and richer symbolically than Bierce's, as evidenced by the figure of the girl on shipboard who appears to be a multiple symbol, standing variously for Mr. Arcularis' mother, love, beauty, and finally his soul. She cannot go with him on his dream voyage and can only embrace him in a vain effort to give him her warmth and life before he takes his final voyage "to the little blue star which pointed the way to the Unknown," where he will become one with the Absolute.

Although Mr. Arcularis' dreams have Freudian implications, his story seems less obviously founded on psychoanalytic theory than "Silent Snow, Secret Snow," which portrays the development of schizophrenia in a young boy. At the beginning of the story the boy's withdrawal is well advanced, although the causes of it are never made explicit. To the boy it is a mystery, albeit a delightful one, that he should have a sense of snow falling about him, making a screen between himself and the world. Each day he feels the snow grow heavier, making for him a new world of incalculable wonder and beauty. It is from the generalized ugliness of the real world, symbolized by the heavy steps of the postman, which are deadened increasingly by the snow, that the boy seems to be withdrawing. He is aware that his mother and father are worried by his "daydreaming," but though he likes them and wants to be kind to them, he feels that he must keep his precious secret no matter what the cost. After he is examined by the family doctor, who can find nothing wrong with him physically, the boy escapes to his room. Followed by his mother, who seems to him alien and hostile, the boy cries out to her to go away, and he is left alone, his withdrawal complete, enveloped in a world which is "a vast moving screen of snow—but even now it said peace, it said remoteness, it said cold, it said sleep."

Aiken has written other notable, if less-well-known stories. "Strange Moonlight" is a memorable story of a small boy confronting the mystery of experience and of how he comes to realize the presence of mutability and death in life. "Gehenna" and "State of Mind" vividly portray the sensations of men who feel themselves on the verge of losing their sanity. "Bow Down, Isaac!" is a powerful story in which religious fanaticism and fear result in a gruesome murder. "Your Obituary, Well Written" is the best of several stories with Jamesian nuances. A number of Aiken's stories have as their principal theme the meaninglessness and futility of life, but as a rule they are less impressive than the stories which treat, often with great perceptiveness and in a highly artistic prose, the hidden springs of man's behavior.

LIKE WILBUR DANIEL STEELE, Stephen Vincent Benét (1896–1943) was one of the most prolific and widely read short-story writers of the 1920's and later. As a recipient of O. Henry memorial awards he rivaled Steele, winning first prize three times: "An End to Dreams" (1932), "The Devil and Daniel Webster" (1937), and "Freedom's a Hard-Bought Thing" (1940). Benét's first love, however, was poetry. He published three volumes of verse before he was twenty and at thirty won a Pulitzer Prize for the long poem *John Brown's Body* (1928). But like Fitzgerald he found that he could make more money by writing short stories for mass-circulation magazines, and he early developed a facility for turning out contrived stories of the sort that the big magazine editors wanted.

Throughout a career of a little more than two decades of short-story writing, Benét remained primarily a writer of commercial stories. A considerable number of the more than 120 stories he published have traditional or stock themes, but they are often developed very skillfully. Among the better stories of this kind are "An End to Dreams," the point of which is that riches do not necessarily make for happiness; "Blossom and Fruit," the most important thing in life is love; "The Story About the Anteater,"—husbands and wives can find happiness if they can only learn to tolerate and accept each other's faults; "The Bishop's Beggar,"—men become good when their better instincts are appealed to; and "The Bright Dream," life has its ups and downs, but things usually turn out all right in the end.

Benét also wrote a number of stories which read like the works of various other writers. "The Treasure of Vasco Gomez," a romantic piece about greed and pirate gold, is in the manner of Robert Louis Stevenson. "A Story by Angela Poe," although more melodramatic, is reminiscent of James's "Greville Fane," in both situation and manner of telling. "The King of the Cats" calls to mind Saki (H. H. Munro), although it has less wit and humor than a good Saki story. "A Gentleman of Fortune" is almost pure O. Henry. "Too Early Spring," a poignant story of a teen-age love affair, may well have derived its tone from Sherwood Anderson's "I'm a Fool." There are also several stories which recall Ring Lardner and Dorothy Parker. In "Everybody Was Very Nice," a Lardner-like first-person narrator tells a friend how he happened to get a divorce. He is now remarried and so is his former wife, but he is not happy, and he is still puzzled why he and his first wife separated. Perhaps the story is too much a tract against easy divorce, but the reader is moved by the narrator's account and feels more sympathy for him than he is usually able to summon for a Lardner character. Two very effective Dorothy Parker-like monologues are "Life at Angelo's" and "Among Those Present." Both portray shallow, self-centered people, a man who is an alcoholic and going downhill fast although he does not realize it, and a woman who has driven her first husband to drink and who seems to be in the process of doing the same to her present one.

Benét's considerable knowledge of and keen interest in American history is reflected in a substantial number of stories which he began writing as early as the middle 1920's. "The Sobbin' Women," published in 1926, in which Benét turned the legend of the Sabine women (to which the story alludes) into an American folk tale, is typical of the earlier stories of this sort. During the early 1930's, Benét turned to glorifying certain aspects of the American past, as in "The Captives," which praises the spirit of the pioneers of the time of the French and Indian wars. And in 1936, with great artistry, he infused the combined themes of the greatness of America and the essential goodness and dignity of the individual into a tall tale which is one of the most famous of contemporary American short stories, "The Devil and Daniel Webster." Webster, who is made larger than life, a kind of Paul Bunyan, Davy Crockett, or John Henry figure, through his eloquence saves a New Hampshire farmer who has sold his soul to the devil. Everything about the story is superbly handled: the oral tone, the New England setting, the background of folklore and superstition, the humor, and the picturesque exaggeration. The faith that the story affirms is expressed without sentimentality and flag waving, and its main statement in the form of Webster's speech to the devil's jury of renegades is, despite the fact that it is given in paraphrase, a magnificent piece of rhetoric.

Benét wrote other stories in the same vein, but he never

achieved quite the same success as in "The Devil and Daniel Webster." Two other Webster stories, "Daniel Webster and the Sea Serpent" and "Daniel Webster and the Ides of March," are amusing, but they do not come to life the way their predecessor does and are no more than competent imitations. In them, as in certain other stories, like "The Angel Was a Yankee," a story about P. T. Barnum, the tall-tale elements at times seem forced and mechanical. Still another weakness, and a more important one artistically, in some of Benét's later stories of American history and legend is that he tried too hard to give them topical undertones for his own day—to make them say, for instance, that American ideals and traditions provide a strong defense for democracy against subversive forces both within and without America. In "William Riley and the Fates," for example, a boy living about 1910 has a vision of the future and wonders how he and his contemporaries will be able to endure what they will have to live through, but he comes to realize that he must do his job, whatever it is, and never believe that the country is finished. But Benét's "propaganda" is usually more restrained than this, and some of the late folk tales are notable achievements, among them being "Johnny Pye and the Fool-Killer," "O'Halloran's Luck," "A Tooth for Paul Revere," and "Freedom's a Hard-Bought Thing." The artistry with which Benét combined folk humor and idiom, fantasy, and homely realism in these tales, and, of course, in "The Devil and Daniel Webster," is such that, without their existence, American short fiction would indeed be considerably diminished."

## 11
## The Discovery of a Style

## Ernest Hemingway

ALTHOUGH not as prolific in either the short story or the novel as some of his contemporaries, no other American writer of fiction in the twentieth century, except perhaps William Faulkner, has won as great a measure of distinction as a literary artist or exerted as much influence on other writers as has Ernest Hemingway (1899–1961). His subject matter, attitudes, narrative techniques, and style have been reflected in the works of many lesser writers of the past several decades. Like Chekhov, Joyce, and Katherine Mansfield, he has been a major force in shaping the form and character of the modern short story.

The son of a doctor, Hemingway was born in the Chicago suburb of Oak Park. After graduating from high school there in 1917, he worked for several months as a cub reporter on the Kansas City *Star* before volunteering for service as a Red Cross ambulance driver in Italy, where he was badly wounded in the leg and decorated for valor by the Italian government. After the war he returned to journalism, working in Toronto and also in Chicago, where he met Sherwood Anderson. In his spare time he wrote stories and poems, and while headquartered in Paris as a foreign correspondent for a Toronto newspaper, he received valuable encouragement and assistance from Gertrude Stein and Ezra Pound in his efforts to make himself a creative writer.

Hemingway's first book was *Three Stories and Ten Poems* (1923), the three stories being "Up in Michigan," "Out of Season," and "My Old Man." The first of these stories indicates the kinds of "simple things" Hemingway said that he began with in teaching himself to write and shows him already employing a manner of expression which was to develop into the famous Hemingway style. Liz Coates, a country girl, is attracted to a blacksmith, Jim Gilmore, who boards at the house where she works. One evening when they go for a walk he roughly seduces her and afterward falls asleep. Unable to rouse him, Liz, who is cold and miserable and for whom "everything felt gone," covers him with her coat and goes home. Everything in the story is sharp, clear, and vivid, and although Hemingway's use of colloquial and very simple diction, short declarative sentences, and much repetition may have been influenced by Sherwood Anderson, passages like the following suggest that he was consciously following not only Gertrude Stein's advice really to see and feel and not merely to report, but also her stylistic practice:

> Liz liked Jim very much. She liked it the way he walked over from the shop and often went to the kitchen door to watch for him to start down the road. She liked it about his mustache. She liked it about how white his teeth were when he smiled. She liked it how much D. J. Smith and Mrs. Smith liked Jim. One day she found that she liked it the way the hair was black on his arms and how white they were above the tanned line when he washed up in the washbasin outside the house. Liking that made her feel funny.

"Out of Season" is the first of a number of stories dealing with unhappy or strained relations between married couples or lovers, others being "Mr. and Mrs. Elliot," "Cat in the Rain," "Hills Like White Elephants," and "The Sea Change." They are all noteworthy for the artistry with which they render obliquely through action and dialogue a full sense of the incompatibility of the characters and the resulting feelings of boredom, hostility, frustration, and futility it entails. In this kind of story particularly Hemingway has had many imitators.

"My Old Man" bears a close resemblance to Sherwood Anderson's "I Want to Know Why," which appeared two years before the publication of Hemingway's story, although Hemingway said that it was written before he had read any of Anderson's work. Both stories have racetrack backgrounds and are told in the first person by adolescent boys who love horses and the world of nature but who have another kind of world—one of corruption and evil—revealed to them when each comes to realize the true character of the older man whom each boy loves and admires. Both stories render the vernacular of their narrators convincingly and are moving studies of disillusionment, but Hemingway's seems more poignant and meaningful, mainly because its father-son relationship is much closer and more important insofar as the boy is concerned than is the relationship between the boy and the horse trainer in Anderson's story.

In 1925, Hemingway published a second, larger volume of stories entitled *In Our Time*, and in 1926, two novels, *The Torrents of Spring*, an extravagant but often hilariously funny burlesque satirizing Sherwood Anderson, Gertrude Stein, and other writers, and *The Sun Also Rises*, a brilliant portrayal of "lost generation" expatriate society in postwar Paris and Spain, which has become one of the classics of modern American fiction. Seven of the fifteen stories in *In Our Time* and several more in Hemingway's two other story collections, *Men Without Women* (1927) and *Winner Take Nothing* (1933), are episodes in the life of a character named Nick Adams, who to a considerable extent appears to be an autobiographical hero. Nick has a doctor-father; some of his earlier experiences take place in northern Michigan, where Hemingway summered as a boy and revisited after the war; Nick sees something of the seamier side of life "on the road," as no doubt did the youthful Hemingway, who ran away from home twice; and a good deal of what we learn about Nick's war and postwar experiences and resulting attitudes seems clearly to be modeled on those of Hemingway's. Most if not all of these stories, as Philip Young and other critics have pointed out, deal with Nick's discovery of some of the less pleasant realities of existence: the perversity in human nature, the complexities of human relationships, suffering, violence, and evil.

In "Indian Camp," Nick, a small boy, accompanies his father to the Indian reservation across the lake from their summer cabin, where a young woman has been in labor two days. Her husband, who has cut his foot badly with an ax, has had to lie in the bunk above her and listen to her screams. After carefully explaining the situation to Nick and encouraging him to watch, his father performs a successful Caesarean operation with a jackknife and sews up the incision with fishing gut. Proud of his performance, he then remembers the husband. " 'Ought to have a look at the proud father. They're usually the worst sufferers in these little affairs,' the doctor said. 'I must say he took it all pretty quietly.' " Pulling the blanket back, he discovers that the Indian has slit his throat from ear to ear with a razor. Nick gets a good look at the body before his father sends him out of the shanty. Nick thus comes in contact with pain and death, but the ending of the story implies that, since he is only a child, he remains more or less impervious to the experience. He is more curious than horrified. On the way home he asks his father whether it is always so difficult to have a baby, why the husband killed himself, whether dying is hard; and "in the early morning on the lake sitting in the stern of the boat with his father rowing, he felt quite sure that he would never die."

"The Doctor and the Doctor's Wife" portrays Nick's mother and father; Nick, still a small boy, appears only at the very end of the story. Nick's father has a row with an Indian whom he has hired to cut up some logs that have drifted onto the beach after having been lost from a steamer towing them down the lake to the lumber mill. The Indian refers to the logs as stolen, and, angered by this comment, the father tells him to get out and later explains to Nick's mother that the Indian probably provoked the argument so that he would not have to take out in work the considerable sum he owes the doctor for medical services. Nick's mother, a Christian Scientist, says that she can not believe anyone could be capable of such an action, to which her husband merely replies, "No?" The doctor goes for a walk and, finding Nick reading in the woods, tells him that his mother wants to see him. Nick replies, "I want to go with you," and his father assents. That a boy should prefer his father's company to his mother's is natural enough, but Nick's statement can perhaps be regarded as suggesting something more: an instinctive rejection of his mother's attitudes and conception of reality.

In "The End of Something," Nick, now in his middle or late teens, apparently having decided beforehand to break off an affair he has been having with a girl named Marjorie, does so one evening while they are out trout fishing. But one must read another story, "The Three-Day Blow," to learn the reason for his action and the effect it has on him. Nick goes to visit his friend Bill one stormy autumn afternoon, and the two have a long talk about baseball, books, and other matters while getting drunk on Bill's father's whisky. Then Bill brings up the subject of Marjorie. " 'You came out of it damned well,' Bill said. 'Now she can marry somebody of her own sort and settle down and be happy. You can't mix oil and water and you can't mix that sort of thing any more than if I'd marry Ida that works for the Strattons.' " Nick does not disagree, but he is puzzled and feels a great sense of loss. " 'All of a sudden everything was over,' Nick said. 'I don't know why it was. I couldn't help it. Just like when the three-day blows come now and rip all the leaves off the trees.' " But when Bill counsels him not to think about it any more, since to do so might result in his getting together with Marjorie again, Nick realizes that he can resume the affair if he wants to. Nothing is irrevocable, and he is happy in the thought. He and Bill decide to go hunting: "Outside now the Marge business was no longer so tragic. It was not even very important. The wind blew everything like that away." Its psychological penetration and its masterful handling of dialogue and imagery make "The Three-Day Blow" one of Hemingway's richest and most artistic stories.

Three other stories of Nick's adolescence and young manhood are of a grimmer kind and are similar in that each shows him being introduced to some unadmirable specimens of humanity. Each also avoids explicit statement, the meaning and emotion being conveyed almost entirely through dialogue and action. In "The Battler," Nick hops a moving freight train, but is soon knocked off by a brakeman, and in the woods alongside the track he has an encounter with two hobos, a former prizefighter named Ad Francis, whose face is badly mutilated and who admits that he is "not quite right," and his Negro companion, Bugs. While the three are eating ham and eggs prepared by Bugs, Ad suddenly turns ugly and threatens to beat up Nick. The Negro thereupon taps Ad on the head with a blackjack and knocks him out, explaining to Nick that he has been obliged to do this before. He goes on to tell Nick something of the prizefighter's unsavory history and then suggests that it would be well for Nick to leave before Ad comes to. The Negro is smooth-spoken and excessively polite, but there is clearly something unusual and sinister about him and his attachment to Ad.

The seventeen-year-old boy in "The Light of the World," who is not named but who could very well be Nick, and his friend Tommy almost have a fight with a belligerent bartender and also have an encounter with some whores and a male homosexual in a railroad station. These boys and the Nick of "The Battler" are hardened and tough enough not to be particularly affected by their experiences, but that is not true of the Nick of "The Killers." In this famous story Nick is in a lunchroom talking to the waiter, George, when two men dressed in tight overcoats and derby hats enter. After tying up Nick and the Negro cook in the kitchen, they announce that they have come to kill a boxer named Ole Andreson, who usually comes to the lunchroom for his evening meal. The gangsters finally leave when Andreson does not appear, and Nick goes to Andreson's rooming house to warn him. Andreson tells Nick that he "got in wrong" and that there is nothing that can be done about it. "Couldn't you get out of town?" Nick asks. "No," Andreson replies. "I'm through with all that running around." Nick goes back to the lunchroom and tells George that he is going to leave town because he cannot bear to think about Andreson waiting in his room, knowing he is going to be killed. George can only reply, "You better not think about it."

Separating the stories in *In Our Time* are thirteen short interchapter vignettes, originally published in a pamphlet entitled *in our time* in Paris in 1924. Precise and vivid in their execution, they seem to be in the nature of practice exercises which Hemingway set himself. Most of them are about war or bullfighting, and one of the former describes Nick just after he has been wounded in the spine, presumably while fighting with the Italians against the Austrians. The story "A Way You'll Never Be" also shows us a wounded Nick. Riding a bicycle through country where many have been left dead from a recent battle, he comes to an Italian battalion headquarters and there tells his friend Captain Paravicini that he has been ordered to circulate among the Italian troops wearing an American uniform so that they will think American troops are coming to their support. In their conversation the captain and Nick allude to the fear they felt during attacks, "a subject," the latter says, "I know too much about to want to think about it any more." Nick assures the captain that he is now all right except that he "can't sleep without a light of some sort." But a moment later he says, "It's a hell of a nuisance once they've had you certified as nutty."

At the suggestion of the captain Nick lies down to take a nap, and mixed-up recollections of war and other experiences race through his head. When some soldiers enter the dugout he ironically explains his mission, refers to having been wounded "in various places," and gives them a dissertation on the use of grasshoppers in fishing. Before he leaves, Nick suffers a brief nightmarish spell in which he is shot at by a bearded man with a rifle, after which he tells the captain: "I'm all right now for quite a while. I had one then but it was easy. They're getting much better. I can tell when I'm going to have one because I talk so much." Although the story leaves unexplained some of the particulars of Nick's situation, it makes dramatically evident the fact that he has been wounded psychologically as well as physically.

"Now I Lay Me" gives us another view of the Nick who has been "blown up" and who lies awake in the dark, fearing that if he closes his eyes his soul will go out of his body. He occupies himself by thinking of the trout streams he has fished and by saying prayers "for all the people I had ever known." At the end of the story Nick refers to having been hospitalized in Milan, and this is where we find the unnamed but Nick-like narrator of "In Another Country," undergoing therapeutic treatments on a machine to restore the use of his wounded leg. This superb story deals with bravery—about which some passages in Hemingway's novel of love and war, *A Farewell to Arms* (1929), also have a good deal to say—and the Italian major whose young wife dies reminds us of Frederick Henry and his loss of Catherine Barkley in the novel. Both the major and Henry know that a man, as the major puts it, "should not place himself in a position to lose. He should find things he cannot lose." Like the narrator, who is honest enough to admit to himself that he is not brave, even though he has been decorated, and like the three young Italian officers also undergoing treatment who have been very brave in battle ("hunting-hawks," the narrator calls them), the major has medals, but he does not believe in bravery. Once the greatest fencer in Italy, he does not believe that therapy will ever restore his right hand, withered from a wound, but in him we see courage manifested in the face of irreparable loss. The highest form of bravery, says Hemingway with marvelous restraint and understatement, is a rigid self-discipline which enables us to endure stoically whatever calamity may befall us.

Five stories give us glimpses of the postwar Nick Adams. In "An Alpine Idyll," the narrator, again unnamed, seems clearly to be Nick. After skiing in the mountains, he and his friend John come down to the Swiss inn in the valley which is their headquarters and hear a story about a peasant who has kept his dead wife's frozen body propped in his woodshed over the winter and has hung a lantern from her mouth so that he could see to cut wood, distorting it terribly. Although the story may appear to depend too much on this rather gruesome anecdote, the point is that it evokes no horrified reaction from John or Nick. Nick, it appears, may be somewhat affected, but on the whole he would seem to have developed considerable immunity insofar as such incidents are concerned. In another skiing story, "Cross-Country Snow," we learn that Nick is married, that his wife Helen is expecting a baby the following summer, and that she and Nick will be going back to the United States. Nick and his friend George, although they never explicitly say so, would clearly like to avoid life's complexities and problems by prolonging their skiing holiday indefinitely, but this is impossible, nor can they be sure they will ever be able to ski together again.

That the only "good place" is away from society and civilization is also suggested in the very good two-part "Big Two-Hearted River," in which we find Nick back in northern Michigan on a solitary fishing trip. Nick's actions—making camp, cooking meals, catching fish—are rendered with exact and vivid detail, but there is much more to the story. Nick is making an escape, and his actions are a kind of ritual intended to ward off unnamed hostile forces in life. "He felt he had left everything behind, the need for thinking, the need to write, other needs." In his camp, where nothing can touch him, he can feel happy and secure. That Nick ultimately manages to come to terms with life is suggested by the final view we have of him in "Fathers and Sons," in which he is thirty-eight, a writer, and the father of a young son. The story is mainly one of reminiscence. Nick thinks back to his boyhood summers in Michigan and particularly to the Indian girl Trudy with whom he learned about sex and to his father who taught him to hunt and fish and who later committed suicide. He will eventually rid himself of the pain of his father's death, about which he has thought many times, by writing about it: "He had gotten rid of many things by writing them."

Of Hemingway's non–Nick Adams stories, two of the most frequently anthologized and highly regarded are "The Undefeated" and "Fifty Grand," both of which, like "In Another Country," are studies in courage. For Manuel Garcia, the has-been bullfighter of "The Undefeated," bullfighting is his life, and he cannot bring himself to quit and do some other kind of work. Just out of the hospital after recovering from an injury, he prevails on a reluctant manager to arrange another bullfight for him and also on his disapproving friend Zurito, a semiretired picador, to work with him on Zurito's condition that if Manuel is not a big success he will give up bullfighting. Manuel shows flashes of his one-time brilliance in his passes at the bull, but after several unsuccessful attempts to kill the bull he is gored. Despite his humiliation and serious injury he refuses to quit and manages to kill the bull, thus achieving a victory in defeat through his courage and endurance.

Jack Brennan, the welterweight boxing champion of "Fifty Grand," is a less admirable character than Manuel, but he too shows courage in a crucial situation. Money is very important to Jack, and when gamblers come to his training camp to bribe him to throw his fight with Walcott, the challenger, he agrees and bets fifty thousand dollars on his opponent. To Jack this is "just business," since he is past his prime and is convinced that he cannot win anyway. During the early rounds Jack has the advantage because of his superior skill, but when he tires he begins to take a beating from his younger, stronger opponent. Although Jack is now sure of his money and is suffering terrible punishment, he is determined not to be knocked out. This indicates that Jack has pride, but he also turns out to have something more, namely, the quick-wittedness and guts to protect his investment when Walcott deliberately hits him below the belt. Despite the excruciating pain, he refuses to take the foul, hangs on, and fouls Walcott in turn, thereby foiling the doublecross the gamblers had planned to pull on him. It does not seem to be a matter of Jack's maintaining his honor, as one critic has suggested, but simply that Jack is out for himself. He can agree to throw a fight, and he can be a "dirty" fighter if it is to his advantage, but he is also, as his manager says, in a tone that seems both admiring and ironic, "some boy."

Another of Hemingway's best stories is "The Gambler, the Nun and the Radio." Mr. Frazer, one of the three principal characters, has some of the characteristics of the adult Nick Adams. He is a writer, he has insomnia and "tricky" nerves, and he tries to avoid thinking as much as possible. While he is recovering in a Montana hospital from a broken leg suffered in a fall from a horse, a Mexican named Cayetano Ruiz is brought in with two bullet wounds in the abdomen. Although Cayetano is a crooked small-town gambler, he has a sense of honor which will not let him reveal to the police that it is another Mexican from whom he had won at cards who has shot him, suffering his great pain in silence so that he will not disturb the other patients in his ward. It develops that Cayetano will recover, although one of his legs will be paralyzed. One day while visiting with Mr. Frazer he confides that when he has money he likes to gamble without cheating, but that he always loses. He has, in fact, bad luck in everything, but he still hopes for good luck. When Mr. Frazer asks what one can do in such circumstances, Ruiz replies, "Continue, slowly, and wait for luck to change." Mr. Frazer says, "I wish you luck, truly, and with all my heart," but, unlike Cayetano and the young nurse, Sister Cecilia, who naïvely hopes to become a saint, he remains without hope or belief. He can admire their conduct and attitudes, but he has no faith to live by as they do. For him not only is religion "an opium of the people," as another character in the story misquotes, but many other things are also—music, economics, patriotism, drink, gambling, ambition, bread, and much more —and he plays his radio to help keep himself from thinking.

This nihilistic view of life is given even stronger expression in the brief but marvelously written "A Clean Well-Lighted Place," with its unforgettable figure of an old man whose nightly habit it is to drink brandy in a cafe until it closes. The story is carried on mainly by the conversation of the two waiters in the cafe, one old and one young, in which it is brought out that the old man had made an unsuccessful suicide attempt. The older waiter, unlike the younger one, understands what is wrong with the old man. He and the old man know that life is a "nothing" and that man is a "nothing" also, and that when one is aware of this he needs a clean well-lighted place to keep off the darkness of the thought. But it is also significant that, although the old man has nothing to live for, he conducts himself well. He is clean, and he drinks without spilling even when he is drunk, as the older waiter points out, and when he leaves the cafe he walks "unsteadily but with dignity." In his way the old man exemplifies what has often been called the Hemingway "code," as do the major in "In Another Country," Manuel Garcia, Cayetano Ruiz, and other Hemingway protagonists. In the face of despair, disaster, or defeat, a man should endeavor to act well instead of badly—to manifest the qualities of stoicism, resignation, courage, and endurance and to maintain his dignity and honor.

Even in stories that may seem slight or otherwise fall short of his best achievement, Hemingway was usually able through his narrative artistry to produce effects which leave a strong impression on the reader. An example is "A Very Short Story," the first half of which is virtually a synopsis of the earlier stages of the Frederic Henry–Catherine Barkley love affair in *A Farewell to Arms*, but which has an outcome very different from that of the novel. The soldier in the story returns to America after the armistice to find a job so that he can send for his nurse-sweetheart. Later she writes that she has fallen in love with an Italian major who will marry her in the spring. But the major does not marry her then or later, and the soldier back in Chicago has a sordid affair with a department-store salesgirl from whom he contracts a venereal disease.

"Soldier's Home," another story of a soldier back from war, gives the reader a full sense of the malaise and sense of futility which is likely to afflict the former soldier in his readjustment to civilian life. Young Krebs gets back to his Oklahoma hometown too long after the war is over to be accorded a hero's welcome, and no one is interested in hearing about his war experiences. He slips into a routine of sleeping late, reading, and playing pool in the evenings. He feels vaguely that he would like to have a girl but does not have the energy to court one. Like everything else, the effort would be too much trouble and make for consequences. "He did not want any consequences ever again. He wanted to live along without consequences." Concerned because he seems to have no ambition, Krebs's mother insists on having a heart-to-heart talk with him, which embarrasses Krebs and causes him to speak to her unkindly. He cannot explain to her that he only wants his life to go smoothly and be uncomplicated. "He would go to Kansas City and get a job and she would feel all right about it."

There is much verbal irony in Hemingway's fiction, and certain of his less well-known stories show the great skill with which he could utilize it. In "Che Ti Dice La Patria?" the narrator and his friend take a ten-day trip through Italy. "Naturally in such a short trip," remarks the narrator, "we had no opportunity to see how things were with the country or the people." Actually there is ample opportunity, and the narrator's seemingly disarming account is really a bitingly sardonic and damning portrait of a country which Hemingway implies was formerly very different. A Fascist hitchhiker—of whom the narrator says to his friend, "That's a young man who will go a long way in Italy"—forces himself upon them; two waitresses in a restaurant are obviously ready for a price to serve the travelers in another way besides bringing them food; and a Fascist policeman, taking advantage of the muddy roads, extorts a fine from them on the grounds that he cannot read their license.

Some of Hemingway's harshest irony is to be found in the first part of "A Natural History of the Dead," with its detached description of women killed in the explosion of an Italian munitions factory and of a battlefield with many dead soldiers left unburied. Horror involving death is also evoked in the ironic monologue "After the Storm." The narrator, a tough Florida east-coast fisherman who resembles Harry Morgan, the hero of Hemingway's novel *To Have and Have Not* (1937), stabs another man in a barroom brawl and hides out in his boat. There has been a hurricane, and the next day, looking for ships which may have foundered, he comes upon a liner not too far under water. Birds are circling the floating debris. His repeated attempts to break through a porthole to get the money and other treasure he knows must be inside are unsuccessful, and when he returns to the ship later after another storm, he discovers that other scavengers, whom he calls "the Greeks," have beaten him to the plunder. Then he speculates about how the liner went down and how the passengers and crew felt and acted. The boilers must have burst when the ship struck quicksand, which explains "the pieces of things" floating out of the ship that the birds were after. Now fish weighing three to four hundred pounds live inside the ship. The narrator gives no indication that there is any horror in all this for him, and concludes, "First there was the birds, then me, then the Greeks, and even the birds got more out of her than I did."

In 1938, Hemingway issued an omnibus collection of his short fiction entitled *The Fifth Column and the First Forty-nine Stories*. In addition to including the contents of his three earlier story collections, the volume contains *The Fifth Column*, a play written in 1937 about the Spanish Civil War, and four previously uncollected stories: "The Capital of the World," a moving account of the accidental death of a young Spaniard who aspires to be a bullfighter; "Old Man at the Bridge," an equally poignant sketch of the Spanish Civil War, in which a man of seventy-six who is "without politics" is compelled to evacuate his native village because of the imminence of enemy shelling; and two of Hemingway's longest and best stories, "The Short Happy Life of Francis Macomber" and "The Snows of Kilimanjaro," both of which were first published in 1936 and derive their backgrounds from an East African hunting safari which Hemingway made during the winter of 1933–34, and of which he gave an account in *Green Hills of Africa* (1935).

"The Short Happy Life of Francis Macomber" is the story of a coward who becomes a brave man. The first part describes the reactions of each of the three principal characters

—Macomber, a rich American sportsman of thirty-five; his wife, Margot; and the English hunting guide Robert Wilson—to Macomber's conduct on a lion hunt from which they have just returned. Macomber speaks of having "bolted like a rabbit," and he is very ashamed and apologetic. His wife is not only ashamed of him, but also bitter and cruel in what she says. Wilson, although he tries to pretend that what has happened is unimportant, feels contempt for Macomber for having acted badly and for asking him not to tell anyone else about his behavior; yet he can also feel some respect for Macomber, hoping that he can redeem himself when they go hunting for buffalo.

That night Macomber goes to bed feeling, even more than shame, "the cold hollow fear in him" and relives the whole experience which culminated in his fleeing before the charge of the wounded lion, leaving Wilson to hold his ground and kill the beast. Withholding "the story of the lion" until this point makes for suspense, and Hemingway, besides making the account vivid and dramatic, compresses a great deal in it: Macomber's nervousness and fear before and during the hunt; the explicit instructions he receives from Wilson about what one does and does not do as a hunter, especially when one has a wounded lion to cope with; the hunt itself; and its outcome. Macomber remembers too how after Wilson killed the lion Margot kissed Wilson on the mouth, and when he wakes later in the night he discovers that she is missing from her cot. She has gone to sleep with the guide. If she is a bitch, as Macomber tells her when she returns, he, she reminds him, is a coward. Margot may be too beautiful for Macomber to divorce, and Macomber may have too much money for Margot to be able to leave him, but she is certain now that she can make him take anything from her. As for Wilson, who "carried a double size cot on safari to accommodate any windfalls he might receive," his standards are those of his clients—except for the hunting. "He had his own standards about the killing and they could live up to them or get some one else to hunt them."

The next day the three hunt buffaloes, and one of the animals Macomber shoots is wounded and escapes into the bush. As had been the case with the lion, it is necessary to go into the bush and finish him. In some mysterious way the buffalo hunt has made Macomber lose all his fear: "It had taken a strange chance of hunting, a sudden precipitation into action without opportunity for worrying beforehand, to bring this about with Macomber, but regardless of how it had happened it had most certainly happened." He is eager, elated, and happy, and Margot becomes aware that she can no longer be secure in her domination of him. When Macomber stands his ground before the charging buffalo, Margot also shoots from the car behind Macomber and kills him. Although Wilson assures Margot that he will report Macomber's death as accidental, he remarks ironically, "That was a pretty thing to do," and adds, "He *would* have left you too." Macomber, Wilson knows, had become a man.

As both Philip Young and Carlos Baker have demonstrated, a convincing case can be made that "The Snows of Kilimanjaro" reflects Hemingway's dissatisfaction with himself for having gone for a period of several years without producing the kind of writing he felt himself capable of. The writer called Harry in the story, who lies dying of gangrene from an infected scratch on his leg, is filled with self-condemnation because he has not written the things he should have written. He has come back to Africa with the rich, devoted wife he has never really loved because he had been happy there in the past and had felt this might be a prelude to a new start. Now that it is too late, he admits to himself that he has sold out for the wealth, comfort, and security his wife stands for but that it is unfair of him to blame her for destroying his talent: "He had destroyed his talent by not using it, by betrayals of himself and what he believed in, by drinking so much that he blunted the edge of his perceptions, by laziness, by sloth, and by snobbery, by pride and by prejudice, by hook and by crook."

Interspersed throughout the story are vividly rendered retrospective vignettes in which Harry recalls experiences he had "saved" to write about, as well as various images of death: vultures who squat "obscenely" on the plain, of which Harry says ironically, "I watched the way they sailed very carefully at first in case I ever wanted to use them in a story"; a hyena that howls in the night; and the shapeless, foul-breathed thing that moves closer and closer to Harry, finally crouching on his chest. The principal image, however, is the snow-covered Kilimanjaro with the dried and frozen carcass of a leopard close to its western summit, known to the natives as "the House of God." It is to the top of Kilimanjaro, "great, high and unbelievably white in the sun," that Harry knows he is going when he "dreams" just before he dies that he is aboard the plane which has been sent for to fly him to the hospital. The possible meanings of the mountain and the leopard and the question of their symbolic appropriateness have evoked much critical discussion. The mountaintop can be viewed as standing simply for death, but it seems more likely that Hemingway meant it to be a symbol of immortality, of Harry's vision of achieving a new life in death. Although all we are told of the leopard is that "no one has explained what the leopard was seeking at that altitude," the statement clearly suggests aspiration. It had striven for some ideal just as Harry once had and knows he should have continued to do. The leopard did not achieve its goal, but its effort is both manifested and immortalized in its frozen, imperishable carcass.

Hemingway's last two novels, *For Whom the Bell Tolls* and *Across the River and into the Trees* were published in 1940 and 1950, respectively, and in 1952 he published his last significant work of short fiction, *The Old Man and the Sea,* a novelette of 140 pages that won the Pulitzer Prize in Fiction for that year. The old man of the story is a Cuban fisherman named Santiago who has fished for eighty-four days without making a catch. At first he was accompanied by the boy Manolin, but after forty days the boy's parents made him go out with another fisherman because the old man was so unlucky. A great love exists between the old man and the boy, who still brings Santiago food and bait and the newspapers so that he can read about the American big-league baseball games.

Early on the morning of the eighty-fifth day Santiago says good-by to the boy and rows far out into the ocean, resolved that he will "fish the day well." At noon Santiago hooks a fish, one so heavy that Santiago cannot raise him. The rest of the day and all that night, Santiago, holding the line against his back, is pulled out to sea by the submerged fish. Santiago feels pity for the great fish. He wishes the boy were with him, both to help and to experience this contest with so wonderful and strange a creature. Santiago's hands are cut and cramped from the pull of the line, and the fish too is feeling the struggle. "I wish I could feed the fish," he thinks, as he eats a raw tuna he caught earlier. "He is my brother. But I must kill him and keep strong to do it."

During the morning of the second day, almost, it seems to Santiago, as if to show how big he is, the fish jumps, revealing himself to be two feet longer than the skiff—biggest fish Santiago has ever seen or heard of. The fish submerges again and continues to pull the boat throughout the rest of the day and all that night. At sunrise on the third morning he begins to jump repeatedly, and the work of catching him begins. Santiago battles the fish until noon before harpooning him, and then, spent almost beyond the limit of his endurance, he lashes the fish alongside the skiff and sets sail for home. An hour later a great shark attacks, tearing off forty pounds of the fish before Santiago kills him. More sharks come at intervals during the day and that night, until finally there is nothing left but the skeleton of the fish. Sailing into the harbor, Santiago beaches his boat and, carrying his mast on his shoulder, makes his way to his shack with several pauses for rest. When he wakes in the morning, he finds Manolin there to minister to him. "You must get well fast," the boy says, "for there is much that I can learn and you can teach me everything. How much did you suffer?" "Plenty," replies the old man. That afternoon he again sleeps peacefully with the boy beside him, dreaming his favorite dream of the young lions he watched play on the African beach when he was a boy.

Hemingway once said of *The Old Man and the Sea* that if he had succeeded in his attempt to make everything in it as real as possible—the old man, the boy, the sea, the fish, and the sharks—then the story would mean many things. Despite its surface simplicity, different meanings can indeed be found in the story. Santiago knows that he was beaten because he went out too far, but he also knows that "a man can be destroyed but not defeated." If a man acts honestly and well, and if he has courage, will, and endurance, he can achieve a moral triumph in defeat. Santiago, however, represents something more than a courageous and tragic individualism in a naturalistic universe. He has a sense of the unity and solidarity of the universe which helps sustain him, and in his humility and capacity for suffering he is Christlike. Santiago's cry of "*Ay*," when the first shark appears, says Hemingway, was perhaps "just a noise such as a man might make, involuntarily, feeling the nail go through his hands and into the wood." This and other suggestions of Christian symbolism do not necessarily make the story a religious allegory but do serve to enhance Santiago's heroic stature.

Although *The Old Man and the Sea* has generally been regarded as a masterpiece, the reader may have some reservations about its artistic achievement. He may feel that some of Hemingway's stylistic mannerisms, especially his repeated use of the word "truly" are a little irritating, that the Christian symbolism is too obtrusive, and that sometimes, instead of dramatic representations of Santiago's qualities of pride, humility, endurance, and compassion, there is too much generalizing and explicit statement about them. He may feel, in short, recalling the iceberg metaphor Hemingway used in *Death in the Afternoon* to describe good writing, that there is a little too much of *The Old Man and the Sea* that is above water:

> If a writer of prose knows enough about what he is writing about he may omit things that he knows and the reader, if the writer is writing truly enough, will have a feeling of those things as strongly as though the writer had stated them. The dignity of movement of an iceberg is due to only one-eighth of it being above water.

But it must be acknowledged that *The Old Man and the Sea* has much poetic beauty and richness of meaning. In its conception and execution it is certainly the most ambitious and in many respects the most impressive short narrative of a great master of the short story. It may well be, as some critics have suggested, that in writing it Hemingway had in mind his own efforts to achieve mastery of his art. It makes a fitting symbol of that struggle.

## 12
## Virtuoso Storyteller William Faulkner

Some of William Faulkner's critics, among them Edmund Wilson and Irving Howe, have expressed the view that Faulkner's great talent for portraying experience in all its density and complexity did not lend itself to achieving the effect of intensity so important in the short story. Whether or not this is so, Faulkner's many fine short stories, no less than his best novels, impress the reader in the breadth and range of the fictional world they create, their deeply felt themes, their moral implications, and the variety and artistry of their style and narrative techniques. No other American short-story writer, it seems safe to say, has accomplished so much in the form, with the possible exception of Henry James.

William Faulkner (1897–1962) lived most of his life in Oxford, Mississippi. His great-grandfather, who was born in Tennessee in 1825, went to Mississippi as a youth, practiced law, fought in the Mexican and Civil wars, built a railroad, and wrote a best-selling novel, *The White Rose of Memphis*. This larger-than-life ancestor, who was shot and killed in 1899 by a former business associate, appears, along with other members of Faulkner's family—often somewhat romanticized—in the early novel *Sartoris* and in the interrelated stories which make up *The Unvanquished*. Much of Faulkner's other fiction is also laid in Oxford and the surrounding region, to which he gave the fictional names Jefferson and Yoknapatawpha County, respectively.

Faulkner did not take kindly to formal education. He left high school without graduating and later dropped out of the University of Mississippi after a year as a special student. During World War I, Faulkner was not, as some of his biographical accounts would have it, shot down and wounded while flying for the Royal Air Force. He did complete flying cadet training in Toronto, but the war ended before he could be sent overseas. Returning to Oxford and finding university studies uncongenial, Faulkner worked at various jobs and resumed the wide though miscellaneous reading and the poetry writing he had begun before the war.

Faulkner's first book was a small volume of poems with a distinct flavor of Keats and Swinburne, entitled *The Marble Faun*, the publication of which in 1924 was subsidized by his friend and mentor, Phil Stone, a cultivated Oxford lawyer. During the following year Faulkner spent several months in New Orleans, where he associated with Sherwood Anderson and other writers. Their friendship and encouragement led him to write a group of prose sketches for the *Double Dealer*, a New Orleans literary magazine, and sixteen stories and sketches which were published in the *New Orleans Times-Picayune*. Although several of these pieces foreshadow certain of the themes and other elements in Faulkner's more mature fiction, they give little hint of what he was later to accomplish in the short story. Several are primarily human-interest portraits of odd or unusual characters—a beggar, a racetrack tout, a jockey, a cobbler, a young hobo, an idiot.

Others have more story quality, but, except for some occasional effective touches, as in "The Liar," which has something of the comic flavor of *The Hamlet*, and in "Sunset," a moving story of an ignorant and misguided Negro, they are essentially amateur work; Faulkner's power of invention and ear for reproducing the speech of his characters were still largely undeveloped.

Faulkner's first novel, *Soldier's Pay*, also written in New Orleans and published in 1926, and his second, *Mosquitoes*, published in the following year, are also apprentice work, but within the next three years he emerged as a major novelist with the publication of *Sartoris* (1929), *The Sound and the Fury* (1929), and *As I Lay Dying* (1930). Although Faulkner did not begin publishing short stories in national magazines until 1930, it is likely that he had been writing short fiction for some time previously. There is an apparently well-authenticated story that after the popular success of *Sanctuary*, published early in 1931, Faulkner was able to sell at more or less his own price a number of stories to magazines which had earlier rejected them. That may explain why Faulkner published a surprisingly large number of stories during the latter part of 1931 and in 1932: four in *Harper's*, three in the *Saturday Evening Post*, three in *Scribner's*, and one in *American Mercury*. *These Thirteen*, Faulkner's first volume of stories, was published in 1931, and a second, *Doctor Martino and Other Stories*, in 1934.

The stories in these volumes, along with other early ones which were not collected, have a wide variety of settings, situations, and themes. There are some rather unsuccessful fantasies: "Carcassonne," a dialogue between a man and his skeleton; "Black Music," whose main character believes he has turned into a faun; and "Beyond," in which an old judge dies and goes to heaven, where he searches for his long-dead son. There are thin, conventional, or otherwise undistinguished stories, among them "Pennsylvania Station," "Elly," and "Artist at Home," as well as others with more substance but whose conflicts do not seem sufficiently realized, such as "Dr. Martino," a study of a girl who struggles unsuccessfully to become independent of her dominating mother; "Mistral," a skillful if somewhat artificial story whose characters and Italian setting are reminiscent of Hemingway; and "Fox Hunt," which makes too-obvious use of the analogy of a fox caught and destroyed by hunters to suggest the heroine's fate.

Of several stories about World War I, the best is "Ad Astra," which portrays a group of pilots, most of them American, in the Royal Flying Corps just after the war has ended. Their hopelessness and war-weariness is excellently evoked, and an effective contrast is drawn between the drunken pilots on the one hand and a captured German aviator and a turbaned Indian major in the British army on the other, although the story makes somewhat too obtrusive its two main themes that all men are brothers and that, as the major puts it, "all this generation which fought in the war are dead tonight. But we do not know it yet." One of the pilots is Bayard Sartoris, and the story serves to make somewhat more understandable his behavior as the lost-generation hero of *Sartoris*. So too does "All the Dead Pilots," a somewhat incongruous mixture of the farcical and the serious, about Bayard's twin brother, the madcap John Sartoris, who is shot down in action and to whose death Bayard, in the novel, is never able to reconcile himself.

Faulkner's other portrayals of men at war are for the most part of British soldiers and are compounded principally of stock characters, melodrama, and sentiment. "Turnabout," published in the *Saturday Evening Post* and later made into a motion picture, in particular reveals a facility in utilizing commercial-story techniques which Faulkner may have learned during periods spent in Hollywood as a scriptwriter. On the same order as these stories is "Honor," a variation on the love-triangle formula, in which Monaghan, one of the American pilots in "Ad Astra," returns to civilian life to become an airplane-wing walker and later an ineffectual automobile salesman. "Death Drag," on the other hand, another story of barnstorming stunt flyers, is devoid of contrivance and, with Hemingway-like objectivity and directness, makes fully convincing the atmosphere of futility and frustration in which its characters move.

With a few exceptions Faulkner's best early stories are those in which he wrote of his own region: stories of the Indians who lived in Mississippi in the old days, episodes in the life of the town of Jefferson in which both past and present often figure, and stories of the country folk living in the vicinity of Jefferson. "Red Leaves," one of Faulkner's masterpieces of short fiction, and "A Justice," one of his finest comic stories, portray the Chickasaw Indians during the early 1800's in Yoknapatawpha County. In the former story, Issetibbeha, the chief, has died, and, according to custom, his horse, dog, and Negro bodyservant must be buried with him. When the slave tries to escape, he is patiently and implacably pursued. The task is a nuisance to the placid, easygoing Indians, but it is one they are resigned to undertake, since the same thing had happened when Issetibbeha's father died.

It is not just this one slave, however, who is a trial to the Indians. They regard all their slaves as burdens. "This world is going to the dogs," one old Indian observes. "It is being ruined by white men. We got along fine for years and years, before the white men foisted their Negroes upon us." This theme lends irony to the dramatic, tragic spectacle of the Negro's futile struggle for survival, and there is powerful irony also in the contrast between the slave who exerts himself to the limit of his endurance and the almost lifeless Moketubbe, who has become the chief, or "Man," a mountain of almost inert flesh, whose only interest lies in wearing the red slippers which his grandfather Ikkemotubbe, the first "Man," had brought back from Paris. Moketubbe, according to tradition, should have actively directed the pursuit, but he leads it only nominally, being carried on a litter. But if he is impervious to what is happening and to its import, that is not true of the pursuers. They may regard their Negroes as little better than animals, but they can nevertheless recognize and pay tribute to the humanity and endurance of the manservant after he is finally captured, saying to him: "You ran well. Do not be ashamed."

"A Justice" tells how the part-Chickasaw, part-Negro Sam Fathers got his name. Sam also figures prominently as a hunter in two of Faulkner's later stories, "The Bear" and "The Old People" (in which a somewhat different version of his paternity is given), but here he is a carpenter-handyman on a farm owned by the Compsons (who appear in *The Sound and the Fury* and elsewhere in Faulkner's fiction) to which the family drives from Jefferson every Saturday. On one of these visits Sam tells the twelve-year-old Quentin Compson, who is too young to comprehend the point of the tale, much less its humor and irony, how Ikkemotubbe, who had become the "Man" by poisoning the old chief and the chief's son, tried unsuccessfully to prevent Craw-ford, Sam's father, from consorting with the wife of one of Ikkemotubbe's slaves. When the woman bears a yellow instead of a black child, her Negro husband protests to Ikkemotubbe and asks for justice. "I don't see that justice can darken him any," the latter replies, but after dubbing the child Had-Two Fathers, he compels Craw-ford and another Indian, Herman Basket, (who later tells all this to Sam), to build a fence around the Negro's cabin too high for Craw-ford to scale. On the day it is completed some months later, the Negro pronounces it a good fence and proudly brings from his cabin a newborn child of the proper color.

Faulkner's two other Indian stories are also humorous ones. In giving "Lo!" its title Faulkner was undoubtedly referring ironically to the lines in Pope's *Essay on Man* beginning "Lo, the poor Indian! whose untutored mind/Sees God in clouds, or hears him in the wind." Frank Weddel, a Chickasaw chieftain, and his entire clan make the fifteen-hundred-mile trip in the middle of winter from Mississippi to Washington, where they cause President Andrew Jackson no end of annoyance and frustration by camping on the doorsteps of the White House. Weddel's nephew has been accused of killing a white man, and Weddel, who professes to be an ignorant and simple man, maintains that justice can only be done if the Great White Father himself determines the nephew's guilt or innocence. Although all this is highly amusing, the story has an underlying significance in that the actions of the Indians are really an indirect protest of the fact that the white man comes into the territory of the Indians not to hunt in peace but to exploit them.

No such theme informs "A Courtship," its humor being that of the tall tale. Ikkemotubbe, here a young man in the days before he becomes chief, and his good friend David Hogganbeck, a young Bunyanesque white man, fall in love with the same Indian girl. The descriptions of the whisky-drinking match, the eating contest, and the epic footrace in which the two engage to determine which will have a clear field in his courting employ exaggeration in the best tall-tale tradition.

In order of publication Faulkner's stories which have Jefferson settings begin with "A Rose for Emily," probably his most famous story, and also his first to be published in a national magazine. It has been variously called a horror story, a story showing the conflict between the values of the Old and New South, a portrait of a woman who retreats from reality and goes mad because she is denied normal human relationships, and simply a tour de force. Miss Emily is enough like Miss Havisham of *Great Expectations* to cause one to wonder whether Faulkner may have modeled her on Dickens' character. Both women, after they are jilted by their lovers, isolate themselves in decaying houses and refuse to recognize the passage of time, and both are proud, haughty, and independent. Miss Havisham's lover, who was showy and not a gentleman, may also be compared with Miss Emily's lover, Homer Barron, a rather flashy man and a Yankee, who in Jefferson's eyes appeared little better than a day laborer. And like Miss Havisham's room, Miss Emily's, which must be forced open after her death, is "decked and furnished as for a bridal." But unlike Miss Havisham, who tries to avenge her jilting by bringing up the girl Estella to make men her victims and ultimately repents having done so, Miss Emily, in effect, refuses to be jilted. She will have her lover in death if she cannot have him in life, as the shocking but nevertheless carefully prepared-for ending reveals. Miss Emily is seen throughout as the townspeople of Jefferson see her, and their ambivalent attitude—compounded of respect, awe, condescension, and pity, along with the recognition that her perversity and madness had its roots in her pride and dignity, in her father's repressive treatment, and in her betrayal by Homer Barron—serves to explain if not mitigate her monstrous action and makes her story something more than a mere horror story or pathological case history.

Horror of another kind is evoked in "That Evening Sun,"

which portrays Nancy, the Negro woman who does the Compson family's washing, and her fear that her husband, in whom she has "woke up the devil" because of her relations with white men, is going to kill her. The pathos and horror of her situation are accentuated not only by the obvious inevitability of her fate but also by the attitudes of the Compsons toward her. Mrs. Compson asserts somewhat querulously that the police should be expected to protect Nancy, while Mr. Compson, although he shows some sympathy for Nancy, is inclined to regard her fears as nonsense and can only suggest that she keep her door locked. As for the three children, although Quentin (who is telling the story fifteen years later) is only nine, his "Who will do our washing now, Father?" indicates his realization that Nancy is doomed; the other two, Caddy and Jason, being only seven and five, naturally do not achieve this awareness. Caddy's unsatisfied curiosity about Nancy's unusual behavior, Jason's small-boy bravado and prattling that he is not a "nigger," and the childish bickering of the two lend a terrible irony to the spectacle of Nancy's suffering, particularly in the scene in which she induces the children to come home with her, hoping to rely on them for temporary protection against the man she knows is lying in wait in the ditch.

In two other Jefferson stories the barber Hawkshaw is a principal character. "Hair" effectively sustains the reader's interest as it unravels the mystery of his background, but it lacks the power of "Dry September," in which Hawkshaw, courageously insisting that they first find out the "facts," tries to deter a group of his fellow townsmen from lynching a Negro accused by a neurotic spinster of attacking her. The lynching itself is left to the reader's imagination, two sections of the story being concerned with Hawkshaw and his ineffectual efforts to prevent it, two with the spinster, and one with the leader of the lynchers. Faulkner handles this difficult method of narration brilliantly, not only bringing the three characters and their involvement in the situation vividly to life but also revealing the horror implicit in the triumph of unreason over reason.

Several of Faulkner's early stories of the country folk of Yoknapatawpha County were later republished, usually extensively revised, as episodes in longer works. The main outlines of "Wash," another of his best stories, reappear in *Absalom, Absalom!* (1936). Colonel Sutpen, the protagonist of that novel, is, in the story, reduced after the Civil War from being the master of a plantation to running a country store. Now sixty, he has a child by Milly, the daughter of Wash Jones, a poor white who lives on the Sutpen place. Wash has built up in his mind an ideal image of Sutpen as the epitome of bravery and honor, and from this image has derived a pride of his own and sense of what is right. When Sutpen callously shows no more concern for Milly and her baby than he does for his mare and its newborn colt, Wash, rudely stripped of the illusion which has given life meaning and purpose for him, brings about the destruction of Sutpen, his daughter, and himself.

Four episodes in *The Hamlet* (1940), the first of Faulkner's three novels about the Snopes clan (the others are *The Town*, 1957, and *The Mansion*, 1959), are expanded versions of previously published short stories. "Fool About a Horse" is a tall-tale anecdote about a horse trade; "The Hound" develops the familiar story formula of the murderer whose fear betrays him into actions which result in his exposure; and "Lizards in Jamshyd's Courtyard" and "Spotted Horses" are accounts of how the virtually inhuman Flem Snopes exploits the greed and gullibility of certain of his fellow country folk of Frenchman's Bend to his material advantage. The first three of these stories, in both their original and their revised forms, are competently executed but not particularly distinctive. Both versions of "Spotted Horses," on the other hand, render brilliantly, but quite differently, essentially the same incident. The short story is a remarkable achievement in conveying serious moral implications through the medium of what ostensibly is a humorous tall tale. Told in the first person by the sewing-machine salesman Ratliff, it is a broadly sketched, graphic, fast-moving narrative illustrating the point made at both the beginning and the end that no one can get ahead of Flem Snopes. Ratliff has a keen eye for detail, and his country idiom is racy and vivid, but he is content mainly with reporting events and letting them speak for themselves. A sharp trader himself, he must give credit to Flem for being a better one, but we feel sure that he is not inhumane, even though he does not expressly say whether or not he shares the indignation toward Flem and the pity for poor Mrs. Armstid, whom Flem cruelly cheats, which his narrative so strongly arouses in us. But artistic though the short story is, the episode in *The Hamlet*, which is almost three times as long, overshadows it qualitatively as well as quantitatively. In shifting to the third-person point of view, employing a more perceptive Ratliff as a kind of chorus instead of as a garrulous yarn spinner, Faulkner established a perspective which made possible an even fuller realization of almost all the elements in the story, the tall tale giving way to narration which has been markedly enriched and is characteristic of Faulkner at his very best.

It is most satisfying for the reader to see one of Flem Snopes's schemes backfire on him, as it does in "Centaur in Brass," and to see also in "Mule in the Yard" the thwarting of one of Flem's conniving relatives, the mule dealer I. O. Snopes. Both stories, somewhat revised and expanded, appear as the first and sixteenth chapters of *The Town*, but the revised versions, if anything, lack some of the spontaneity and unity of the originals, both of which belong among Faulkner's best humorous stories. The vivid depiction of the wild ponies breaking loose in "Spotted Horses," with all its ensuing violence and confusion, is duplicated on a smaller scale, with a mule, a cow, and two old women as the chief actors, in "Mule in the Yard," while one of the most comic situations in all of Faulkner results from Flem Snopes, as superintendent of the Jefferson power plant, scheming to play his two Negro firemen against each other to his advantage in "Centaur in Brass." Only a little inferior to these two stories in manifesting Faulkner's genius as a humorous storyteller are "A Bear Hunt," in which Ratliff is again the narrator, and "That Will be Fine," in which a seven-year-old boy assists his irresponsible, philandering uncle to do "business" with women, without, of course, the boy realizing what is happening.

There is humor also in the earlier episodes of *The Unvanquished* (1938), a group of seven interrelated stories about the Sartoris family during the Civil War and Reconstruction. Bayard Sartoris (the elderly banker of *Sartoris*) is the first-person narrator in all these stories. In the first, "Ambuscade," we find him as a boy of twelve, along with Ringo, his Negro playmate, involved in an amusing adventure with Yankee soldiers. In the absence of Bayard's father, the dashing Colonel John Sartoris, they decide that they must defend the Sartoris plantation. Lying in wait beside the road one day with an antiquated musket they shoot the horse from under a passing Yankee. Fleeing to the house, they find sanctuary under Bayard's grandmother's voluminous skirts, and when she denies to a Yankee officer any knowledge of the boys, his gallantry compels him to take the word of a southern lady, although he obviously knows better.

"Retreat" is concerned largely with Colonel John, who leads a band of irregular cavalry and whose daring exploits and hairbreadth escapes from the enemy are in the best romantic tradition. No less remarkable than the colonel but a somewhat more convincing character is Granny Rosa Millard, Bayard's grandmother, who refuses to accept the loss of her chest of silver and two mules which have been stolen by the Yankees, as well as two slaves who have run off after being told by the Yankees that they are free. In "Raid" she determinedly sets out by wagon, accompanied by Bayard and Ringo, to get her possessions back. There are serious elements in the story: the Yankees have destroyed the railroad, burned down plantations, including the Sartorises', and liberated the Negroes, who tramp the roads at night toward the Mississippi, believing that they are on their way to Jordan; and Bayard, now fourteen, gains some insight into the ravages the war has wrought from his Cousin Drusilla, whose fiancé has been killed at Shiloh. Nevertheless, all these incidents are conveyed more romantically than realistically, and the story eventually turns comic when, through a misunderstanding, the single-minded Granny is rewarded for her persistence by receiving a Yankee general's order that puts her in possession of ten chests of silver, 110 mules, and 110 Negroes. Subsequently, in "Riposte in Tertio," Granny, to assist her poor-white and Negro neighbors, defrauds the Yankees of additional mules and money. Her humanitarianism, which overcomes her moral scruples, is admirable, but her ability to carry out her scheme with so much success, the circumstances of her death at the hands of the desperadoes known as Grumby's Independents, and finally the exciting and suspenseful account, in "Vendee," of how Bayard and Ringo avenge her murder put something of a strain on the reader's credulity.

Colonel John Sartoris returns to view in "Skirmish in Sartoris," with its account of how he keeps a Negro from being elected as marshal of Jefferson shortly after the war. How he accomplishes it by killing two carpetbaggers and forcibly preventing the Negroes they have organized from voting forms a serious and significant action in the story, but its implications are ignored in favor of another, essentially trivial, action, in which Faulkner represents humorously and satirically the efforts of Bayard's Aunt Louisa, Cousin Drusilla's mother, to make her daughter an "honest" woman by bringing about Drusilla's marriage to Colonel John. In "An Odor of Verbena," however, Faulkner questions Colonel John's use of force in the "Skirmish in Sartoris" days, as well as later, to realize his dream of rehabilitating, as Drusilla puts it to Bayard, "this whole country which he is trying to raise by its bootstraps." Eventually Colonel John, believing that he has accomplished his aim, renounces the killing of men, and, refusing to defend himself, is killed by a former business partner he had alienated. This event occurs when Bayard is twenty-four, in his last year of law school, and it is taken for granted by Drusilla and the people of Jefferson that he will avenge his father. But Bayard, too, has decided that there has been enough killing in his family and, unarmed, confronts his father's killer and outfaces him.

Although eminently readable and important for the background they provide for *Sartoris* and other Faulkner tales, the episodes in *The Unvanquished* leave something to be desired as meaningful representations of the effects of the Civil War and Reconstruction on southern life and character. A more significant collection of Faulkner's stories, and one which shows his talent at its finest, is *Go Down Moses* (1942). In general, the seven stories in the volume are less closely related than are the stories in *The Unvanquished*, but often one story is illuminated by another, since they are for the most part about the progeny through several generations, both white and black, of Carothers McCaslin and since they are all concerned with the relationship of the white man and the Negro, both to the land and to each other.

"Was," as its title suggests, is a story of "the old days." The time is 1859, before the birth of Isaac McCaslin, the central figure in several of the other stories. Isaac, or Uncle Ike, now almost eighty, heard the story long before from his older cousin McCaslin Edmonds, who as a boy of nine participated in its action. Living with his great uncles, Buck and Buddy McCaslin, who are sixty-year-old bachelor twins, the boy goes with Uncle Buck to recover the Negro slave Tomey's Turl, who periodically runs off to see the girl Tennie on the plantation of Mr. Hubert Beauchamp. Mr. Hubert has a spinster sister, Miss Sophonsiba, whom he has long endeavored to foist off, with her full approval, on either Uncle Buck or Uncle Buddy, and the story turns into a comic double manhunt. Events so conspire against Uncle Buck that brother and sister seem sure of their quarry, until Uncle Buddy effects his rescue by beating Mr. Hubert at a hand of poker. "Was" is "situation comedy," the pursuit of a bachelor being a stock subject of this genre, but it is nevertheless a masterpiece of humorous storytelling.

Although "Was" is a prelude to the stories about Isaac McCaslin, it indicates little of a specific nature about his antecedents. These forebears, as well as details of his own history, are alluded to in "The Fire and the Hearth" and are recounted at length in the fourth section of "The Bear." "The Fire and the Hearth," however, focuses on proud, stubborn old Lucas Beauchamp, also a principal character in the novel *Intruder in the Dust* (1948). Lucas is a Negro with white blood, being, like Isaac, the grandson of old Carothers Mc-

Caslin and at sixty-seven the oldest McCaslin descendant on the plantation, which is now in the possession of Carothers Edmonds, old Carothers' great-grandson. Several flashback episodes are devoted to showing how Lucas, acutely conscious of his ancestry, asserts his manhood and maintains his integrity and independence despite his Negro blood. These episodes have meaningful implications for the relationship between white and Negro, but unfortunately they are incongruously combined with a comic and somewhat contrived main plot involving Lucas's dogged and unrelenting search for money that he is convinced is buried on the plantation.

"Pantaloon in Black" is a grimmer treatment of the relationship between white and black. The Negro sawmill worker, whose grief over the death of his young wife is more than he can bear and who is lynched after he kills a white man who has cheated him in a dice game, is one of Faulkner's best characterizations, and, although Faulkner's indictment of callous Southern whites, like the sheriff's deputy and his wife, who take it for granted that "niggers" aren't human, may seem too heavily ironic, the story is one of his most powerful and poignant ones.

"The Old People," "The Bear," and "Delta Autumn" are companion pieces dealing with the history and more particularly the moral education of Isaac McCaslin. In "The Old People," Sam Fathers, half-Negro but also the son of a Chickasaw chief, prepares young Isaac for the time when he will go into the big woods to hunt bear and deer and become a man. At age twelve he kills a deer, and Sam marks his face with the animal's hot blood. The boy becomes "one with the wilderness which had accepted him."

In "The Bear," Isaac serves "his novitiate to the true wilderness" under the further tutelage of Sam Fathers on the annual trips his cousin, McCaslin Edmonds, and other grown hunters make into the big woods to hunt the indomitable and seemingly immortal bear called Old Ben, who is the "epitome and apotheosis" of a "doomed wilderness whose edges were being constantly and punily gnawed at by men with plows and axes who feared it because it was wilderness." Isaac learns woodsmanship, as well as humility, patience, and endurance, but he must also learn to conquer fear by going out into the wilderness alone, without gun, watch, or compass. Only then, although he has earlier sensed its presence, does he see the bear. Soon afterward, when Isaac is fourteen, Sam Fathers finally finds a dog he believes is capable of baying and holding Old Ben. Isaac does not hate the fierce, implacable dog Lion because he comes to recognize, during the two years before the dog and the half-blood Indian Boon Hogganbeck bring about the end of Old Ben, "that there was a fatality in it. . . . It was like the last act on a set stage. It was the beginning of the end of something, he didn't know what except that he would not grieve." The killing of Old Ben and the death soon afterward of Sam Fathers foreshadow for Isaac the eventual end of the wilderness. Two years later he goes back to the hunting camp for the last time, ostensibly to hunt with Boon Hogganbeck but actually to commune with the wilderness at the graves of Sam and Lion.

There is more, however, to "The Bear" than a hunting story, in which Isaac learns values from Sam Fathers, the wilderness, and Old Ben. The influence of these values is indicated in the long fourth part of the narrative. There Isaac, now twenty-one, asserts in the course of an extended dialogue with his cousin, McCaslin Edmonds, that the land belonging to Isaac's grandfather is not Issac's to inherit. Isaac holds that God

> created man to be His overseer on the earth and the animals

> on it in His name, not to hold for himself and his descendants inviolable title forever, generation after generation, to the oblongs and squares of the earth, but to hold the earth mutual and intact in the communal anonymity of brotherhood, and all the fee He asked was pity and humility and sufferance and endurance and the sweat of his face for bread.

Old Carothers' land is tainted and accursed like that of all other men who have held land for themselves. Isaac repudiates the land and becomes a propertyless carpenter to make restitution as best he can for the unhappy history of the Negro descendants of the evil and unregenerate old Carothers, a "chronicle which was a whole land in miniature, which multiplied and compounded was the entire South."

Although "The Bear" has been generally admired and often highly praised by the critics, to a good many of them its structure has seemed faulty and its meanings somewhat unclear. In composing "The Bear," Faulkner reworked and expanded the materials of two earlier hunting stories ("Lion," *Harper's*, December, 1935, and a much shorter "The Bear," *Saturday Evening Post*, May 9, 1942) and to their development of the wilderness theme added the fourth section with its theme of the Negro. It seems fair to say that he was not altogether successful in welding the two parts into a fully unified whole. We may also feel that Faulkner unduly extols a romantic primitivism, that his logic in seeming morally to equate man's exploitation of nature and the exploitation of one's fellow man through slavery is questionable, and that Isaac's repudiation of his inheritance does nothing to atone for the wrong done the Negro but is merely a means by which he can appease his conscience. But even with these reservations we can still feel the power, vividness, and originality of the story.

The time of "Delta Autumn," about 1940, is more than

half a century after that of "The Bear." Isaac, now almost eighty, makes his annual hunting trips with the sons and grandsons of the men he hunted with in his youth. The territory where game still exists is now two hundred miles from Jefferson, where once it had been thirty. But the wilderness is still Uncle Ike's true home, and he goes to it, though too old to hunt, to achieve peace and tranquillity. Peace is destroyed, however, by the appearance of the young woman with whom his kinsman Roth Edmonds, without being aware of her ancestry, has had an affair. She is a mulatto descendant of old Carothers McCaslin, and Uncle Ike can do nothing for her and her child except insist that she take the conscience money Roth has left for her, and say: "Go back North. Marry: a man in your own race. That's the only salvation for you—for a while yet, maybe a long while yet. We will have to wait. Marry a black man."

The title story, and the final one, of *Go Down, Moses* is the brief, unhappy history of yet another Negro McCaslin descendant. Reared by his grandparents Lucas and Mollie Beauchamp, the boy Butch Beauchamp at nineteen is ordered off the Carothers Edmonds farm after he is caught breaking into the commissary store. It is the beginning of a criminal career which ends at twenty-six, when Butch is executed for killing a Chicago policeman. Although Beauchamp is "a bad son of a bad father," the lawyer Gavin Stevens raises the money from the citizens of Jefferson to bring him home and give him a proper funeral for the sake of Aunt Mollie, who laments that her grandson was sold to Pharaoh in Egypt.

In addition to his role as Lucas Beauchamp's lawyer, as well as pontifical and dogmatic apologist for the South in *Intruder in the Dust*, Gavin Stevens also figures in the six stories, written at intervals during the 1930's and 1940's, which Faulkner collected in *Knight's Gambit* (1949). In them Stevens plays the role of a Sherlock Holmes, his Dr. Watson being his young nephew, Charles Mallison. A Harvard Phi Beta Kappa and Heidelberg Ph.D., whose hobbies are playing chess and translating the New Testament into classical Greek, Stevens has most of the familiar attributes of the acutely perceptive and eccentric fictional supersleuth. For the most part the stories also employ familiar detective-fiction formulas. In "Smoke," for instance, Stevens uses a clever ruse to make the killer expose himself, and "Tomorrow" employs the device of the murderer who impersonates his victims but who makes an inadvertent misstep which gives him away—although, as Stevens emphasizes, it is really the murderer's pride and vanity in his gift of impersonation which is his undoing. The *Knight's Gambit* stories are skillful in their creation of mystery and suspense, they have something of the flavor of Faulkner's serious fiction with Jefferson and Yoknapatawpha County settings, and they purport to have, on occasion, significant psychological and moral implications; but they are essentially contrived stories, none too plausible, and for Faulkner, as Edmund Wilson has said, a little cheap.

Most of Faulkner's short stories were written during the 1930's and early 1940's, and his career as a short-story writer virtually came to a close several years before the publication of *Collected Stories of William Faulkner* in 1950, a volume of forty-two stories which includes most of his work outside the *The Unvanquished, Go Down, Moses*, and *Knight's Gambit* collections. These four volumes contain a richly abundant variety of stories, some of which give testimony, as do certain of his novels, to Faulkner's weaknesses—his tendency on occasion to be obscure, overrhetorical, didactic, and fuzzy and confused in his philosophy—but almost all of them show to some extent Faulkner's magnificent powers of invention and his superlative storytelling genius.

# 13
# Social Protest and Other Themes in the Short Story, 1930 to 1940

## Erskine Caldwell
## James T. Farrell
## John Steinbeck
## William Saroyan
## John O'Hara
## Dorothy Parker
## James Thurber
## and
## Kay Boyle

For more than a century the social and economic evils of an ever-growing, increasingly industrialized nation have provided subject matter and themes for American short stories and novels. Herman Melville, in "The Tartarus of Maids and the Paradise of Bachelors," included in his *Piazza Tales* (1856), condemned the cruel exploitation of women "operatives" employed in the paper mills of New England; Rebecca Harding Davis published a grimly realistic story, "Life in the Iron Mills," in the April, 1861, issue of *Atlantic*; and around the turn of the century Stephen Crane condemned social and economic injustice in his Bowery stories, as did Hamlin Garland in *Main-Travelled Roads* and Jack London in a number of proletarian stories. In no other period, however, was there so much social protest reflected in American fiction as in the Depression years of the 1930's. The short story was a vehicle for protest in the work of Erskine Caldwell, James T. Farrell, John Steinbeck, and William Saroyan, as well as in that of many lesser writers, many of whom, because of their Marxist or other doctrinaire attitudes, produced mostly propaganda and little art.

ERSKINE CALDWELL (1903– ), the son of a Presbyterian minister, was born in Georgia and grew up in various regions of the South, although during most of his writing career he has lived in New England and elsewhere in the North. One of America's most prolific and widely read writers, he has published more than twenty-five volumes of fiction and nonfiction, and his novels, notably *Tobacco Road* (1932), *God's Little Acre* (1935), and *Trouble in July* (1940), have sold in the millions.

Although Caldwell's output of short fiction is extensive and varied, his most impressive stories, with a few exceptions, are those which portray Negroes and poor whites in rural areas of the Deep South as victims of social and economic injustice. These tenant farmers and sharecroppers, especially the Negroes, have no recourse when they are treated oppressively by their landlords, as is the case with the old Negro in "The People v. Abe Lathan, Colored," who, because in his landlord's opinion he has outlived his usefulness, is with his family evicted from the farm where he has been a faithful and hardworking tenant for forty years. Far worse, however, is the fate of the Negro sharecropper in "The End of Christy Tucker." He is a "biggity nigger" because he couples ambition and initiative with hard work and is too resourceful and thrifty to go into debt to his plantation owner, as he is expected to do. As a result he must be taught a lesson; when he resists being beaten, he is shot dead. The sin of being "too damned good for a Negro" is also that of Will Maxie, in "Saturday Afternoon," who is lynched on the trumped-up grounds that he has "said something" to a white girl. Caldwell makes effective if somewhat obvious use of irony, contrast, and repetition in the story to evoke horror and revulsion in the reader. For the narrator and his fellow townsmen there is nothing unnatural, or apparently even unusual, in the lynching of a Negro like Will, whom nobody liked because he had always been too smart and hardworking, and for them the lynching is no more than a mildly diverting interruption in the routine of their Saturday afternoon.

Another, even more powerful, lynching story is "Kneel to the Rising Sun," which is not so much about a lynching as it is a study of a white sharecropper who is so afraid of his landlord that he cannot bring himself to ask for adequate rations for himself and his always-hungry family. The system under which he exists has robbed him of whatever manhood he may once have had, and he betrays to a lynch mob his best friend, a Negro sharecropper who courageously stands up for him to the landlord. In its structure and tone this is one of Caldwell's most artistic stories. Each of its three scenes culminates in a violent and horrifying action—the brutal and sadistic cutting off of Lonnie's dog's tail by the landlord Arch; the discovery by Lonnie and the Negro Clem of Lonnie's old father, who has wandered out in the night to look for something to eat, chewed to death by Arch's fattening hogs; and the betrayal of Clem by Lonnie and Clem's consequent lynching—but despite their sensational character these actions are made credible and meaningful as well as highly dramatic.

In "Daughter" we are shown another oppressed white sharecropper, Jim Carlisle, who has been jailed after killing his daughter with a shotgun. Jim had no food for her because his landlord, Colonel Maxwell, had a month before taken Jim's share of the crop when one of the colonel's mules used by Jim dropped dead. In reply to his fellow townsmen gathered outside the jail, who want to know how the daughter's death occurred and who keep asking whether it was an accident, Jim can only repeat: "Daughter said she was hungry, and I just couldn't stand it no longer. I just couldn't stand to hear her say it." Repetition and understatement are effectively utilized to make us feel the poignancy of Jim's situation and the indignant reaction of the men outside the jail, who proceed to take the law in their own hands and let him out because, in the words of one of them, "it ain't right for him to be in there."

Caldwell has written a number of stories with New England settings. In these stories, on the whole, he seems less interesting and convincing in representing regional speech and character than in his southern stories. Most of the New England stories, such as "Country Full of Swedes" (awarded a one-thousand dollar prize by the *Yale Review*), "Over the Green Mountains," "The Grass Fire," and "Balm of Gilead," are humorous and often ironic anecdotes revealing crotchets and prejudices of Maine and Vermont farmers. There are also stories in which Caldwell appears to strive overmuch to achieve pathos and poignancy; among these are "Man and Woman" and "Wild Flowers," both stories of homeless, hungry young married couples, and "Masses of Men," a Dreiser-like naturalistic story in which a destitute widow, desperate for food for her young children, sells the favors of her ten-year-old daughter to a stranger. There are also stories in which cruelty and violence are introduced much too gratuitously. In "The Growing Season" a father, crazed by heat and fear of losing his cotton crop, kills his moronic child. "Savannah River Payday" depicts the horrifying acts of two degenerate sawmill workers who transport the body of a dead Negro from the country to the undertaker in town. In "Thunderstorm" two suitors of a girl fight to the death with pitchforks.

Many of Caldwell's stories tend to a somewhat repetitious air because of his proclivity for writing variations on the same situation or theme. There are, for example, several versions of the "courting" story, which figures prominently in the tradition of American folk literature. For example, in "The Courting of Susie Brown," "An Autumn Courtship," and "A Day's Wooing" the complications involving bashful suitors, unwilling ladies, and so on, are at most only mildly amusing. But the taming-of-the-shrew situation in "Big Buck" is infused with a wonderful tall-tale humor which makes it one of Caldwell's best stories. For Big Buck's Bunyanesque counterpart, however, in "Candy-Man Beechum," another very good story, the outcome is less happy. Innocently passing through town on Saturday night to court his girl, he is accosted by the night policeman, who has decided that it will save trouble to lock up "niggers," and when Candy-Man resists, he is callously shot down.

Although Caldwell has remained a widely read author, his work receives very little serious critical attention today, and he is no longer regarded, as he once was in some quarters, as one of America's foremost writers of fiction. The novels and stories with which he established his reputation in the 1930's came to seem less impressive in their art and in the force of their social protest than they once did, and the fiction he has written in recent years has been regarded, with considerable justification, as clearly inferior to his best earlier work. Yet there is at least a small body of Caldwell's fiction that manifests considerable literary artistry, psychological insight, and moral significance. Few writers have been endowed with more humorous imagination or with more gift for oral tall-tale storytelling, and few have excelled him in providing examples of social injustice which can move us to sympathy, pity, and indignation.

An even more pronounced sociological emphasis is apparent in the novels and short stories of James T. Farrell (1904–     ). Born in Chicago, Farrell used the South Side of that city as the setting for his best-known work, the *Studs Lonigan* trilogy (*Young Lonigan: A Boyhood in Chicago Streets*, 1932; *The Young Manhood of Studs Lonigan*, 1934; *Judgment Day*, 1935), and for many of his other novels and short stories. The world of most of Farrell's fiction is not a pretty or happy one, since, as he has said, much of his writing has been concerned with portraying "conditions which brutalize human beings and produce spiritual and material poverty."

Farrell has been one of the most prolific contemporary short-story writers, having published during the past thirty years over two hundred stories, collected in twelve volumes. Most of his best stories are case studies of individuals—stories like "The Benefits of American Life," an objectively told but highly ironic and moving account of the "success" of a Greek immigrant boy who wins fame and money by competing in dance marathons in Chicago but contracts tuberculosis in the process. Other noteworthy stories of this kind include "A Front-Page Story," "A Jazz-Age Clerk," "The Fastest Runner on Sixty-first Street," "Father Timothy Joyce," and "They Don't Know What Time It Is."

It must be admitted that a number of Farrell's stories, especially his earlier ones, show the influence of other writers. Farrell has frequently treated everyday characters and the emptiness, vulgarity, or sordidness of their lives in the manner of Chekhov and Joyce; he has written some Sherwood Anderson–like studies of repression and frustration; and he has written still other stories which are reminiscent in various ways of Hemingway, Lardner, and Dreiser. It must also be admitted that Farrell's work in the short story is very uneven, that at times it is overly doctrinaire, and that it is sometimes undistinguished in form and style. But though one does not find the artistry of a Chekhov or Joyce in Farrell, one does find intensity and moral seriousness and, in at least a few of the stories, an ability to powerfully affect the reader.

LIKE CALDWELL AND Farrell, John Steinbeck (1902–68) is better known to the general reader as a novelist than as a short-story writer, and particularly as the author of *The Grapes of Wrath* (1939). His published work in the short story is much less extensive than that of either Caldwell or Farrell, consisting of fifteen stories in *The Long Valley* (1938) and a few uncollected stories. In a sense, however, Steinbeck has shown more of an inclination for the short-story form than his output suggests, since several of his ostensible novels —*The Pastures of Heaven* (1932), *Tortilla Flat* (1935), *Cannery Row* (1945), and *Sweet Thursday* (1954)—employ an episodic structure in which the episodes are sometimes virtually independent stories.

Most of the stories in *The Long Valley* are laid in Steinbeck's native Salinas Valley in central California. For the most part, however, they do not have a great deal of regional flavor, and their implications are more often psychological than sociological. "The Raid," for instance, although concerned with the same subject as *In Dubious Battle* (1936), Steinbeck's novel of striking California apple pickers, is more a psychological study than a proletarian story, concentrating on the reactions of a neophyte labor organizer who knows that the meeting at which he is to speak will be raided and that he will suffer a beating. "Vigilante," likewise, although it is about the lynching of a Negro, restricts itself mainly to analyzing the feelings of one of the participants. After the event he feels tired but satisfied, and when he returns home, his manner prompts his wife to say, "You been with a woman." Although he denies it, he admits to himself, "That's just exactly how I do feel."

Of three stories dealing with the relations of husband and wives—"The Murder," "The Harness," and "The White Quail"—the first makes clear that the brutal punishment a rancher inflicts on his young Slavic wife is the only effectual way to cure her of infidelity, and the last two are studies in repression in which the husband is dominated by the wife. In "The Harness," after his wife's death, a Monterey County farmer throws off the harness she had him wear for twenty-one years to hold in his stomach and maintain an erect posture, but this symbolic gesture is of no avail, since he finds that he cannot bring himself to change the old habits she had instilled in him. To Mary Teller's husband in "The White Quail" his wife is, like her carefully ordered garden, inscrutable and untouchable. In it she is aloof and secure from the outside world, symbolized by the thicket on the far side of the garden. Mary identifies herself with a beautiful albino quail that comes to drink at the garden pool. Her husband promises her that he will frighten away a cat that is stalking the bird, but he kills the quail instead. He tells himself that he had intended only to frighten it away and reproaches himself, but his action is really prompted by his need to get back at the wife who has condemned him to loneliness.

In contrast to Mary Teller, Eliza Allen, in "The Chrysanthemums," has energies and drives that make her want to act as a man would. Confident that she could live the life of the itinerant pot mender who repairs her pans and do his work as well as he, she is deflated when she discovers that he has thrown away the chrysanthemums he has asked her for merely to ingratiate himself with her, flowers that have a symbolical significance because she so vigorously and successfully cultivates them. We see her at the end of the story, overcome by frustration, "crying weakly—like an old woman."

In a very different vein is "St. Katy, the Virgin," an amusing parody of a medieval saint's legend, in which an evil pig is converted to Christianity and performs many good deeds and miracles of healing which result in its canonization. The story satirizes materialism and expediency in religion, but its tone is more playful than serious.

Although less unconventional in their technique and subject matter, "Johnny Bear" and "Flight" further demonstrate Steinbeck's versatility as a storyteller. A half-witted young man in a small Salinas Valley town who can "photograph words and voices," Johnny Bear eavesdrops on the private, intimate conversations and actions of the townspeople and without understanding their meaning reproduces them in the saloon, which serves as a kind of social club for the town's male citizens. It is with the attitudes of these men rather than the freakish Johnny Bear that the story is mainly concerned. They pay him to make his disclosures by buying him the whiskey he craves, their curiosity being stronger than their sense of embarrassment and shame, until on one occasion Johnny reveals that an artistocratic spinster in the town had become pregnant by a Chinese laborer and to conceal her condition had committed suicide with the tacit assent of her sister. These two women symbolized for the town everything that was good and respectable, and their secret must at all costs be kept from the town at large so that its sense of order and decorum will not be destroyed.

"Flight" tells of the nineteen-year-old Pepé who changes

from boy into man after he kills a man for calling him insulting names and who is then pursued into the mountains and finally shot to death by a mysterious avenger. At the end of his flight, having lost his horse and gun and suffering from a festering wound and lack of water, he is reduced to the state of an animal, but yet remains a man in stoically accepting his fate. Although the story as a whole may seem to be somewhat of a tour de force and its ending melodramatic, its symbolism is effective, and its setting and action are vividly realized.

The last four stories in *The Long Valley* are also about a boy, the ten-year-old Jody Tiflin, whose father is a Salinas Valley rancher. These stories, which also appear by themselves under the title *The Red Pony* (1945) and constitute perhaps Steinbeck's most artistic accomplishment, show Jody moving toward adulthood through learning about birth, old age, and death. In "The Gift," his father, a stern disciplinarian, gives him a pony, and Jody learns to accept responsibility in breaking and training it under the tutelage of the ranch hand Billy Buck. But the pony is inadvertently left out in a cold rain one day and contracts a sickness which Billy Buck, despite all his skillful and unremitting ministrations, is unable to cure. The dying pony breaks out of the barn one night, and Jody finds him the next morning just as buzzards are beginning to prey on the carcass. In a blind rage Jody catches and brutally kills one of the birds, giving vent to his feelings over his great loss, even though he knows, of course, as his father says, that the buzzards did not kill the pony.

"The Great Mountains" is somewhat similar in situation to Robert Frost's poem "The Death of the Hired Man." Rather than going to his relatives in Monterey, as Jody's father declares he should, a Mexican laborer who has grown too old to work comes back to the Tiflin ranch to die because he was born on its land. Told by Jody's father that he can only stay overnight, by the next morning he has disappeared, along with an old worn-out horse to which Jody's father had callously compared him. Later it is reported that the old man has been seen riding the horse toward the great mountains to the west. Jody knows that the old man has left behind his scanty possessions, except for a beautiful rapier inherited from his father, and remembers that the old man has described these, to Jody, mysterious and awesome mountains as "quiet" and "nice." The boy senses the old man's purpose and is filled with "a nameless sorrow."

In "The Promise," to compensate Jody for the loss of his pony, his father has a mare bred so that the boy can have a colt, which, however, Jody must earn by working all summer to pay for the breeding and by patiently taking care of the mare during the eleven months' pregnancy. Jody has small-boy dreams of glory of his colt turning into a magnificent black horse named Demon, but he also fears that something may go wrong when the mare's time comes, a dread that is borne out when Jody gets his colt as Billy Buck has promised him, but only after Billy is obliged to kill the mother in order to deliver it.

The last story, "The Leader of the People," focuses on Jody's maternal grandfather, who comes to the ranch for a visit. The old man had led a wagon train westward to the coast during pioneer days, and it is his habit to tell the same stories over and over again about his great adventure. Jody loves to hear the stories, involving Indians and crossing the plains, but his father is bored and irritated by them and says so to Jody's mother, not realizing that the old man is within earshot. The son-in-law is embarrassed and apologetic, but the grandfather tells him not to be. "Maybe you're right," he says. "The crossing is finished. Maybe it should be forgotten, now it's done." Later he says to Jody: "I tell those old stories, but they're not what I want to tell. I only know how I want people to feel when I tell them."

What the grandfather wants to communicate is the spirit that animated him and the people he led: "When we saw the mountains at last, we cried—all of us. But it wasn't getting here that mattered, it was movement and westering. . . . Then we came down to the sea, and it was done. That's what I should be telling instead of stories." Jody only vaguely understands all of this, but he can sense the pathos in his grandfather's situation, the implications of which are clear to the reader. Steinbeck's work as a whole may have philosophic shortcomings—notably a mysticism which attempts, not altogether successfully, to reconcile human ideals and aspirations with a universe of natural and social forces indifferent and even hostile to the individual—and he may also have suffered an artistic decline in most of the fiction he published after *The Grapes of Wrath*, as most critics have asserted, but through his art in that work and in his other earlier novels and in *The Long Valley*, he enables us to understand one another better, as he said he wanted all his writing to do.

LIKE STEINBECK, William Saroyan (1908– ) was born and grew up in California (his father was an Armenian immigrant who settled in Fresno). Although he had only a public-school education, he began early to read widely and write stories. By the age of twenty-five he had achieved considerable fame with the publication of *The Daring Young Man on the Flying Trapeze* (1934), a collection of highly subjective narratives, mostly about himself and his fellow Armenians.

The title story of this volume is made up of the thoughts of a young, unpublished writer during the Depression era who cannot find work to support himself and as a consequence starves to death. On the last day of his life he wakes in the morning from a dream reverie in which are jumbled together thoughts of those things that have given life interest and richness for him—the past, art, music, the beauty of nature, great writers. Dressing and shaving, he puts on his only tie and goes out in the street. There he finds a penny in the gutter and thinks ironically that, though it will not buy the food he needs, he can use it to weigh himself. There is no work at the department stores he visits, but he does not feel displeased or even personally involved: "It was purely an abstract problem which he wished for the last time to solve. Now he was pleased that the matter was closed." He spends an hour in the Y.M.C.A., where he helps himself to paper and ink, composing "An Application for Permission to Live," and another hour reading in the public library, all the while sustaining himself by drinking water copiously. Returning to his bare room, he amuses himself by polishing the penny he found until he begins to feel drowsy and ill, and, falling upon the bed, "swiftly, neatly, with the grace of the young man on the trapeze, he was gone from his body. For an eternal moment he was all things at once: the bird, the fish, the rodent, the reptile, and man. . . . The earth circled away, and knowing that he did so, he turned his lost face to the empty sky and became dreamless, unalive, perfect."

An impoverished, struggling young writer—clearly Saroyan himself—is found in several other stories in *The Daring Young Man*, the most notable being "Seventy Thousand Assyrians," "Myself upon the Earth," and "A Cold Day." In these pieces there is little if any story line. Instead, the young writer talks about himself and his love of man and of life, his deliberate disregard of the rules and conventions of story writing, his desire to be humble and to use language honestly, his poverty and other inhibiting forces on his writing. Because they repeat the same formula, the stories become a little tedious, but Saroyan on occasion compensates for intruding himself overmuch, as, for instance, in "Seventy Thousand Assyrians," which contains a genuinely moving portrait of a young San Francisco barber of Assyrian descent who is saddened because his once-great race is oppressed and dying out.

Besides some obviously Whitmanesque passages in these "writer" stories, the influence of other writers can be discerned elsewhere in the volume. "And Man" portrays the thoughts and emotions of an adolescent boy in a manner somewhat reminiscent of Sherwood Anderson. "The Earth, Day, Night, Self" employs a Joycean stream-of-consciousness technique to convey the impact of his surroundings on a small boy. The heavily ironic discussion of Hollywood melodrama and the terse, clipped dialogue of the two young men who visit a whorehouse in "Love" call Hemingway to mind. "Harry" is an ultrarealistic portrait in the manner of James T. Farrell of a go-getting young man who "could turn anything into money."

More to be deprecated than this occasional imitativeness are Saroyan's tendency to write too impressionistically and rhetorically and his inclination to seem to have little concern for form or coherence. Furthermore, his flippant and brash tone in declaring himself in the Preface to be a young rebel who has deliberately flouted all the conventional rules for writing fiction is not likely to ingratiate him with the reader. Nevertheless, *The Daring Young Man* is an interesting and individual book. Some of its attempts at experimental prose are very effective; setting and atmosphere are sometimes vividly evoked, especially in those stories whose backgrounds are the Depression years of the 1930's; and in some passages dealing with loneliness, suffering, and sorrow Saroyan, writing with uncharacteristic restraint, achieves great poignancy.

During the five years after the appearance of this first book, Saroyan published six more volumes of short stories, but none of them comes up to the quality of *The Daring Young Man.* His subjects and style remained much the same. He continued to write about unemployment, poverty, and other effects of the depression, although he was by no means Marxist in his social protest. He portrayed lonely, unfulfilled characters striving, usually unsuccessfully, to achieve satisfactory relationships with others. Sometimes, more optimistically, he wrote of love as providing compensation for the pain and boredom of existence and as giving beauty and meaning to life. Occasionally he wrote a humorous or even a farcical story—and frequently a trivial or banal one. Although he continued to be more or less indifferent to form, as time went on he did less declaiming and exhorting and exhibited a tendency to suggest his themes through character and action.

Having written about five hundred stories, according to his own account, during the period from 1934 to 1940, Saroyan, as a culmination to his efforts, published in the latter year what is probably his best volume of short fiction. Written out of his boyhood memories and experiences in Fresno, the stories in *My Name Is Aram* are concerned with the adventures and escapades of a boy of Armenian extraction named Aram Garoghlanian and with Aram's unusual relatives and his experiences with them. In "The Summer of the Beautiful White Horse," Aram and his cousin Mourad spend several glorious weeks taking surreptitious early-morning rides on a neighbor's horse, which they keep hidden in a deserted barn. In "The Circus," Aram and a friend play hooky from school, and an understanding principal gives them only token punishment. In "Locomotive 38, the Objibway," although he has never driven an automobile before, Aram is employed as a chauffeur by an eccentric Indian oil millionaire from Oklahoma. In each of these stories Saroyan seems to be making the point—with unusual restraint and unsentimentally—that adults can be sympathetic with children and want to contribute to their happiness. The neighbor who owns the horse pretends not to recognize the animal when he encounters the boys with it one morning and after its return declares that its condition is improved. The school principal knows that small boys cannot withstand the lure of the circus, and the Indian, it turns out, did not really need a chauffeur since he is himself an excellent driver.

Aram is an interesting and amusing character, but the most memorable portraits in *My Name Is Aram* are those of other members of the Garoghlanian tribe, notably three of Aram's uncles. There is the irresponsible idler, Uncle Jorgi, in "The Journey to Hanford," who is incapable of performing gainful labor but whose singing and beautiful zither playing are more important to the family than any money he might earn. There is Uncle Gyko, in "The Fifty-Yard Dash," who studies Oriental philosophy and lives like an ascetic, but who fails to achieve the transcendent state he hopes for. And finally there is Uncle Melik, in "The Pomegranate Trees," who spends thousands of dollars in an unsuccessful effort to make a garden bloom in the desert. Uncle Melik is such an impractical idealist that he may seem scarcely credible, but his story is a charming little fable of man's desire for beauty in a world that does not always provide it. Saroyan frankly exploits whimsy and eccentricity in these stories, but they are also utilized for something more than their own sake.

Although Saroyan has continued to be a prolific writer during the past quarter-century, he has devoted himself mostly to writing plays, novels, and autobiographical works. Only three of his many books published after 1940 are short-story collections. One of these, *Dear Baby* (1944), is a slender volume of stories and sketches written during the 1930's. The most interesting of the eleven pieces in *The Assyrian and Other Stories* (1950) are the title story, a novelette of seventy-four pages, and "The Cocktail Party." Neither is like the earlier Saroyan in tone and style, and both appear to be autobiographical, at least insofar as they reveal Saroyan's state of mind and attitudes at the time of their writing. The protagonists of both stories are writers, and of the first Saroyan said that it was a story he needed to write, presumably to purge himself of disillusionment and weariness. Life seems meaningless and full of unhappiness to the writer-hero, but he will keep on doing the best he can. The writer in the second story also feels apathetic about things generally, but he is not quite as pessimistic. The old Saroyan exuberance and optimism are also missing in most of the twenty-two pieces in *The Whole Voyald and Other Stories* (1956). Both this volume ("Voyald," according to Saroyan, was "a way of saying Void, Voyage, and World at the same time") and *The Assyrian and Other Stories*, it is worth noting, have prefaces, the first entitled "A Writer's Declaration," and the second, "The Writer on Writing," respectively. Both have the tone of apologia. Saroyan defends himself for being careless in his writing. To be careful, he felt, was to block creation, which he believed must always be impromptu and spontaneous. If his work appeared to lack form, he wrote, that was because he did not want to write short stories or essays but rather something in between. He disagreed with the critics who considered his writing unrealistic and sentimental; he was proud that his writing had always been free of inhibitions and restrictions; and he believed that he had always written his best. The familiar Saroyan personality, less flamboyant certainly than it once was but still recognizable, is reflected in these prefaces, and it is this personality which is responsible in large measure for giving the best of Saroyan's short fiction its vitality and individuality.

THE SHORT STORIES of Sherwood Anderson, Fitzgerald, and Faulkner written during the 1930's, as well as Katherine Anne Porter's, do not have the "social significance" of either Caldwell or Farrell, and this is also true of other good, if lesser, writers, such as John O'Hara (1905–70), who was concerned mainly with depicting manners and customs in the tradition of Sinclair Lewis, Lardner, and Fitzgerald. The first, and probably the best, of O'Hara's many novels, *Appointment in Samarra* (1934) provides both a sociological study of a small Pennsylvania city called Gibbsville (modeled on O'Hara's home town, Pottsville) and an account of the collapse and tragic end of Julian English, one of its prominent younger businessmen, who rebels against the monotonous, respectable life of his social milieu.

O'Hara's first collection of short fiction, *The Doctor's Son and Other Stories* (1935), also has some stories laid in Gibbsville and its environs, notably the long and, to a considerable extent, autobiographical title story, which portrays a boy of fifteen whose maturing is furthered by certain insights he gains into some of the less attractive aspects of life and sexual love. Most of the stories, however, as well as those in O'Hara's second volume, *Files on Parade* (1939), are of a different sort. Some are monologues or narratives in letter form in the vein of Lardner and Dorothy Parker, in which a character is made to expose his stupidity, insensitivity, or vanity. A good example is "Invite," a letter written by a sophomoric middle western undergraduate for the purpose of persuading a girl at Smith to come to his university (Northwestern?) for a weekend of fraternity parties and the like. Brief though it is, the piece is a masterfully ironic exposé of the inanity and vacuity that characterized much of the college social life of the thirties.

Other stories depict dull, crude, and sometimes vicious individuals, among them Broadway and Hollywood show-business people and jazz musicians. There is much irony in most of these pieces, as there is in the relatively few stories in which O'Hara treats his characters sympathetically. One of the best of the latter is "Over the River and Through the Wood," which portrays with genuine pathos an old man who is an alien in a young person's world. Several stories have themes or situations reminiscent of Fitzgerald. In "Price's Always Open" a town boy in a New England village is only apparently but not really accepted by the young people who summer there, and in "Trouble in 1949" a man and a woman who were once lovers try unsuccessfully to recapture the past.

Four of the sketches or stories in *Files on Parade* are letters written by a young night-club entertainer to his friend Ted. These stories, along with ten more like them, make up *Pal Joey* (1940). With their malapropisms, bad grammar and spelling, and slang, they would seem to derive most immediately from Ring Lardner's *You Know Me Al.* Joey, although perhaps somewhat more sophisticated, possesses much the same quality of egotism, vulgarity, brashness, and naïveté, despite a certain shrewdness in small things, as Lardner's baseball protagonist. If he is not an altogether admirable character, Joey is not contemptible either. His encounters with other people, his efforts to improve his situation, and his numerous contretemps make interesting and amusing reading, and it is not surprising they should have provided the basis for a highly successful musical comedy with music and lyrics by Richard Rodgers and Lorenz Hart.

After publishing two more collections of stories during the 1940's, *Pipe Night* (1945) and *Hellbox* (1947), which, though they contain some good pieces, show no real development over his earlier work, O'Hara did no more short-story writing for eleven years. Having produced in this interval *A Rage to Live* (1949), *Ten North Frederick* (1955), and several more novels, he found himself during the summer of 1960, according to his account, with what seemed to be an almost unlimited supply of story ideas and "an inexhaustible urge" to express them. As a result he wrote most of the twenty-six stories included in *Assembly* (1961). O'Hara's creative urge continued unabated, almost unbelievably so, and he published a volume of twenty-odd stories almost every year up to the time of his death in 1970. (*The Cape Cod Lighter*, 1962; *The Hat on the Bed*, 1963; *The Horse Knows the Way*, 1964; *Waiting for Winter*, 1966; *And Other Stories*, 1968).

Most readers are likely to find the later stories better, on the whole, than O'Hara's earlier ones. By and large they have more substance, more story quality, more interesting characters, more penetrating social observation, and more significant implications. Like O'Hara himself, many of his characters have grown older, and there is more concern than in the earlier short fiction with rendering thoughts and feelings, particularly those having to do with how a character has lived his life and what he has or has not made of it. These characters, especially when they suffer disappointment, deterioration, or defeat, are often presented so as to evoke our sympathy, but it should also be noted that the number of unsympathetic characters in the later stories is not inconsiderable—vulgarians, scoundrels, degenerates, unfaithful wives, philanderers, and worse are held up to view, though usually with more subtlety and restraint than in the often heavily ironic and sardonic exposés of such persons in the earlier stories.

In a number of the later stories O'Hara goes back to the Gibbsville of the twenties and thirties and even earlier, and some of them are very good. O'Hara had not forgotten what it was like in the old days, and he gave the reader much of the flavor of the times. Some of the stories, as well as a good many others whose time is more or less contemporary, treat of domestic relations, the characters often being married and divorced and involved in illicit love affairs. But O'Hara is probably consistently at his best in the many stories which are primarily character studies. These include "The Pioneer Hepcat," the history of a young jazz singer who dies of drink; "Claude Emerson," the story of a Gibbsville journalist; "John Barton Rosedale, Actor's Actor," the portrait of an actor who lives on his memories of the old days, being too proud to take smaller parts and lower salaries than those he once commanded; "Natica Jackson," a long case study of a young Hollywood actress; and several other particularly noteworthy stories. O'Hara obviously knew many different kinds of people and subjected them to close and careful scrutiny. He made them seem very real, not so much through penetrating deeply into them psychologically as by showing in detail what they say and do in relation to their environments and to the other characters in the story. Often he restricted himself to a minimum of auctorial exposition and description, relying heavily on dialogue instead to tell his story, and he had a very fine ear for the speech of his characters, no matter what their occupation or social station.

Throughout his long career O'Hara usually was not regarded as favorably by critics as he obviously believed he deserved to be, and he more than returned the compliment, carrying on his "war" against them with great gusto and not a little rancor. The general indictment against O'Hara contains a number of specific charges: He was a limited writer who was too flatly and literally realistic, too preoccupied with social distinctions and trivial details, and unable to transcend his realism as Hemingway did. He wrote too rapidly and discursively; he ought to have revised and polished more. His stories need more of such elements as humor, emotion, symbolism, imagery, mystery, and often they need more point. They may be vivid and plausible, but they do not have enough truth. There is undoubtedly some basis for these criticisms, but taken altogether they perhaps asked O'Hara to be something he was not, and they were sometimes urged more strongly than seems justifiable, perhaps because O'Hara deliberately provoked unsympathetic treatment and because his stories are not, as Granville Hicks suggested, the kind most influential critics prefer. Actually, O'Hara has also come in for a good deal of praise, albeit usually qualified, more for short stories than for his novels, in recognition that he possessed not inconsiderable talents, developed through long and assiduous practice, as a literary craftsman and social historian.

O'Hara was one of the early regular contributors to the *New Yorker*, and a good many of his stories of the sixties also appeared in that magazine. The question of the *New Yorker*'s influence, whether for good or ill, on the development of the short story during the past four decades is somewhat moot. It has been accused, though less so of late than formerly, of forcing its contributors into a Procrustean bed of writing stories to certain formulas, such as those of the brief, repertorial anecdote and the autobiographical reminiscence, with technique and a mannered, sophisticated style too often being emphasized at the expense of truly significant meaning.

To a certain extent this charge may be true, but more than a few *New Yorker* stories over the years give evidence of having been affected very little or not at all by these inhibitions, as a reading of the three omnibus collections of *New Yorker* stories published in 1940, 1950, and 1960 will show. Furthermore, the roster of its contributors is a remarkably diverse one, as is indicated by such names as Sherwood Anderson, Erskine Caldwell, Thomas Wolfe, Dorothy Parker, Irwin Shaw, Albert Maltz, George Milburn, Kay Boyle, E. B. White, J. D. Salinger, Shirley Jackson, Mary McCarthy, Frank O'Connor, Jean Stafford, Carson McCullers, Marjorie Kinnan Rawlings, J. F. Powers, Eudora Welty, Philip Roth, Saul Bellow, and John Updike.

CLOSELY IDENTIFIED with the *New Yorker*, especially during its early years, was Dorothy Parker (1893–1967). Of the writers of the 1920's and 1930's who produced stories on the order of those of Ring Lardner, only she came close to matching his telling irony and satire and his ear for recording common speech. Narrower in range than Lardner, she excelled in witty and humorous monologue and dialogue rather than in storytelling, as attested to by most of the pieces in her two collections of sketches and stories, *Laments for the Living* (1930) and *After Such Pleasures* (1933). Undoubtedly her finest story is "Big Blonde," a trenchant portrait of a shallow woman devoid of any inner resources who becomes an alcoholic.

JAMES THURBER (1894–1961), a long-term *New Yorker* staff member and regular contributor, but primarily a humorous and satirical essayist and cartoonist, deserves particular notice as the author of two classics of contemporary short fiction and of a third story which, though less well known, rivals them in originality and skillful execution. "The Secret Life of Walter Mitty" portrays a typical Thurber ineffectual, wife-dominated male who escapes the boredom and frustrations of reality by daydreaming himself into various adventurous, heroic roles. "The Catbird Seat" is a wonderfully droll and clever variation on the "perfect crime" story. Its mild-mannered, old-maidish, filing-clerk protagonist puts to rout with an inspired ruse an obnoxious female efficiency expert who threatens his job security. Likewise ironic, but grimly instead of humorously so, is "The Whip-poor-will," a chilling but effectively developed story of a neurotic husband driven to committing multiple murder and suicide when his condition is exacerbated by the maddeningly recurrent cry of a whippoorwill and gains no sympathy or understanding of his state from his wife and servants.

ANOTHER CONTEMPORARY of O'Hara and Thurber who has often appeared in the *New Yorker* is Kay Boyle (1903– ). Versatile and prolific, she published in 1931 the first of her many novels, *Plagued by the Nightingale*, the story of an American girl who marries into a French family. Her first volume of short stories, *Wedding Day and Other Stories* (1930), was followed by two other collections, *First Lover and Other Stories* (1933) and *The White Horses of Vienna and Other Stories* (1936). *Crazy Hunter* (1940), a short novel which portrays a seventeen-year-old girl striving to find herself and to achieve satisfactory relationships with her mother and father, is probably her most notable piece of longer fiction.

Miss Boyle's early stories, published mostly in avant-garde magazines, sometimes seem attentuated and show her straining perhaps too much for experimental effects and for too poetic a style, but at least several are very effective, for example, "Rest Cure," which pictures a writer dying on the Riviera, clearly modeled on D. H. Lawrence. She lived in Europe for almost twenty years (1922–41), and many of her stories are laid in France, Austria, and Germany. In some of them the European background is not important, as in the poignant "Natives Don't Cry," in which an English governess traveling with a family in Austria compensates for her dull life by inventing an imaginary lover. More often, however, these stories reflect European political and social tensions of the period between the two world wars. Examples are "Effigy of War," a 1940 *New Yorker* story, which, though its specific situation has to do with an ugly, vicious manifestation of wartime superpatriotism in a French resort town, is so treated as to give it a universal application; and her two first-prize O. Henry Memorial Award stories, "The White Horses of Vienna" (1935), laid in Austria at a time not long before the Anschluss, and "Defeat" (1941), which portrays dramatically and with bitter irony the humiliation of the French after their country fell to the Germans. "Winter Night," one of Miss Boyle's most moving and artistic stories, although laid in New York City, also has political implications, saying as it does that the world can be a terrible place for children, both for a little American girl denied mother love and for European children in a time of war and concentration camps.

Miss Boyle's most notable work was published during the 1930's and 1940's. Her most significant fiction since then consists of a volume of short stories, *The Smoking Mountain: Stories of Germany during the Occupation* (1951), and a novel, *Generation Without Farewell* (1959), both dealing with the relationship between conquerors and conquered in the American Zone of Germany and demonstrating the theme that although they can superficially adjust they cannot really relate to each other. Here, as earlier, Miss Boyle has written in the belief, as she has put it, that "an equally important responsibility [in addition to holding the reader's interest] of the short story is that it attempt to speak with honesty of the conditions and conflicts of its time." That she has accomplished this aim, often with high art, is attested to by the praise she has been given by, among others, another distinguished stylist and literary craftsman, Katherine Anne Porter.

# 14
# Symbolism and Sensibility

## Katherine Anne Porter

THE BEST OF the short-story writers to come into prominence in the 1930's is Katherine Anne Porter (1894– ). Miss Porter was born at Indian Creek, near San Antonio, Texas. Although she has written during a long literary career only a relatively small body of fiction—three volumes of stories and one novel, *Ship of Fools* (1962)—she is, as Robert Penn Warren has remarked, a writer whose artistry and originality entitle her to rank with such acknowledged masters of the modern short story as Joyce, Katherine Mansfield, Anderson, and Hemingway.

*Flowering Judas* (1930, enlarged edition, 1935), Miss Porter's first volume, contains ten stories, four of which have their settings in Mexico, where the author lived and traveled at various times between 1920 and 1931. "María Concepción," her first published story (1923), which, she said, was "rewritten fifteen or sixteen times," portrays the primitive attitudes of Mexican villagers toward revenge, love, and death. Juan Villegas, a likable but irresponsible young peasant, to escape the boredom of digging for an American archaeologist as well as the responsibilities of family life, deserts his wife, María Concepción, and goes off to join the army, taking his mistress, María Rosa, with him. When he returns, tired of military life, he is saved from being shot as a deserter through the intervention of his archaeologist employer. At the same time Juan becomes a father when the pregnant María Rosa gives birth to a son, but his felicity is short-lived since the betrayed María Concepción murders María Rosa with the knife she uses to kill the chickens she raises for a livelihood. His primitive code demands that Juan protect his wife, and he rises to the occasion, seeing that all the incriminating evidence is destroyed and fabricating stories for them to tell the police, which, along with testimony by the villagers in behalf of María Concepción, absolves her of any apparent part in the crime.

Everything in the story is artistically treated, particularly the characterizations of both Juan and María Concepción. Juan, who might well have been a stereotype, is entirely convincing, and María Concepción is portrayed with great psychological insight. Her suffering, caused by the loss of her husband and the death of her child shortly after its birth, is made very real, and the motivation for her deed is so presented as to make it seem inevitable as well as justifiable in terms of her primitive nature and heritage. Juan, ironically, will suffer a kind of death in having to go back to his dull labor of digging, whereas María Concepción will find a new life and happiness in taking over as the mother of María Rosa's newborn son.

Two other early Mexican stories (which Miss Porter did not include in the *Flowering Judas* volume but which appear in her *Collected Stories*), "The Martyr" and "Virgin Violeta," are slight and not particularly distinguished. "That Tree," however, is notable for the self-revelation it provides in its monologue of a man who had wanted to live an easy, carefree life in Mexico as a Bohemian poet but who became, through an ironic set of circumstances, a highly successful journalist. "Hacienda," a short novel concerned with a Russian film director making a movie on a Mexican plantation, while not as successful as the three stories of comparable length in Miss Porter's *Pale Horse, Pale Rider*, has some good character portraits and some fine humor and irony.

The best of the Mexican stories, and beyond doubt one of the finest of all twentieth-century stories, is "Flowering Judas." This character study of a young woman who "cannot help feeling that she has been betrayed irreparably by the disunion between her way of living and her feeling of what life should be," despite the fact that its conflict and theme are made very explicit, conveys its meaning and achieves its comic, ironic, and poignant effects through style and symbol in a highly subtle and complex way. Laura, an idealistic American girl with a Roman Catholic background, has gone to Mexico to work for the Marxist revolutionary cause, but her experience has disillusioned her. The well-fed, self-loving, expedient, yet skillful and courageous revolutionary leader Braggioni symbolizes for Laura her many disillusions, since she has believed, though she now knows better, that "a revolutionist should be lean, animated by heroic faith, a vessel of abstract virtues." But Laura's predicament is clearly caused by something more. "She is not at home in the world" because she cannot love. Having lost her religious faith, she cannot feel divine love. She cannot respond to Braggioni or the handsome army captain or the romantic youth who stands in her patio and sings to her, all of whom offer her erotic love. She cannot love humanity in the sense of feeling a genuine commitment to her revolutionary activity, and even the children she teaches daily and who write on the blackboard,

"We lov ar ticher," are as strangers to her. To everyone and everything she must say no.

What is responsible for Laura's inhibitions? Are they the result of her early religious training, her seeming inability to believe in anything, a temperament which somehow makes her incapable of all normal feeling, or perhaps a combination of all these? Laura is a betrayer not only of others but also of herself, as she comes to realize in her nightmarish dream at the end of the story. In it the figure of Eugenio, the young revolutionary who took an overdose of the drugs she brought to him in prison, appears to her and offers her the flowers of the Judas tree, the symbol of betrayal, and when she greedily eats them, he calls her "Murderer" and "Cannibal." This vividly rendered and, for Laura, terrifying vision is the culmination of Miss Porter's use of much subtle yet powerfully ironic religious imagery and symbolism in a story which seems perfect in its every aspect.

Although none of the six remaining stories in the *Flowering Judas* volume can match the superlative title story, they are all very good ones and, furthermore, exhibit a remarkable variety of subject and narrative style. "Magic" is a brief but subtle sketch of the brutal treatment a young prostitute suffers in a New Orleans brothel. "He" portrays the attitude of a mother toward her mentally defective son. Although the mother protests too much for us to believe her when she says she loves him, and although we may deplore her false pride and hypocrisy where he is concerned, the achievement of the story is such that it evokes an ambivalent attitude toward her in the reader, making it impossible to condemn her altogether. "Rope," although something of a tour de force, is a wonderfully comic and satirical depiction of a domestic altercation a young married couple have over a trivial matter, catching the essence of how a foolish quarrel can come about and be needlessly prolonged. "Theft," although ostensibly about the theft of a purse from a woman by the janitress of her apartment building, implies that life involves all sorts of theft and loss, the heroine learning that the greatest deprivation of all comes when one steals from oneself through being unable to love others.

Loss is also the subject of "The Jilting of Granny Weatherall," one of Miss Porter's best stories. In this case the heroine is a woman of almost eighty who lies on her deathbed on the last day of her life. Only at intervals is she aware, and then mostly in a confused way, of the relatives, the doctor, and the priest at her bedside, her past life as it unfolds in her memory being much clearer to her. Widowed early, she had been left with children to rear, and through fortitude, business sense, and hard work she had provided them and herself with a good life. But though she had had this satisfaction and had loved her husband, she has not been able in sixty years to forget that her first lover jilted her the day they were to be married. No matter how much she protests to herself, the hurt is still there, and it is compounded when she realizes at the end of the story that she is dying and that in again being taken by surprise she has suffered a second jilting.

One of the great virtues of "The Jilting of Granny Weatherall" is that Miss Porter avoids sentimentalizing her character, and this is true also of the portraits of Dennis and Rosaleen, the Irish couple in "The Cracked Looking-Glass." Dennis O'Toole, a widower and headwaiter in a New York hotel, had married Rosaleen, a young chambermaid thirty years his junior. Now, twenty-five years later, they are living in retirement on a little farm in Connecticut, and he has been for some time too old to be a husband to her. What the story brings out with both humor and pathos is that Rosaleen, despite her romantic temperament, which causes her to gloss over her situation in various ways, is able finally to abandon her illusions and self-deceptions and to accommodate herself to her life as it is, and even to find a measure of happiness in it.

Every bit as memorable as the characterizations of Rosaleen, Granny Weatherall, and Laura is that of Mr. Thompson in "Noon Wine," first published in 1937 and reprinted in Miss Porter's second collection of short fiction, *Pale Horse, Pale Rider* (1939). A small south Texas dairy farmer, who is incompetent and lacking in energy and who considers much of the farm work he has to do beneath his dignity, he finds it increasingly difficult to make even a bare living and has about resigned himself to failure. But although he is "a noisy proud man," much too concerned about the way he appears to others, he is by no means an altogether unsympathetic character. In his blustering way he is very fond of his wife, who returns his affection, and he does not hold it against her, but rather against his fate, that her ill health requires him to do the churning and other chores he regards as woman's work.

Thompson's situation is greatly improved, however, after he hires Olaf Helton, an itinerant Swedish farm worker from North Dakota. Although he is taciturn and likes to keep to himself and his demeanor is sometimes strange, Helton is capable and hardworking, and before long, through his efforts, the farm begins to pay. As a result Thompson can enjoy a feeling of importance and a measure of prosperity he had never supposed possible. But after nine years all this is lost when Homer T. Hatch comes to the farm to apprehend and return Helton to North Dakota, revealing to Thompson that the Swede is a murderer who had escaped from an insane asylum. Thompson feels an immediate dislike for the far-from-ingratiating Hatch, which turns into resentment and anger when Hatch, after conversing about other matters, finally reveals his purpose. The encounter ends with Thompson suffering what appears to be a hallucination. Thinking he sees Hatch stabbing Helton in the stomach when the latter suddenly appears on the scene, Thompson smashes Hatch's skull with an ax.

Thompson is tried and acquitted, but he does not feel absolved of guilt. He was not able to say what he wanted to in his defense, his lawyer's strategy preventing him from either telling fully or altogether truthfully what happened. He feels compelled every day for a week following the trial to visit all his neighbors, taking his wife with him, to protest his innocence and to call on her to testify, though she had not been a witness to the deed, that he did not deliberately kill Hatch. No one, however, will believe him, and, what is worse, his wife and his sons also believe him guilty. Knowing no other way to prove that he is innocent of premeditated murder, and not really certain that he is, he commits suicide.

Whether Thompson should be regarded as a tragic and heroic figure, as one critic has suggested, is questionable. Certainly he is made vulnerable by his pride to the circumstances which engulf him by, as we are told early in the story, "the appearance of things, his own appearance in the sight of God and man." Besides the portrait of this simple man, who yet has a complexity which gives him and his situation a universal application, there are other aspects of this very good story which deserve mention: its detailed and convincing description of farm life, its warmly human yet unsentimentalized depiction of the family life of the Thompsons, the convincing rendering of the idiom of the characters, and the long, brilliantly written scene between Thompson and Hatch. Although the characters and setting are not the kind Miss Porter treats in her other stories, she writes about them with complete authority.

The other two novelettes in the *Pale Horse, Pale Rider* volume, "Old Mortality" and the title story, are concerned with the history of the girl Miranda, a semiautobiographical character who figures in Miss Porter's short fiction in much the same way, if not as extensively, as Nick Adams does in Hemingway's. Like Miss Porter, Miranda belongs to a southern family, very conscious of its heritage, which had moved from Kentucky to Louisiana to Texas. She goes to a convent school, runs away from school to get married, works as a reporter on a Denver newspaper, and almost dies during the World War I influenza epidemic. These correspondences in themselves are not, however, important qualities in the stories. What is important is what Miranda's experiences teach her about herself and about life, as she comes to maturity.

In Part I, dated 1885–1902, of "Old Mortality," we see Miranda and her sister Maria, little girls of eight and twelve, eagerly absorbing details about persons and episodes in the family past which have taken on a legendary quality, particularly the sad but romantic story of their Aunt Amy. Beautiful, willful, and consumptive, she had been known to dance all night three times in one week, had capriciously broken two engagements, had very nearly had a duel fought over her, had suddenly married Gabriel, her devoted second cousin, after making him wait five years, and had died, in New Orleans only six weeks later from an overdose of medicine. Uncle Gabriel's heart was broken, of course, even though he had married again quite soon.

Young though they are, Miranda and Maria can perceive certain discrepancies between what they are told and what appear to be the facts, the result of love of legend which causes their father and other grown-ups around them to romanticize the past. The disillusionment the little girls experience when legend and reality are juxtaposed is dramatically illustrated in Part II, dated 1904, when their father visits them at their convent school in New Orleans and takes them to the races, where their Uncle Gabriel is running a mare. This presumably handsome, romantic husband of Aunt Amy, whom they now see for the first time, is "a shabby fat man with bloodshot blue eyes, sad beaten eyes, and a big melancholy laugh, like a groan," who weeps drunken tears after his horse, a hundred-to-one shot, wins her race. Gabriel insists on taking the girls and their father to his rundown hotel for an uncomfortable meeting with his second wife, the taciturn and bitter Miss Honey, and Miranda knows that the woman loathes and despises them all, including her husband.

In Part III, dated 1912, we find Miranda, now eighteen and married after running off from school and eloping, returning home for Uncle Gabriel's funeral. On the train she encounters a middle-aged woman who turns out to be Cousin Eva Parrington. A contemporary of her father, a former teacher, and an ardent feminist who knew Miranda when she was little and who read Latin to her, Cousin Eva, as Miranda's family had often reminded her, had been a homely girl with a receding chin and no belle like Aunt Amy—about whom she has much to say. Cousin Eva, in effect, demolishes the legend of Aunt Amy, maintaining that in reality she was spoiled, made people suffer, had a bad if undeserved reputation, treated Gabriel badly on their honeymoon, and in all probability committed suicide because she was in some kind of trouble. Girls like Amy, Cousin Eva avers, conducted themselves as they did—rode too hard, danced too much, contested with each other for the attention of suitors—because they were driven by sex, even though they pretended not to know about it. To Miranda, Cousin Eva's interpretation of the past seems no truer or more objective than the romantic version she had been given earlier. She resolves to free herself from family and past, to make her own way, and, "in her hopefulness, her ignorance," assures herself that she "can know the truth about what happens to me."

In "Pale Horse, Pale Rider" we find Miranda, now twenty-four, working as a reporter on a Denver newspaper. World War I is in progress, and Miranda feels emotional insecurity and a sense of malaise brought on by the confusions and tensions of the time. She has also developed an affection for Adam, a young army officer from Texas, but their brief period of happiness together is terminated when she succumbs to influenza during the 1918 epidemic. Given up for lost by the doctors, she recovers just as the armistice comes, but Adam in the meantime had caught the disease from her and died, and it will be a while before Miranda can feel that it is good to be alive.

This very moving story is one of Miss Porter's most subtle and complex, as well as most artistically structured. It begins with Miranda having a nightmarish dream and ends with her delirium brought on by illness, in both of which she is confronted by Death, and much of Miranda's waking experience as well is like a bad dream. Everything, except perhaps her love for Adam, seems unreal: the pressure put on her by super patriots to buy a Liberty Bond when she cannot afford one out of her meager salary, her feeling of ineffectuality when she visits the hospital to cheer wounded soldiers, her dreary job of reviewing plays and vaudeville performances, her premonition that Adam will die. In her premonitory dream she rejects the lank, greenish stranger who rides with her, recognizing him as Death and ordering him to ride on without her. But she and Adam admit his power and presence when they sing the old Negro spiritual "Pale horse, pale rider, done taken my love away." Miranda adds that death always leaves one lover to mourn, and there is the poignant spectacle, after she has won her victory over death, of Miranda being that bereaved lover.

Six of the nine stories in *The Leaning Tower and Other Stories* (1944) have to do directly or indirectly with Miranda as a little girl. "The Source" and "The Witness" are brief, beautifully executed vignettes dealing, respectively, with her grandmother and with the old former family slave Uncle Jimbilly. "The Old Order" adds substantially to the portrait of the grandmother and provides in addition a warm and touching but unsentimentalized picture of the close relationship, which began when the two were little girls, between her and the old former slave Aunt Nannie. After the death of the grandmother, which occurs at the end of this story, we find Aunt Nannie, in "The Last Leaf," leaving the family house to pass her remaining days in a little cabin given her by Miranda's father, having earned the right, she believes, to independence, rest, and solitude.

Miranda is the central figure in "The Circus" and "The Grave," as well as in "The Fig Tree," which was accidentally omitted from *The Leaning Tower* but later included in the *Collected Stories*. All three stories are incidents in which Miranda, although she is only nine or so, makes important discoveries about human nature and experience, the most meaningful having to do with birth and death in "The Grave."

The remaining three stories in *The Leaning Tower* are very different from the Miranda stories and from one another as well, except that their view of life in each case is by no means a happy one. In "The Downward Path to Wisdom," given no real affection by his quarreling parents and never sure how they and the other insensitive adults around him will react to what he does, the four-year-old boy Stephen decides that he hates them all. "A Day's Work" is another

Irish story like "The Cracked Looking-Glass," dealing with a husband-wife relationship, but here there is no underlying affectionate feeling to compensate for the incompatibility of the middle-aged couple, he blaming his wife for his failure to achieve success as a ward politician, she shrewish and puritanical, and both filled with vindictiveness and recrimination.

Another unpleasant view of humanity is provided in "The Leaning Tower," which describes the experiences of Charles Upton, a young painter from western Texas who goes to Berlin to study and practice his art. The year is 1931, the season is winter, and the city turns out not to be the wonderful romantic place he had been led to expect from the descriptions of a childhood friend of German descent. Germany is in the depths of a depression, and there is much misery and suffering; he finds something porcine and otherwise inhuman about the fat older Germans he sees in the streets; there is a general suspicion and distrust of foreigners; and Charles feels himself very much an alien. He does achieve a certain rapport with his Viennese landlady, who had once known better days, and with her three other young men lodgers, a Heidelberg student who has come to Berlin to have an infected dueling wound treated, a young Polish pianist, and a University of Berlin mathematics student of peasant stock. Yet even with them Charles remains conscious that there is a gulf between nationalities and that the Germans feel animosity toward America for having contributed to their defeat in the war. Briefly, during the New Year's Eve drinking party which they attend at the conclusion of the story, Charles feels that he likes all the people there, but later, back in his room, still a little drunk, he senses something threatening around him and experiences "an infernal desolation of the spirit, the chill and knowledge of death in him."

"The Leaning Tower" effectively renders Charles's complex, ambivalent reactions to his surroundings, even if, as some of Miss Porter's critics have felt, its symbolism is perhaps too obtrusive, its characters somewhat wooden, and its view too strongly anti-German—this last criticism also later to be made of *Ship of Fools*. Such weaknesses are not to be found, however, in "Holiday," a well-nigh perfect short novel. It was not published until 1960, although it was composed, except for minor revisions, about twenty-five years earlier. Its first-person narrator, a young woman who is not identified but who might well be Miranda, having troubles she feels she must get away from, goes to stay for a month's "holiday" with the Müllers, a "family of real old-fashioned German peasants, in the deep blackland Texas farm country, a household in real patriarchal style." Besides Father and Mother Müller, there are two married daughters and their husbands, three sons, several grandchildren, and a third son-in-law who joins the family late in the story when he marries the youngest daughter. Although we are shown the father, mother, and youngest daughter more fully than the others, the members of the family are so much alike in appearance and attitudes that the only person who seems an individual to the narrator is the crippled, dumb servant girl Ottilie, who performs the Herculean labor of cooking and serving the three enormous meals needed daily to sustain the family. Although ignored and unacknowledged as a member of the family, Ottilie, whose condition was caused by a physical accident in childhood, is actually another daughter, and after the narrator inadvertently learns this, she sees that logically it is right and necessary for the others to have forgotten their relationship to Ottilie, that "they with a deep right instinct had learned to live with her disaster on its own terms, and hers; they had accepted and then made use of what was for them only one more painful event in a world full of troubles, many of them much worse than this." Emotionally, however, the narrator is unable to accept this truth fully until the end of the story, when a further experience with Ottilie dramatically and ironically indicates, as her family knows, that she is beyond the reach of others. The story becomes profoundly meaningful because of the narrator involvement in it, but its richness is the result of several elements, all given their due emphasis but kept in perfect balance: the pictorial detail, the regional flavor, emotion without sentimentality, an eloquent yet simple style, philosophical significance without moralizing or sententiousness. "Holiday" is the work of a very great artist in fiction.

# 15
# The Short Story Since 1940

## Eudora Welty
## Mary McCarthy
## Jean Stafford
## J. F. Powers
## J. D. Salinger
## Bernard Malamud
## and
## Flannery O'Connor

WHAT HAS BEEN the development of the short story in America since 1940? Does it constitute perhaps, as William Peden, its principal historian during the period, has averred, the major literary achievement of the decades since World War II? Or has it, on the other hand, become mainly a reflector of middle-class attitudes and emotions, as Alfred Kazin has argued, with its practitioners, though often talented and skillful, depending perforce on style and manner to interest the reader, since their subjects deal with the problems we already know all too well and have thought and felt much about—relations between parents and children, relations between husbands and wives, loneliness and disillusionment, the search for identity, and so on. Several contemporary short-story writers of talent have treated these knotty issues with originality and individuality. Among them—all writers of firmly established reputation—are Eudora Welty, Mary McCarthy, Jean Stafford, J. F. Powers, J. D. Salinger, Bernard Malamud, and Flannery O'Connor (for commentary on other contemporary short-story writers, see the Appendix.)

UNDOUBTEDLY ONE of the most gifted short story writers of our time is Eudora Welty (1909– ). Except for relatively brief periods Miss Welty has lived in Jackson, Mississippi, where she was born, and her home region of southern Mississippi has supplied her with the settings for most of her stories. She has published four volumes of short stories; a novelette, *The Ponder Heart* (1954); and two novels, *The Robber Bridegroom* (1942) and *Delta Wedding* (1946).

Miss Welty's first collection of stories, *A Curtain of Green* (1941), is an impressive one. Its seventeen stories exhibit, as Katherine Anne Porter has said, "an extraordinary range of mood, pace, tone, and variety of material." There is whimsical humor in "Lily Daw and the Three Ladies," in which three middle-aged women abandon their designs to send a half-witted girl who is their protegée to the state institution for the feeble-minded and instead marry her off to a traveling xylophone player. More ironic than humorous is "A Visit of Charity," in which a campfire girl takes flowers to an old ladies' home to earn points and finds her visit with two elderly women a gruesome experience. Humor, irony, and caustic satire are blended in "Petrified Man," a dialogue between a garrulous beauty-shop operator named Leota and her customer, Mrs. Fletcher. The main subject of their conversation is Leota's friend Mrs. Pike, who sees a picture in one of Leota's magazines of a man wanted for raping "four women in California, all in the month of August" and who collects a $500 reward for recognizing him as the petrified man in a freak show running in town. The purpose of the story, however, is not to exploit this grotesque situation, but rather to lay bare the shallow, vulgar natures of the women. Like Ring Lardner, Miss Welty has a marvelous ear for the natural idiom of her vulgar characters, a talent exhibited again in "Why I Live at the P.O.," a comic monologue of a girl whose jealousy of her younger sister amounts to a persecution complex. Obsessed with the idea that all the members of her family are against her too—her mother, her uncle, and her grandfather, who got her a job as the local postmistress—she avenges herself by carrying off all her belongings and taking up residence in the post office.

Some of Miss Welty's presentations of abnormal or grotesque characters and situations have little or no leavening of humor and satire. In "Keela, the Outcast Indian Maiden" a young man named Steve, who has been a carnival barker, tells how Little Lee Roy, a clubfooted Negro, was abducted by the operator of the carnival, who painted and dressed him to look like an Indian woman and made him eat live chickens as an act in the show. He was also forced to growl and threaten to strike the spectators with an iron bar if they came too close. But one day in Texas a compassionate stranger took pity on this poor debased creature and notified the authorities, who freed him and sent him back to his home in Mississippi.

Steve is haunted by the fact that he did not realize until the stranger came along that he had been an accessory to a horrible exploitation of another human being. He feels compelled to seek out Little Lee Roy. At the beginning of the story we find him, with the Negro and Steve's guide, the cafe owner Max, as listeners, recounting what had happened and how it has affected him. Max listens patiently, but he is puzzled, and when the frustrated Steve finally hits Max in the jaw because he cannot make him understand, the tolerant Max does not take umbrage but instead asks: "What you want to transact with Keela? You come a long way to see him." Little Lee Roy's reaction is even more ironic. He smiles and at intervals laughs with amusement at some of the most horrible points in the anguished Steve's story. In answer to Max's question he says that he does not remember the stranger who came to his rescue, but he adds happily that he "remembas" Steve. Childlike and without moral feeling, he can recall his experience, but he has never comprehended his degradation. It is not, however, the degradation, terrible though it is, that affects us most in the story. Instead it is the fact, terrible for Steve, that he is denied the understanding and forgiveness he needs to free him from his sense of responsibility and guilt.

Three of Miss Welty's stories exhibit her ability to use very different moods and tones in conveying the essence of a character. Bordering on fantasy and narrated in the style of a children's story is "Old Mr. Marblehall," which portrays a sixty-year-old man who tries to counteract the dullness and insignificance of his existence by leading a double life (or imagining he does?) with two different wives in different parts of town. "Powerhouse," a vivid portrait of a Negro jazz pianist, is written in a somewhat similar style, but its tone and pace are very different, the story throughout exuding the vitality, energy, and dynamic quality of the man. The story also has great poignancy, as does "Clytie," a study of an eccentric spinster. A drudge for her family—consisting of a dominating older sister, a paralyzed father, an alcoholic brother, and another brother who shoots himself—and long denied the love and normal human relationships she needs, she finally has a revelation of the grotesque she has become and drowns herself.

Pathos and tragedy appear in several other stories—in "A Curtain of Green," with its bereaved widow who works all day in her garden, in order, as it were, to raise a curtain that will blot out the memory of the accident which killed her husband; in "Flowers for Marjorie," with its young husband who in desperation and despair because he cannot find work kills his young wife; and in "Death of a Traveling Salesman," to whose protagonist illness brings heightened sensibility and a resulting awareness of his isolation from others. There is pathos also, but without tragedy, in the beautifully written "A Worn Path," with its very old and feeble Negro woman who walks to town from the country at regular intervals to get "soothing medicine" for her little grandson.

In the eight stories in *The Wide Net* (1943), Miss Welty managed to achieve even further range and variety in subject and style. Two remarkable stories utilize American history and legend: "First Love," in which Aaron Burr is a principal character; and "A Still Moment," which pictures the naturalist John James Audubon. Fantasy, the supernatural, mythology, allegory, and realism are blended variously, although not always entirely successfully, in "Asphodel," "The Winds," "At the Landing," and "The Purple Hat." "The Wide Net," awarded the first-prize O. Henry Memorial Award in 1942, subtly combines mythical, tall-tale, and realistic elements in relating how a young husband and his friends drag the river for his pregnant wife, mistakenly believing that she has drowned herself because he got drunk and stayed out all night. An even better story is "Livvie," also an O. Henry first-prize winner in 1943, in which an old Negro marries a young girl and takes her to live deep in the country up the Natchez Trace. He has a "nice" house, but it is isolated, and Solomon, Livvie's husband is so old that he "wears out" and has to stay in bed. The outside world comes to Livvie when a cosmetic saleswoman calls, but Livvie has no money to buy her wares. Soon afterward comes Cash, formerly Solomon's fieldhand, but now decked out in flashy city clothes. Solomon dies, but not before asking God for forgiveness in taking too young a girl for a wife and keeping her prisoner, and Livvie goes off with Cash. Its fairy-tale tone, brilliant descriptive detail, symbolism, and poetic prose make the story an outstanding example of the kind of modern short fiction which is more akin to the lyric poem than to the traditional short story.

The seven stories in *The Golden Apples* (1949) deal mostly with various members of the "main families" in the fictitious small town Morgana, Mississippi. Perhaps the most interesting character is King MacLain, whose history is recounted in the first story, "Shower of Gold," by the garrulous Mrs. Katie Rainey. Like Zeus, to whom he is comically likened, King is a great lover and seducer. A traveling salesman who sells tea and spices and is a figure of mystery and legend, he appears also in "Sir Rabbit" and in the epiloguelike final story "The Wanderers," having come home finally in his sixties to stay after many years of wandering. Two other stories are about King's twin sons, Randall and Eugene, after they have reached manhood. "The Whole World Knows" is a soliloquy by Randall revealing the effect on him of his young wife's infidelity. "Miser from Spain" recounts an unusual episode in Eugene's life while he is living in San Francisco and has to do also with his relationship with his wife. The longest and best of the stories, "June Recital," is concerned mostly with the unhappy history of the town's piano teacher, Miss Eckhart. Somewhere between a novel and a collection of stories, *The Golden Apples*, although less impressive than her two earlier story volumes, again reflects Miss Welty's virtuosity and originality in its use of myth, symbolism, and literary allusiveness to give varied representations of what it means to be lonely, to love, and to become aware.

*The Ponder Heart* is the most broadly comic of Miss Welty's stories, and it fully succeeds in achieving its apparent aim of being simply an amusing entertainment. The longest of Miss Welty's several monologues, its narrator is the garrulous, shrewd Miss Edna Earle Ponder, who tells about her Uncle Daniel Ponder, a lovable eccentric who loves people and loves giving away everything he has. Generally regarded as simple minded and committed for a time to an asylum by his father, Uncle Daniel later has a prolonged and complicated trial marriage with a stupid, vain young girl he found working in a ten-cent store, a relationship concluded by her accidental death and his trial for her murder. But the love he epitomizes prevails, and he is acquitted.

*The Bride of the Innisfallen* (1955) differs from Miss Welty's earlier short-story volumes in that four of its seven stories have settings outside Mississippi. It differs, too, in that, taken as a whole, the stories display even more of the elaborateness, subtlety, and density so frequently encountered in the earlier stories. One of the three Mississippi stories, "The Burning," a Civil War tale of rape, arson, murder, and suicide, is as brilliantly experimental in narrative technique, especially in use of point of view, as anything in Faulkner, but the result is that the reader is left a bit confused about some of the happenings in the story. Less deliberately experimental but still hardly conventional in narrative method are the non-Mississippi stories, three of which—"The Bride of the Innisfallen," "Going to Naples," and "No Place For You, My Love"—are about voyages or trips, with the relationships between travelers and strangers brought together for a time by chance delineated with what would seem to be almost too much indirection. Miss Welty's great talent is, to be sure, reflected intermittently in *The Bride of the Innisfallen*, but for its fullest manifestation one must go back to *A Curtain of Green* and *A Wide Net*, and to such perfect or nearly perfect stories as "Livvie" and "A Worn Path."

WHEREAS EUDORA Welty's stories are characterized by poetical, mythical, and symbolical qualities, those of Mary McCarthy are distinguished by their intellectual and analytical ones. Although Miss Welty delves deeply into character, she leaves us with the feeling that the human personality is ultimately a mystery, while Miss McCarthy subjects her characters to microscopic analysis, stripping them bare of their façades and exposing them for the fallible and usually somewhat less than admirable beings that they inevitably are. Miss Welty is concerned with states of mind and the nuances and ambiguities of human relationships; Miss McCarthy, with the individual as a social and a political being, not only in her literary and other essays and in such novels as *The Groves of Academe* (1952) and *The Group* (1963) but also in her short stories.

Orphaned by the influenza epidemic of 1918, Mary McCarthy (1912– ) grew up in Minneapolis and Seattle under the care of grandparents. After graduating from Vassar, she wrote book and drama reviews for the *New Republic*, the *Nation* and the *Partisan Review*. Encouraged by her second husband, the critic Edmund Wilson, to try writing fiction, she published *The Company She Keeps* in 1942, a collection of stories in which an admittedly autobiographical heroine named Margaret Sargent, a young divorcée living in New York City, appears sometimes as a secondary and sometimes as the main character. In "Rogue's Gallery" she is employed as a private secretary to an unscrupulous art dealer, whose "essence," to use the author's expression, Miss McCarthy tried to convey. Two other masculine characters are microscopically analyzed in "The Genial Host" and "Portrait of the Intellectual as a Yale Man," the protagonist of the latter being modeled on the one-time socialist writer John Chamberlain.

In the three remaining stories in the volume the focus is on Margaret. "Cruel and Barbarous Treatment" is a brilliant piece of irony and satire depicting the several stages she goes through and the continuous self-examination she engages in as she plays the role of putative divorcée, culminating in her eventual realization that in her vanity and egotism she has made herself the victim of an "imposture." In "The Man in the Brooks Brothers Shirt," Margaret again displays hyper-self-conscious introspection and a compulsion to act in such a way as to cause herself embarrassment and mortification. She gets drunk with a male passenger in his compartment on a train bound for the West Coast and wakes up the next morning with at first a vague and then an all-too-clear recollection of having made love with him. But Margaret conducts her most acute self-examination in "Ghostly Father, I Confess," in which, after a session with an analyst, in which he counsels her to liberate herself from her domineering architect-husband, she realizes her difficulty: "that it was some failure in self-love that obliged her to snatch blindly at the love of others, hoping to love herself through them, borrowing their feelings, as the moon borrowed light." But she is consoled by the knowledge that she can still see herself as she is—all her frauds and her equivocal personality—and she prays to be allowed to retain this spiritual insight.

Margaret Sargent as architect's wife reappears in "The Weeds," the first story in Miss McCarthy's second volume of short fiction, *Cast a Cold Eye* (1950). She has left her husband, but married life in the country has made her a stranger to the New York City in which she had hoped to reestablish herself, and she is unable to resist when her husband comes to take her home. Like her flower garden, overrun by weeds during her absence, the part of her which had always held faith in life's possibilities has been destroyed, and there is left only the course of hypocritical acceptance of things as they are.

The other stories in *Cast a Cold Eye* are not so bleak and

despairing, but they view "coldly," or at least with dispassionate analysis, persons and character types, as in "The Friend of the Family," more essay than story, which exhibits uncanny insight into urban social relationships, and in an unforgettable portrait of the author's Catholic grandmother, "Yonder Peasant, Who Is He?" The latter piece, described by Miss McCarthy as "an angry indictment of privilege for its treatment of the underprivileged," which was evoked by her wealthy grandparents' lack of concern for Mary and her little brothers, she also used as the opening selection in what is perhaps her best book, *Memories of a Catholic Girlhood* (1957). Although essentially autobiographical, it has some invented details and dialogue and also the heightening, exaggeration, and dramatizing characteristic of fiction. To the sketches which comprise its contents, most of which were first published in the *New Yorker*, Miss McCarthy added a long general introduction and additional commentary in the form of epilogues to each piece in which she attempted to separate as far as possible truth from invention.

Besides the Catholic grandmother, other sketches in *Memories* bring vividly to life Miss McCarthy's Aunt Margaret and her selfish, sadistic husband, who served as the author's guardians; her Protestant maternal grandfather, by whom Mary and her two older brothers were ultimately rescued from the paternal grandparents and removed to Seattle; and her maternal grandmother, who was Jewish. The remaining pieces deal with experiences, usually both comic and ironic, connected with Mary's schooling, a recurring theme being the need for a young person to dissemble in one's relationships with adults.

Mary McCarthy's one other piece of short fiction published in book form is *The Oasis*, a story of some 35,000 words about a group of people, mostly intellectuals, who establish a Utopian community soon after World War II on a mountaintop apparently somewhere in New England. Called by its author a *conte philosophique*, it is also very much of a *roman à clef*. Several of the colonists are modeled on actual persons. Pacifist and idealist Macdougal Macdermott is patterned on Dwight Macdonald, social and literary critic and onetime editor of *Politics*; and Will Taub, the leader of the "realist" group, is modeled on Philip Rahv, also a critic and founder and editor of the *Partisan Review*. The actions of the characters and the conflicts which result from them are presumably imaginary, although altogether plausible and with meaningful implications. Factionalism soon rears its ugly head, crises and contretemps occur, and in the dramatic concluding episode the community finds itself in the position of having violated the nonviolent principles on which it presumably is based in order to protect its property, although the conflict is resolved in a sense in the mind of the idealistic young wife, Katy Morell, who realizes that "her hunger for goodness was an appetency not of this world and not to be satisfied by actions, which would forever cheat its insistencies." Although *The Oasis* has been criticized as lacking in fictional quality and as being too sharply satirical and too mordant in its wit and irony, it can be said that no other work, except perhaps for Lionel Trilling's *The Middle of the Journey*, has provided as searching and meaningful a fictional representation of the attitudes and dilemmas of the liberal intellectual as a political and social being. In it are vividly revealed the qualities of style and of intellect which give Miss McCarthy's writing its special distinction.

JEAN STAFFORD (1915–    ) combines an intellectuality somewhat similar to that of Mary McCarthy with a sensitivity and a style reminiscent of Katherine Anne Porter. She has published three very good novels—*Boston Adventure* (1944), *The Mountain Lion* (1947), and *The Catherine Wheel* (1952) —all of which deal with the theme of the adjustment of children or young people to the adult world, as well as two volumes of short stories—*Children Are Bored on Sunday* (1953) and *Bad Characters* (1964)—which are reprinted, along with some previously uncollected stories, in *The Collected Stories of Jean Stafford* (1969).

Miss Stafford's critics have noted that her stories often portray cruelty and suffering, both mental and physical, and that they reveal a preoccupation with characters whose idiosyncrasies, personal misfortunes, or physical afflictions turn them into lonely, isolated people, and even, in some instances, into freaks or grotesques. The queerest, as well as one of the most vividly drawn, of these characters is Ramona Dunn, the terribly obese girl, at once both contemptible and pathetic, in "The Echo and the Nemesis," who compensates for or rationalizes her gluttony by identifying with the beautiful twin sister she has invented, the image of her former thin self. Like Ramona, the young wife in "A Country Love Story," denied the love and companionship of her convalescent husband, who is preoccupied with his health and scholarly writing, also seeks to compensate for her situation by conjuring up an imaginary lover.

The loneliness and suffering of some of Miss Stafford's characters results when they are unable or refuse to adjust themselves after being removed from a familiar environment into an alien one. In "A Summer Day" an orphan Indian boy, sent from his home in Missouri to an Indian reservation school in Oklahoma, can think only of how he can assert his independence and escape. In "The Bleeding Heart," a young Mexican girl after graduating from college goes to live in New England and feels herself an alien. So, too, does the young woman from the Middle West in "Children Are Bored on

Sunday," who, because she feels inferior and embarrassed, cannot bring herself to feel at home in the world of Manhattan intellectuals. Besides dramatizing her heroine's conflicts, Miss Stafford provides a penetrating satirical exposé of the artificiality and sterility of the intellectual life.

The lonely middle-aged Jewish doctor in the wartime story "The Home Front," a refugee from Hitler Germany in an alien Connecticut industrial community where he ministers to defense workers, finds in the landlady of his rooming house and her family an anti-Semitism and a brutish cruelty worthy of the Nazis. Goaded by their killing of the stray cat he has befriended, the doctor is reduced to their level and in his hatred kills a pet bird belonging to the landlady's son. A similar theme and horrifying outcome is to be found in "In the Zoo" (awarded the O. Henry first prize for 1955), which portrays the relationship of two little orphan girls with their foster mother. The woman sadistically turns the girls' dog from an affectionate animal into a vicious one. When the man who gave the dog to the girl tries to right the matter with the mother, she incites the animal to kill his pet monkey, and in retaliation he poisons the dog.

Several of Miss Stafford's best stories deal with the loss of love: "A Country Love Story," already mentioned; "The Liberation," the story of a young woman's struggle to free herself from her possessive elderly relatives; and "A Winter's Tale," a long story of an American girl in Germany who has a love affair with a young Nazi aviator killed in a plane crash.

Not all Miss Stafford's stories are "unhappy" ones. Several are humorous and entertaining, such as "Bad Characters," with its unforgettable portrait of the little girl shoplifter, Lottie Jump; "A Reasonable Facsimile," with its amusing situation of a retired professor being plagued by a too-zealous admiring disciple; and "Caveat Emptor," which has some good satire about the ways of progressive colleges. But these stories are hardly profound and perhaps even a little contrived. Much more characteristic of Miss Stafford's work are the stories of alienation and loss. If they seem sometimes overly depressing, they have truth, and they are the work of a highly talented and disciplined artist.

ALTHOUGH J. F. POWERS (1917– ) has published to date during a literary career of more than twenty-five years only one novel—*Morte D'Urban* (1962)—and two volumes of short stories—*Prince of Darkness* (1947) and *The Presence of Grace* (1956)—there are probably few if any critics and readers who would deny him a place among the foremost living American writers. His stature would seem to be all the more remarkable since, in addition to being far from prolific, Powers' best work appears limited in range and manner, dealing as it does almost entirely, in a witty and ironic way, with the dilemmas and frustrations of the Catholic clergy. Powers' stories are not, however, parochial but have general and universal applications, and his achievement is also impressive because he is a master of comic satire.

Born in Jacksonville, Illinois, Powers was graduated from the Quincy, Illinois Academy, a Franciscan institution, in 1935, and worked at various jobs in Chicago before publishing his first story. "He Don't Plant Cotton," in 1943 in the small but distinguished literary quarterly *Accent*. One of several early stories which Powers said he wrote out of indignation at the situation of the Negro, it deals with a jazz drummer, a piano player, and a girl singer, all Negroes, who walk out during their performance at a Chicago nightclub, refusing to be humiliated by a group of drunken patrons from the South. Another early story which shows Powers' social consciousness is "Renner," in which an Austrian refugee from fascism discovers that anti-Semitism and other fascist attitudes also exist in the United States. Along with irony, "Renner" has pathos and poignancy, and these latter qualities are to be found in such stories as "The Old Bird, A Love Story," an account of an elderly man's first day in a temporary Christmas-season job in a department store; and in "Jamesie," an initiation story reminiscent of Hemingway and Sherwood Anderson, in which a young boy learns that his baseball hero is corrupt.

The most moving and probably the best of Powers' stories is his first clerical story, "Lions, Harts, Leaping Does," about an old Franciscan priest, Father Didymus, who at the end of his life fears that he has been guilty of the sin of pride and lacking in true piety and thus has not achieved a state of saving grace. The reader, however, is made to feel that he will attain salvation, for the nature of Father Didymus' doubts gives evidence of his piety and humility, all of this being conveyed by Powers in a highly poetic and marvelously restrained way.

More typical of Powers' clerical protagonists is Father Burner in "Prince of Darkness." Unspiritual and materialistic, he is a curate ambitious for a parish of his own but in the end is frustrated in his hopes and schemes. What makes Father Burner "a prince of darkness" would seem to be his pride and his lack of any real religious feeling, but it is to his credit that he at least vaguely senses his deficiencies, realizing that "the mark of the true priest" is not on him. He appears again in two other stories, "Death of a Favorite" and "Defection of a Favorite," but he is not quite the same character, still secular but scarcely satanic. In the latter stories, moreover, the narrator is the rectory cat, Fritz, and though they are highly amusing and ironic, there is something of the tone of situation comedy in the conflicts between cat and priest.

One of Powers' most memorable portrayals of a secular priest is that of Monsignor Sweeney, the materialistic, prudent, politically reactionary, and snobbish pastor in "The Forks," who regards his idealistic and liberal young curate, Father Eudex, as naïve if not stupid for believing he should act as much as possible like a true Christian. Sweeney endeavors to convince Father Eudex of the error of his ways, to realize that it is important to be concerned about appearances and the proper way of doing things, and, above all, always to do the expedient thing. The reader is left feeling that Father Eudex will try to retain his integrity and stand up for spiritual values, but whether he can succeed in doing so is by no means certain.

Powers' later religious stories (in *The Presence of Grace*) deal more kindly on the whole with their clerical characters than do the earlier ones. "The Presence of Grace" again portrays a conflict between curate and pastor, but with the curate able to find goodness in his superior where he had seen none before. The Father Fabre of this story appears also in the very funny "A Losing Game," in which in another contest with his pastor he shows himself magnanimous and without vindictiveness in defeat. Also sympathetically portrayed is the bishop, in the highly ironic but good-humored "Zeal," who is brought to realize his shortcomings and moved to try to atone for them through an encounter with an overzealous busybody of a priest, a man, who, it must be admitted, is always trying to do what he considers God's work.

The theme of a cleric made aware of his moral and spiritual deficiencies is also that of *Morte D'Urban*, which Powers began as a short story and several episodes of which were first published as independent short narratives in the *New Yorker* and other magazines. Somewhat in the nature of a companion piece to this novel is the long short story "Keystone," published a year after the novel in the *New Yorker* (May 18, 1963). In it the assistant to a bishop manipulates his ineffectual and old-fashioned superior into approving various progressive, modern improvements in the diocese, including the building of a new cathedral. But this kind of effort, the story seems to suggest symbolically, will not achieve for the church what it needs most, spiritual improvement. Here again, in both novel and short story, as in Powers' other best fiction, in his depiction of the anomalies and complexities of man's motives and actions are to be seen his penetrating insight and his marvelous wit.

PROBABLY BOTH THE BEST KNOWN and most controversial of contemporary fiction writers is J. D. Salinger (1919– ), whose fame rests in great measure on the popularity of his one novel, *The Catcher in the Rye* (1951) but who must also be regarded as an important short-story writer. Salinger grew up in New York City. He received his secondary education at the Valley Forge Military Academy in Pennsylvania and briefly attended several colleges before going into the army in World War II. Having published his first story in the magazine *Story*, edited by Whit Burnett, with whom Salinger studied writing at Columbia, he found time and opportunity to write stories during his army service, which after D-Day he spent in the European Theater as a military security agent. Although he contributed three more pieces to *Story*, a periodical noted for discovering and encouraging young writers of talent, Salinger soon began to publish also in the *Saturday Evening Post* and *Collier's*, and for several years his work appeared mainly in these and other large-circulation magazines, until the later 1940's, when he began to publish almost entirely in the *New Yorker.*

Salinger's *Story* pieces are not particularly impressive for narrative artistry, but two of them, "The Long Debut of Lois Taggart" and "Elaine," are good character studies. The first tells of a rather vain, shallow daughter of well-to-do, socially prominent parents; the second, of a dull-witted girl dominated by her mother. In his *Post* and other "slick" magazine stories, Salinger demonstrated from the beginning a talent for turning out formula pieces with surprise endings and other contrivances. In a few cases Salinger endeavored to give these stories some substance and genuine significance, as in "The Varioni Brothers," which treats the theme of the artist who must choose between commercial success and true artistic achievement, and "The Inverted Forest," a long story of a poet who leaves his wife and his other social responsibilities to withdraw from the world.

Not surprisingly, several of Salinger's stories reflect his military experience or are in other respects about World War II. "Soft-Boiled Sergeant," published in the *Post*, is a sentimental account of an ugly sergeant with a heart of gold who loses his life saving three of his frightened young soldiers. A much better *Post* story, "The Last Day of the Last Furlough," depicts the feelings and actions of a young soldier on the day before he is to be shipped overseas. This protagonist, named Babe Gladwaller and who in some respects resembles Holden Caulfield in *The Catcher in the Rye*, appears again in two other stories which are less artistically effective. "A Boy in France," shows Babe in a foxhole finding comfort in rereading a letter from his little sister. In "The Stranger," Babe, safely home, denounces war.

Undoubtedly Salinger's best war story, and one of the very

best of all World War II stories, is "For Esmé—with Love and Squalor," in which an American soldier who has suffered a nervous breakdown after V–E Day is given what would appear to be the will and confidence to achieve a recovery when he receives an affectionate letter and the gift of her dead father's' watch from a girl of thirteen he had met in England while he was taking preinvasion training, and who had told him good-bye with the hope he would "return from the war with all your faculties intact." A few critics have regarded the story as overly sentimental and not altogether convincing, but most have felt it genuinely moving and highly artistic.

"For Esmé" is included in *Nine Stories* (1953), the only collection Salinger chose to make from the thirty stories he had published by 1953. The stories in the volume, all but two published, in the *New Yorker* from 1948 on, are arranged in the order of their original publication beginning with "A Perfect Day for Bananafish." It describes the suicide and the immediate events leading up to it of Seymour Glass, about whom Salinger was to have much more to say in his subsequent fiction. In this much-discussed story, the main question would seem to be how Seymour should be viewed. He is recently married to a presumably insensitive, materialistic young woman named Muriel, who is shown, in the first part of the story, in a Florida hotel room talking on the telephone to her mother in New York. Seymour, in the meantime, is seen playing on the beach and floating in the water with a little girl named Sybil, to whom he talks about many things but especially about the curious habits of the bananafish, which swims into a hole full of bananas and eats so many that he is too fat to come out. Later Seymour returns to the hotel room where Muriel is now asleep and puts a bullet through his head. Seymour, we learn from what Muriel tells her mother, has been behaving strangely, and certainly some of the things he says and does while he is with Sybil are strange, but the story provides no explanation for his conduct. Perhaps the best way to regard Seymour, as several critics have suggested, is as suffering from an obsession brought on by a hypersensitivity to physical existence which causes him to feel trapped by life, his only escape being through suicide.

Marital incompatibility also figures, although differently, in the next of the *Nine Stories*, "Uncle Wiggily in Connecticut." The result is depression and frustration for a young suburban housewife, who tries to solace herself with the memory of an old lover who was killed in an ironic freak accident in Japan during the war. Also unhappily married is the husband in the highly ironic, O'Hara-like "Pretty Mouth and Green My Eyes," who tells his law-firm associate over the phone that his wife, whom he had feared might be betraying him after leaving a party without him, has after all come home, when at the time she is actually in bed with the other man.

The remaining stories in the volume have either young children or teen-agers as their principal characters. A four-year-old boy in "Down at the Dinghy" compulsively runs away from home, as he has a habit of doing (although he never gets very far) when he hears someone say something which frightens or disturbs him. He is brought back by his patient, understanding young mother, who tries by giving him her love to cure him of his fear. Through understanding, also, a teen-age girl in "Just Before the War with the Eskimos" is moved to feel compassion for the freakish, mixed-up older brother of her girl friend.

In two other stories, a first-person narrator, possibly the same one to judge from his style and tone, looks back on experiences which had a profound effect on him. In "The Laughing Man" a nine-year-old boy suffers vicariously when the love affair in which the man he idolizes is involved is broken off. In "De Daumier-Smith's Blue Period," a more substantial and meaningful story, the experience is more crucial, consisting of a young man of nineteen achieving, through what appears to be a mystical vision, a new sense of unity with and love for others.

Mysticism is also at the center of "Teddy," a story which puts something of a strain on the credulity of the reader insofar as the characteristics of its protagonist are concerned and may also leave him uncertain about its outcome. Teddy is a ten-year-old child prodigy who has the gift of seeing into the future and who believes in the Vedantic theory of reincarnation. Having been examined by European scientists, he is now returning by ship to America with his mother and father, an unpleasant couple much given to sarcastic bickering, and a hateful brat of a younger sister. Teddy, we learn mostly from a conversation he has with a young man who is a teacher, believes that people cannot develop spiritually as long as they put value on logic and the intellect, and he professes to be able to see into the future, even to predicting that his own death might occur that day. This does presumably happen when his sister pushes him into the ship's dry swimming pool, although the reader cannot be certain. Some critics have found Teddy not only unconvincing but also obnoxious, though certainly Salinger wanted him to be regarded sympathetically.

A continuing interest in the phenomenon of mysticism is manifested in Salinger's most recent stories, all of which are also concerned with members of the Glass family. This fictitious family, which seems to have preoccupied Salinger almost to the point of obsession, consists of a Jewish father and an Irish mother, once a well-known song-and-dance team in vaudeville, and their seven children—five boys and two girls

—who all achieved a certain fame by appearing in their early years on a radio quiz program entitled "It's a Wise Child." In *Franny and Zooey* (1961), which consists of two related stories originally published in the *New Yorker* in 1955 and 1957, we see Franny, the youngest child in the family and a college girl of twenty or so, undergoing and eventually recovering from a kind of spiritual crisis. In "Franny," a story of about ten thousand words, she goes to visit her boy friend for what turns out to be an abortive Ivy League college football weekend. At lunch before the game she tries to explain to her rather imperceptive escort her disturbed, unhappy state, brought on by her overwhelming conviction that people are full of "ego," that everything they do is phony and meaningless, and that she has found comfort in a little book, *The Way of a Pilgrim*, which advocates the continuous saying of the prayer "Lord Jesus Christ, have mercy on me," which can have a "tremendous mystical effect" on a person. Her boy friend, however, is more interested in his lunch and in telling her about a paper he has written on Flaubert. Franny, having previously had to go to the ladies' room to compose herself, faints away, and we see her after regaining consciousness murmuring the Jesus prayer at the end of the story.

"Franny" is unquestionably a very good story. "Zooey," which is three times as long, seems, on the other hand, somewhat forced and in other respects less artistic. Its narrator, Buddy Glass (after Seymour, the oldest of the Glass children), calls it "a sort of prose home movie," and not a short story at all, adding that it "isn't a mystical story, or a religiously mystifying story at all. I say it's a compound, or multiple, love story, pure and complicated." Franny does, by the end of the story, apparently recover from her deep depression through becoming convinced by her brother Zooey that she must learn to love everybody. Whether or not one regards this resolution of Franny's conflict as altogether convincing, the story does seem much too drawn out. Buddy Glass is very long-winded, preoccupied with minutiae, such as cataloguing the contents of a bathroom medicine cabinet and every piece of furniture in the Glass family's New York apartment living room, and given also to writing in the *New Yorker* "casual style," with its Narcissistic prose, folksy phrasing, frequent qualification of what has just been said, and other mannerisms.

Buddy is even more conspicuously the narrator of the two lengthy stories (first published in the *New Yorker* in 1955 and 1959 respectively) in *Raise High the Roof Beam, Carpenters; and Seymour: An Introduction* (1963), since although his older brother, Seymour, is ostensibly the subject of both, Buddy also keeps himself very much in the reader's view. In "Raise High," Buddy tells how as a soldier on leave he had traveled from his Georgia army post to attend Seymour's wedding in New York in 1942. Seymour, to everyone's consternation, stands up his bride-to-be at the church, and then a few hours later prevails on her to elope with him. The story is divided between expressing Buddy's discomfiture when, in company with some of the wedding guests, including the matron of honor, he is discovered to be the groom's brother and his efforts to provide for the reader an apologia for Seymour's strange behavior.

"Seymour: An Introduction," which is more an informal and extremely discursive biographical essay than a story, goes much further in its portrayal of Seymour. Buddy (who sounds very much like Salinger himself) is not only even more omnipresent than previously but appears also to be put under a great strain by the effort to conjure up the real Seymour for the reader. Besides providing many miscellaneous recollections, most of them involving himself as well as Seymour, along with a detailed description of Seymour's physical appearance, Buddy develops the thesis that Seymour was a very great poet-seer, given to expressing his feelings and insights in verse modeled on the Japanese haiku, specimens of which Buddy says he can only paraphrase because Seymour's widow has forbidden their direct quotation. Buddy also seems to suggest that if Seymour behaved strangely at times—even like a fool or an imbecile—it was because he was a saint. In saying this, and all that precedes it, Buddy is, to use his own phrase, "graphically panegyric." This, and what Buddy also calls his "perpetual lust to share top billing" with Seymour, hardly serves to ingratiate Buddy with the reader. Salinger unfortunately seems to have indulged too much in a purging of himself at the reader's expense.

On the dust jacket of *Raise High the Roof Beam, Carpenters; and Seymour: An Introduction*, Salinger is quoted as saying that he had "several new Glass stories coming along," but as of the present writing only one has appeared (in 1965), "Hapworth 16, 1924." Again it is not really a story but a very long letter (about thirty thousand words), briefly introduced by Buddy Glass, written by Seymour to his parents at the age of seven from summer camp. Like Salinger's Teddy, Seymour professes to have knowledge of the past and insight into the future, declaring that he is now in one of several appearances or incarnations and that his death will occur when he is around thirty (as we have seen it does in "A Perfect Day for Bananafish"). This and other aspects of the story (Seymour requests a long list of books to be sent to him, a list so formidable in both quantity and profundity that not even the most proficient of speed readers, let alone a seven-year-old, could reasonably expect to get through more than a small part of it in a summer) are such as to give even the most pro-Salinger sympathizer pause and make him wonder what is happening to Salinger's talent. In the past, Salinger's themes have been significant and his ability to give them artistic expression impressive, but whether he will add meaningfully to his achievement seems at this point problematical.

ALTHOUGH J. D. Salinger is Jewish, he is not a Jewish writer in the sense that Bernard Malamud and such equally talented writers as Philip Roth and Saul Bellow are. All of them write "Jewish fiction," that is, stories concerned with the situation of the Jew in contemporary America. Their protagonists often manifest themselves as distinctively Jewish through their awareness of an unhappy heritage of injustice and persecution, through their assumption, all too frequently confirmed by experience, that suffering, rather than happiness, is the rule in life, and finally through their idiom, a special kind of English flavored with Yiddish expressions and constructions.

Both Roth and Bellow have published some excellent short stories—Roth in *Goodbye, Columbus* (1959) and Bellow in *Mosby's Memoirs and Other Stories* (1969)—but neither has shown as much interest in the form as has Malamud, whose reputation at this point is perhaps even higher as a short-story writer than as a novelist. Born in 1914, Malamud grew up in Brooklyn and after graduating from City College of New York earned a master's degree in English at Columbia. After teaching evening high-school classes for several years, he became an English instructor at Oregon State College in 1949, leaving there in 1961 to join the faculty of Bennington College in Vermont. He began writing short stories in the early 1940's, but it was almost a decade before he began to place his work in such magazines as *Harper's Bazaar* and

*Partisan Review*. His first longer fiction, *The Natural* (1952), a comic allegorical novel about a baseball hero, although original and inventive, is not entirely successful, nor is it typical of his work. In his second novel, *The Assistant* (1957), he began to treat themes which are central in his later work: if man is destined to suffer, he must endure with hope, and if man falls into sin, there is yet the possibility of redemption through repentance.

The best pieces in *The Magic Barrel* (1958), Malamud's first story collection, resemble *The Assistant* in style, situations, and themes. Feld, the shoemaker of "The First Seven Years," like Morris Bober, the grocer in the novel, is ambitious for his daughter, wanting her to marry a materialistic young accounting student and "live a better life," but he is compelled to bow to the power of the love with which his poor Polish refugee assistant has won over the girl. The sense of despair and futility that pervades much of *The Assistant* is also found in several of the stories. In "The Bill," as in the novel, a grocery store is a prison for its owners, as is a candy store in "The Prison," and in both stories acts of kindness ironically have injurious results for those who perform them. In "The Mourners," an elderly retired Jew isolates himself in a filthy little flat from which his landlord, in whom greed and compassion contend, seeks to evict him. And in the equally dismal but very affecting "Take Pity," Rosen, a coffee salesman and a man of much compassion, but also of much pride, commits suicide after the destitute widow of a grocery-store owner refuses to let him help her and her children.

Since we see Rosen after his death in apparently some kind of institution telling his story to a "census-taker" named Davidow, "Take Pity" has something of the quality of a fantasy. The combining of fantasy with drab realism, which was to become a distinctive aspect of Malamud's fiction, is especially conspicuous and artistically exploited in "Angel Levine." Its protagonist is Manischevitz, an East Side tailor who suffers the tribulations of Job when his store burns down, his son is killed in the war, his daughter runs off with a worthless lover, and his wife has an almost fatal illness. Called on to believe that a huge, mysterious Harlem Negro named Levine is a Jewish angel who can save him, he at first doubts. When later he does believe, he has a brief vision of Levine ascending to heaven on great wings, and to his wife, who has miraculously become well, he says: "A wonderful thing, Fanny. Believe me, there are Jews everywhere."

Perhaps the finest piece in *The Magic Barrel* is the title story, whose protagonist is a young rabbinical student about to be ordained. His negotiations with a marriage broker, to whom he has gone to find the wife he needs to gain a congregation, cause him ultimately to believe that his motives and actions throughout the business have proved him unspiritual and that perhaps "he did not love God as well as he might because he had not loved man." Ironically, he falls in love with the picture of the marriage broker's daughter, a prostitute, and with her he sees hope for his "redemption."

Malamud's second volume of stories, *Idiots First* (1963), is a more varied and uneven collection than *The Magic Barrel*. "The Cost of Living" and "The Death of Me" are lesser stories in the vein of *The Assistant*, but the title story is one of Malamud's finest efforts in the use of fantasy. Nightmarish in setting and with allegorical implications, it tells of Mendel, an elderly Jew about to die, who has only one evening to obtain the thirty-five dollars he needs to send his idiot son, Isaac, to an uncle in California. Denied by a wealthy philanthropist, who gives "only to institutions," but assisted by an old rabbi who, over the protests of his wife, gives Mendel his fur-lined caftan to pawn, Mendel is barred at the train gate by his mysterious black-whiskered antagonist, Ginzburg (Death). Clutching him by the throat and denouncing him for not understanding what it means to be human, Mendel, instead of being destroyed, is allowed to put his son on the train.

Fantasy appears also in "The Jewbird," a wonderful little fable, both comic and pathetic, and, except for its protagonist, altogether realistic, of a skinny, crow-like talking bird named Schwartz. One evening, seeking sanctuary from anti-Semites and ill fortune, he flies in the window of a New York East Side apartment inhabited by the family of a frozen-food salesman named Cohen. To Cohen, Schwartz is merely a smelly, spunging opportunist, not an object of compassion, and he is angered also because Schwartz takes his place as father by helping Cohen's ten-year-old son with his lessons and otherwise being a companion to the boy. Cohen begins a program of harassing the bird, including bringing a cat into the apartment. Unsuccessful in forcing the bird to leave, he finally throws him out the window, and some time later the son finds his battered body in a nearby vacant lot, the victim, his mother tells him, of "Anti-Semeets."

In a very different vein from these stories is "A Choice of Profession," about a none-too-happy relationship between a young college teacher recently divorced from an unfaithful wife and an older girl student who is a reformed prostitute. With its academic background and its hero who needs to get away from his past and make a new start in life, the story invites comparison with Malamud's third novel, *A New Life* (1961), but it seems too deliberately naturalistic and is noticeably lacking in the rich humor and pathos of the novel. On the other hand, "Black Is My Favorite Color," a departure from Malamud's usual third-person narration in the *Idiots First* volume, is a touching first-person account by a Jewish liquor-store owner in Harlem of his failure, because of racial antagonism, to win the Negro woman he loved. Even more affecting is "The German Refugee," also told in the first person, in which a twenty-year-old New York college boy gives an account of the unhappy past and present history of a refugee literary critic whom he had tutored in English one summer.

A year spent in Rome and traveling in Europe in 1956, made possible by a *Partisan Review* fellowship, provided Malamud with the background for a group of Italian stories, several being included in *The Magic Barrel* and the rest in *Idiots First*. In the former volume "Behold the Key" portrays with wonderful humor the frustration and futility attendant on the attempts of an American graduate student in Italian studies to find an apartment for his family in Rome. "The Lady of the Lake" relates, with perhaps overmuch local color and an ending that smacks of contrivance, how a young New York Jew loses the romance he hopes for when he misrepresents himself as a gentile to the girl he loves. In the second collection, "Life is Better than Death," a story of a young Italian widow deceived and deserted by her lover, is not as moving as it might be. But "The Maid's Shoes," an account of the relationship between an American law professor and his middle-aged Italian maid, shows Malamud at very nearly his best in treating one of his favorite situations: a character is called upon to exhibit understanding and compassion for a less fortunate person who makes embarrassing and inconvenient demands on him.

"The Last Mohican," a third Italian story in *The Magic Barrel*, introduces one of Malamud's most memorable characters in the person of Arthur Fidelman, would-be painter and picaresque antihero, who learns finally, if not how to paint, at least how to live. Two more stories about Fidelman, "Still Life" and "Naked Nude," are included in *Idiots First*, and later all three were republished, along with three new Fidelman narratives, in *Pictures of Fidelman* (1969).

In his first adventure we see Fidelman, late of the Bronx and "a self-confessed failure as a painter," planning to spend a year in Italy sightseeing and writing a critical study of Giotto. On his arrival he is accosted at the railroad station in Rome by an odd-looking character wearing baggy knickers who identifies himself as a Jewish refugee named Susskind in need of a handout. Although Fidelman gives him money, Susskind continues to hound Fidelman for an old suit which Fidelman refuses to give him, denying Susskind's contention that both as a man and as a Jew Fidelman has a responsibility to help him. When the briefcase containing the first chapter of his Giotto study disappears, Fidelman suspects that Susskind has stolen it in retaliation. After weeks of searching and fruitless attempts to rewrite the chapter, Fidelman finds Susskind in the ghetto and offers him the suit as ransom, only to learn that the manuscript has been burned. "I did you a favor," asserts Susskind. "The words were there but the spirit was missing." And Fidelman, his rage dispelled by "a triumphant insight," cannot help but agree.

In "Still Life" we find Fidelman, who has remained in Rome and is filled with a renewed desire to paint, renting part of a chilly studio from Annamaria, a female abstract painter with whom he falls in love and by whom, in return, he is charged an exorbitant rent and treated with indifference or scorn, despite the chores he does and the meals and presents he buys for her. However, after long endurance, Fidelman's conflict is accidentally and comically resolved when his inamorta, seeing him one day dressed in a cleric'c vestments—Fidelman having had the inspiration to attempt, à la Rembrandt, a self-portrait of the artist as priest—wildly pours out to him the sin she had been unable to reveal in the confessional. When she accedes to the penance he demands for forgiving her, Fidelman is able to nail the guilt-plagued lady "to her cross."

More contrived in its situation, but with some very funny touches, is "Naked Nude," in which Fidelman, who has fallen on evil days, is seen working as a captive janitor in a Milanese whorehouse without money or passport. The criminals who have him in their power offer him his freedom in exchange for counterfeiting a Titian Venus they are plotting to steal. Fidelman at first has great difficulty executing the forgery, but by the time he completes it he has so fallen in love with his handiwork that when the theft is carried out he escapes with his copy rather than the original.

The scene shifts to Florence in "A Pimp's Revenge," where we see Fidelman struggling to execute an artistic project that has come to obsess him, a "mother and son" portrait based on a snapshot of his long-dead mother and himself as a little boy. The painting eventually turns into "Procurer and Prostitute," since he has used as a model the young whore for whom he pimps to support himself. The work is a masterpiece, but Fidelman cannot be content with his achievement, and, foolishly allowing himself to be persuaded by the whore's former pimp (who seeks revenge on his successor) that it needs to be "truer to life," he alters the painting and ruins it.

The highly experimental "Pictures of the Artist," composed of a series of disjointed episodes and utilizing a Joycean stream-of-consciousness to express Fidelman's thoughts and emotions, shows him satisfying his need to create by traveling about Italy digging square holes in the ground, exhibiting them to the public as underground sculpture, and justifying them to those who complain of being defrauded of their ten-lire admission on the grounds that the forms of the empty holes possess aesthetic value and provide the beholder with enjoyment. One dissatisfied customer (who turns out to be the Devil, angered because Fidelman has not "yet learned what is the difference between something and nothing,") hits Fidelman over the head with his own artistic shovel, knocking him into the sculpture hole. From this interment Fidelman can only be resurrected, the rest of the story seems to suggest, through divesting himself of his ambition to win fame through art and by learning to love and have consideration for others rather than being preoccupied with self.

In "Glass Blower of Venice," Fidelman, having given up painting, supports himself by working as a Grand Canal ferryman during a gray, dripping winter. After losing that job, he is reduced to carrying people piggyback across flooded squares for one hundred lire a person, "skinny old men half price." He has an affair with one of his women passengers and later is seduced by her homosexual husband, who tells him, "Think of love . . . you've run from it all your life," and who, when he sees examples of Fidelman's pictures and sculptures, advises him to "burn them all." Fidelman learns that one can invent life if he cannot invent art, and after learning to be a glassblower he goes back to America, where he is happy in this trade "and loved men and women." This conclusion to Fidelman's progress as man and artist may be didactic and moralistic, but it is nevertheless both convincing and satisfying—and Fidelman is a major character creation, made so by Malamud's great inventive gifts, rich comic sense, deep insight into human nature and experience, and compelling belief that it is possible for a man to achieve at least a small victory of the spirit.

ONE OF THE MOST provocative of recent short-story writers is Flannery O'Connor (1925–1964), who, before her untimely death at the age of thirty-nine, produced two volumes of stories as well as two novels—*Wise Blood* (1952) and *The Violent Bear It Away* (1960). In her work violence, the grotesque, and other Southern Gothic elements are so intermingled with the theological implications that her fiction often poses real problems of interpretation and the degree of approbation it merits. Born in Savannah, Miss O'Connor graduated from the Georgia State College for Women, attended the Writers' Workshop of the University of Iowa on a fellowship, and spent most of her writing career—suffering from the bone disease which contributed to her death—on a dairy farm near Milledgeville, Georgia, managed by her mother. Her talent and achievement were recognized by many grants and honors, including three O. Henry first prizes for her stories, in 1957, 1963, and 1964.

A Roman Catholic, Miss O'Connor said of her point of view: "I see from the standpoint of Christian orthodoxy. This means that for me the meaning of life is centered in our Redemption by Christ and that what I see in the world I see in relation to that." She asserted that her theological beliefs enhanced rather than restricted what she saw and assured her "respect for mystery," She explained and justified her use of the grotesque and violent as a means to shock the reader into seeing that what may seem normal is really abnormal and evil.

This strategy is used in full measure in *Wise Blood*, whose Christ-haunted protagonist discovers that there is no salvation through denying Christ and striving for self-justification. It is also used in the title story of *A Good Man Is Hard to Find* (1955), a story which, although it is not really one of her best or most representative, has frequently been reprinted by anthologists, probably because it is such a shocking one. It pictures The Misfit, a criminal psychopath who believes that Jesus "thown everything off balance" and who reasons that "if He did what He said, then it's nothing for you to do but thow away everything and follow Him, and if He didn't, then it's nothing for you to do but enjoy the few minutes you got left the best way you can—by killing somebody or burning down his house or doing some other meanness to him. No pleasure but meanness." The Misfit's pleasure in this instance comes from murdering in cold blood a hapless family of six —an old grandmother, her son and daughter-in-law, and their three children—when he and his fellow escaped convicts came upon them after their automobile has accidentally run off a country road. Presumably through this horrifying denouement Miss O'Connor means to suggest that there is no halfway position in the matter of believing in Christ. Either we do or we are led to substitute ourselves for him with dire consequences.

Equally startling, although its shock does not come from violence, is the outcome of "Good Country People," in which Joy Hopewell, a young woman Ph.D. in philosophy—physically unattractive, intellectually proud and atheistic—is duped by a pious young Bible salesman who makes off with her artificial leg for a souvenir. "We are all damned," she tells him in the barn loft where she expects to seduce him, "but some of us have taken off our blindfolds and see that there's nothing to see. It's a kind of salvation." But it is she not he who turns out to be the fool, and he tells her, as he makes off with his booty, that she is not so smart: "I been believing in nothing ever since I was born."

Like the young Bible salesman, Mr. Shiftlet, the one-armed tramp carpenter of "The Life You Save May Be Your Own," turns out to be a con man. After marrying Lucynell, the idiot daughter of the toothless old Mrs. Crater, in return for the mother's old Ford, he drives off, leaving his new wife behind when she falls asleep in a lunchroom on their honeymoon trip. The story may seem reminiscent of the Dude-Bessie episode in *Tobacco Road*, but it has far more serious implications. Mr. Shiftlet asserts that he has "a moral intelligence" but also likens a man's spirit to an automobile. But in abandoning Lucynell, likened to "an angel of Gawd" by the lunchroom waiter, he saves his life but loses his soul.

This story contains one of Miss O'Connor's frequent satirical hits at modern man's trust in "advancement" or progress. When Mr. Shiftlet tells Mrs. Crater that the monks of old slept in their coffins, she replies, "They wasn't as advanced as we are." A similar remark is made by Mrs. Shortley, in "The Displaced Person," about Europeans: "They never have advanced or reformed. They got the same religion as a thousand years ago. It could only be the devil responsible for that." This remark is occasioned when Mrs. McIntyre, for whom Mrs. Shortley's husband works as a hired hand, is persuaded by the local Catholic priest to employ a Polish refugee, Mr. Guizac, to work on her farm. Because he is a threat to the Shortleys' job security, Mrs. Shortley views him as a devil, and because he is a diligent and competent worker, unlike the poor whites and Negroes she has had to put up with, Mrs. McIntyre views him as her "salvation." The complex action that results in the "displacing" of the Shortleys and Mrs. McIntyre and of Mr. Guizac himself, the mordant irony, and the effective symbolization—all serve to bring out with impressive artistry the theme that there is no salvation unless one loves God and one's neighbor as oneself, a theme which is developed again and very nearly as effectively in "A Circle of Fire," in which the materialistic Mrs. Cope, who might well pass for a sister of Mrs. McIntyre, is so selfishly fearful for the safety of her woods that she is blinded to the truth uttered by the underprivileged boys from the city who invade her farm and set the woods afire that "Gawd owns them woods and her too."

"The Artificial Nigger," probably the finest story in *A Good Man Is Hard to Find*, and the author's favorite, portrays with profound insight in tones varying from comic to poignant, the relationship between a grandfather and his grandson. Between old Mr. Head, a Georgia backwoodsman, and the ten-year-old Nelson there is a natural and intense competition. The grandfather wants Nelson to acknowledge freely and fully the old man's superior wisdom and experience, and the boy does not want to admit his inferiority and ignorance. To convince Nelson that he is not as smart as he thinks he is, Mr. Head takes him to the city for his first visit, which the boy insists is his second by reason of his having been there briefly as a newborn infant. On the train Nelson is sufficiently impressed when Mr. Head shows him the first Negroes he has ever seen, as well as such wonders as the men's washroom and the dining car, and he realizes that "the old man would be his only support in the strange place they were approaching." But Mr. Head proves neither very smart nor a support. His unfamiliarity with the city results in their getting lost, and when Nelson accidentally bumps into a woman, causing her to fall down and spill her groceries and call irately for the police, Mr. Head, struck by panic, denies knowing the boy. But there are no police, and nothing happens to Nelson; the two walk on, Mr. Head's sense of shame and guilt having evoked in him a new love and concern for the boy, while Nelson, his shock at his grandfather's treachery having turned to hatred, keeps at a distance behind. Then on the lawn of a mansion they see a plaster statue of a Negro, which to them seems "some great mystery, some monument to another's victory that brought them together in their common defeat. They could both feel it dissolving their differences like an action of mercy." Mr. Head realizes that God has mercifully hidden from him his true depravity, and "since God loved in proportion as He forgave, he felt ready at that instant to enter Paradise."

Like Mr. Head, the characters in several of Miss O'Connor's later stories included in *Everything That Rises Must Converge* (1965) are brought to greater self-awareness through some kind of vision of the true relation between God and his creation. Unlike Mr. Head, however, in some instances the characters are intellectuals, as in the title story. Julian, a year out of college and a would-be writer who is selling typewriters until he can get started, blames his mother for his disenchantment with life and his feeling that he has no future, and he prides himself that "in spite of growing up dominated by a small mind, he had ended up with a large one," free of her polite racism, preoccupation with status, and general insensitivity. In the end, when Julian indirectly brings about his mother's death, he becomes aware of his lack of love and understanding, and his loss "seemed to sweep him back to her, postponing from moment to moment his entry into the world of guilt and sorrow."

Another would-be southern writer and intellectual is Asbury Fox, in "The Enduring Chill," who has come home to die in the mistaken belief that he is afflicted with a fatal malady. Several wonderfully funny scenes contribute importantly to making this one of Miss O'Connor's best stories, but there are serious implications as well, particularly in Asbury's ultimate awareness of his spiritual debility. Even less sympathetic specimens of this same character type are Thomas, the self-centered local historian in "The Comforts of Home," whose quarrel with his mother is that she has an "excess of virtue" insofar as wanting to help others; and the odious Wesley, the college-teacher son of Mrs. May in "Greenleaf," who hates everyone and everything and tells his mother, "I wouldn't milk a cow to save your soul in hell."

Mrs. May is also a familiar character type, being another of Miss O'Connor's hardworking, proud, materialistic "good Christian" widows plagued with shiftless help, who struggle to make a dairy farm pay. Not so familiar, however, is the shocking and incongruous symbolic action by which both her punishment and her illumination are brought about. The neighbor's scrub bull, which disturbs Mrs. May's tranquility by breaking loose, eating her oats, and threatening to ruin the breeding schedule of her cows, represents two lovers, Zeus for Europa and Christ for man. Gored to death by the bull, which is likened to "some patient god come down to woo her" and to "a wild tormented lover," Mrs. May, who had declared that she would die only when she got "good and ready," has "the look of a person whose sight has been suddenly restored but who finds the light unbearable."

If the reader is left here to imagine the particulars of what Mrs. May sees, such is not the case with Mrs. Turpin, in "Revelation," who is vouchsafed the most explicit as well as profoundly shocking vision of any of Miss O'Connor's characters. Even more than the author's other such women, Mrs. Turpin prides herself on being good, and she is grateful that God has made her a respectable home- and landowner instead of white trash. Her smugness, revealed in a conversation she has with a southern lady in a doctor's waiting room, so enrages the lady's epileptic college-student daughter that the latter, just before being seized by a fit, bites Mrs. Turpin and curses her: "Go back to hell where you come from you old wart hog." Back home, still shaken to the depths of her being, Mrs. Turpin in a fury demands to know of God: "How am I a hog and me both? How am I saved from hell too?"

And as dusk descends she has her answer: that the last may be first and the first last. A purple streak in the sky appears to her as a great bridge on which a vast horde of souls are entering heaven, "with white trash clean for the first time in their lives and bands of black niggers in white robes, and battalions of freaks and lunatics shouting and clapping and leaping like frogs." And bringing up the rear of the procession were people like herself, who "always had a little of everything and the God-given wit to use it right. . . . Yet she could see by their shocked and altered faces that even their virtues were being burned away."

That salvation is not achieved through good works is also the revelation afforded the main character in "The Lame Shall Enter First." The story closely resembles *The Violent Bear It Away* in its character relationships and theme, although it lacks something of the artistry and dramatic power of the longer work. Sheppard, a widower with a ten-year-old son and an unpaid reformatory psychologist who counsels because of "the satisfaction of knowing he was helping boys no one cared about," takes Rufus Johnson, a fourteen-year-old former inmate he has worked with into his home. He believes that through providing Rufus affection and obtaining a corrective shoe for his deformed foot the boy will be cured of his criminal proclivities. But Rufus, who has been raised by a fundamentalist grandfather, maintains that Satan makes him lie and steal and that nobody but Jesus can save him. Sheppard persists in trying to "save" the boy; and his own son, Norton, deprived of his father's love and understanding and believing from what Rufus has told him that if he is dead he will go to heaven where his mother is, commits suicide. Too late there comes to the father the tragic insight that "he had stuffed his own emptiness with good works like a glutton. He had ignored his own child to feed the vision of himself."

"Parker's Back," the last short story Miss O'Connor wrote, is one of her finest and also very different from any of her others. Its protagonist is a young former sailor named O. E. Parker, whose body is covered with tattoos except for his back. This had come about because the satisfaction he felt in a new tattoo never lasted for long, and soon he had to have another one. Finally, because his dissatisfaction is so great and because he thinks it will improve his relations with his shrewish and strongly fundamentalist young wife, he has a large, staring Byzantine Christ tattooed on his back. When the tattoo artist asks him if he has got religion, Parker replies, "I ain't got no use for that. A man can't save himself from whatever it is, he don't deserve no sympathy." But he finds that unaccountably he has become a witness for Jesus and that for the first time he is able freely to admit to his Old Testament prophet names, Obadiah Elihue. Ironically, his wife considers him an idolator. "God don't look like that," she indignantly cries. "He's a spirit. No man shall see his face." Beating him vigorously with a broom, she raises large welts on the face of the tattooed Christ.

"Parker's Back" is a very humorous story and at the same time one of Miss O'Connor's most allegorical presentations of the theme that it is natural for man to believe in Jesus and that he needs desperately to believe. Theological assumptions of this sort may well disturb those readers who do not share Miss O'Connor's Christian concerns. They may also feel that the representatives of secular humanism in her stories, such as Sheppard, deserve more sympathetic portrayal and have the cards stacked too much against them, and that, in addition, the religious visions or insights which end some of her stories are perhaps somewhat gratuitous. V. S. Pritchett, for example, confessed that he was not altogether happy about the sudden entrance of the Holy Ghost at the end of "The Enduring Chill," but he could also say, "If these stories are anti-rationalist propaganda, one does not notice until afterwards." And Irving Howe has said that, even though the reader may have different religious beliefs from Miss O'Connor's they need not affect his judgment of the artistry and psychological truth of her stories and that they must in any event be regarded as standing "securely on their own as renderings and criticisms of human experience." Few, if any, critics have disagreed with this judgment.

THE WRITERS discussed in this chapter (along with other contemporary short-story writers whose work is noted in the Appendix) have had much to say to us about ourselves and our times, and, transcending mere reportage, they have often done so with impressive artistry and insight. Despite such obstacles and handicaps as limited outlets for periodical publication, the notoriously poor sale of collected volumes of short fiction compared with that of novels, and the competition of films, television, and other mass media, the short story shows no decline. The decade of the 1960's produced many good stories and several of unusual distinction, such as Katherine Anne Porter's "Holiday" and Lawrence Sargent Hall's "The Ledge" (originally published in the *Hudson Review*), winner of the 1960 O. Henry award. Some excellent recently published stories by Eudora Welty, Bernard Malamud, and other writers make clear that the rich vein of contemporary short fiction in the tradition of realism is by no means exhausted, as some critics have averred. The potentialities of the experimental short story have been realized brilliantly on occasion in the work of such promising younger writers as Joyce Carol Oates, Donald Barthelme, and Robert Coover. American writers, we can hope, will continue to be challenged by the short-story form, providing us with more than a few good stories in the future and not too infrequently a story which has the marks of greatness.

# APPENDIX: Other Contemporary Short Story Writers

SO THAT THIS BOOK may have additional reference value, and because they are well worthy of the reader's attention, a number of contemporary short-story writers are given mention below.

NELSON ALGREN (1909– )

A product of Chicago's Northwest Side, which provides the setting for much of his fiction, Algren is the author of the widely praised *The Man With the Golden Arm* (1949) and two other naturalistic novels, *Never Come Morning* (1942) and the proletarian *A Walk on the Wild Side* (1956), a revision of Algren's first novel, *Somebody in Boots* (1935), the story of a young hobo during the Depression. The *Neon Wilderness* (1947) is a collection of twenty-four stories (several appear in somewhat different form as episodes in the novels), most of which treat the sordid and brutal aspects of Chicago Division Street life, with its slums, crooked cops, dope addicts, and prostitutes. Drab and deterministic though many of the stories are, they convey Algren's compassion for his defeated characters, as in "A Bottle of Milk for Mother," "Stickman's Laughter," and "Design for Departure," and the volume is also enhanced by the inclusion of several very good broadly humorous stories.

LOUIS AUCHINCLOSS (1917– )

In addition to actively engaging in the practice of law, Louis Auchincloss is a prolific writer of fiction and literary criticism. An expert and interesting novelist of upper-class eastern seaboard society and manners in the tradition of Henry James and Edith Wharton in such novels as *The Great World and Timothy Colt* (1956), *Portrait in Brownstone* (1962), *The Rector of Justin* (1964), and *The Embezzler* (1966), he has also published three collections of short stories. *The Injustice Collectors* (1950) is a collection of eight stories whose characters, in the author's words, are "people who are looking for injustice, even in a friendly world, because they suffer from a hidden need to feel that this world has wronged them." The smooth professional craftsmanship and sardonic wit that characterize these stories are displayed again in the sometimes overly contrived but highly readable stories about the personal and professional problems and conflicts of various members of a Wall Street law firm in *Powers of Attorney* (1963). Most of the twelve stories in *Second Chance: Tales of Two Generations* (1970) are about affluent members of the Establishment who find their old values inadequate when they face crises in their lives.

PAUL BOWLES (1911– )

An expatriate American living in Tangier, which, along with Morocco, Spain, the Sahara, and Central America, has provided the exotic and often sinister settings for his fiction, Bowles is the author of three novels: *The Sheltering Sky*

(1949), *Let It Come Down* (1952), and *Up Above the World* (1966). Like his novels, most of the seventeen pieces in his first and best collection of shorter fiction, *The Delicate Prey and Other Stories* (1950), rival and even outdo Poe, whose influence Bowles has acknowledged. At times Bowles seems to exploit horror and violence for their own sake, but the nightmarish reality and allegorical suggestiveness of "A Distant Episode," the title story, and several others is impressive. Two later volumes of stories in the same vein are *A Hundred Camels in the Courtyard* (1962) and *The Time of Friendship* (1967).

TRUMAN CAPOTE (1924– )

Now famous as the author of *In Cold Blood* (1964), a graphic and exhaustive factual account of a Kansas small-town murder case first printed in the *New Yorker*, Truman Capote achieved considerable reputation as a promising young writer with the publication of his first novel, *Other Voices, Other Rooms* in 1948, and his first volume of short fiction, *A Tree of Night and Other Stories* in 1951. In this early work Capote revealed himself as one of the most "Gothic" of those southern writers, among whom are William Faulkner, Eudora Welty, Carson McCullers, and Flannery O'Connor, who have portrayed, in at least some of their fiction, obsessed, degenerate, mentally or physically defective, or otherwise abnormal characters and have emphasized the macabre and decadent aspects of their southern settings. He showed himself also as perhaps too self-conscious a stylist and given to contrivance to achieve his effects, but in several short stories, again usually about children (for example, "Master Misery," "Children on Their Birthdays," "Miriam"), he treated loneliness, the search for identity, and other important themes of modern fiction with dramatic power and psychological insight. A marked change in subject and style is evident in the short novel and the three short stories included in *Breakfast at Tiffany's* (1958), the title story a delightfully humorous, if somewhat slick, piece about a Manhattan playgirl, and the three shorter pieces dealing with characters and settings more nearly normal and natural than those of the earlier stories. A volume revealing the considerable variety of Capote's writing over the past twenty years, with an introduction by Mark Schorer, is *The Selected Writings of Truman Capote* (1963).

JOHN CHEEVER (1912– )

Winner of the O. Henry Memorial Award first prize in 1956 for his story "The Country Husband," John Cheever has specialized in writing about the tensions, frustrations, futilities, and inanities of life in upper-middle-class suburbia. He has been criticized for being slick and contrived in some of his work and for writing according to the conventions of *New Yorker* fiction; but, though their situations and conflicts are often familiar enough, his best stories have freshness and genuine interest, and his fiction is often saved from being too depressingly realistic by a leavening of humor and fantasy. He has published three novels, *The Wapshot Chronicle* (1957), *The Wapshot Scandal* (1964), and *Bullet Park* (1969), and five collections of stories, *The Way Some People Live* (1943), *The Enormous Radio and Other Stories* (1953), *The Housebreaker of Shady Hill and Other Stories* (1958), *Some People, Places, and Things That Will Not Appear in My Next Novel* (1961), and *The Brigadier and the Golf Widow* (1964).

WALTER VAN TILBURG CLARK (1909–71)

The author of the well-known *The Ox-Bow Incident* (1940)

and other novels, Walter Van Tilburg Clark has also published *The Watchful Gods and Other Stories* (1950). Like his longer works, his short stories usually have western settings and deal with animals and men in their relation to nature. "Hook" makes highly dramatic the life history of a hawk, and the frequently anthologized "The Portable Phonograph" is a powerful parable on the themes of atavism and man's need for the aesthetic nourishment afforded him by literature and the arts.

JOHN COLLIER (1901– )

Although born in England, John Collier has lived for many years in the United States, writing for motion pictures and publishing short stories in the *New Yorker*, *Harper's*, and other American magazines. Even more than the earlier English writer Saki (H. H. Munro), whom he resembles, Collier is fond of the weird and macabre, the droll and ironic, and the twist ending. Highly imaginative in their use of fantasy and the supernatural, Collier's stories also often have a grim moral point, with retribution overtaking characters guilty of evil-doing, pride, or folly. *Fancies and Goodnights* (1951) is an omnibus collection containing fifty stories. Earlier volumes are *Presenting Moonshine* (1941) and *A Touch of Nutmeg* (1943).

CAROLINE GORDON (1895– )

Caroline Gordon, the wife of the poet and critic Allen Tate, may be likened to Katherine Anne Porter in that she is also a distinguished literary craftsman who writes of the influence of the past on her southern characters. Her short stories are collected in *Old Red and Other Stories* (1963), which reprints, along with additional stories, an earlier collection, *The Forest of the South* (1945). In Aleck Maury, the protagonist of "Old Red," and also of the novel *Aleck Maury, Sportsman* (1934) she has created one of the more memorable characters in recent American fiction.

SHIRLEY ANN GRAU (1929– )

The quality, if not the quantity, of Shirley Ann Grau's short fiction justifies ranking her alongside such other southern women writers as Eudora Welty, Carson McCullers, and Flannery O'Connor. She is at her best in the stories about Louisiana Negroes, included in *The Black Prince and Other Stories* (1955), but the other stories in the volume, as well as a number of later uncollected stories in the magazines also display a consistently high degree of narrative artistry. She has also published four novels: *The Hard Blue Sky* (1958), *The House on Coliseum Street* (1961), *The Keepers of the House* (1964), and *The Condor Passes* (1971).

SHIRLEY JACKSON (1919–65)

Shirley Jackson's story "The Lottery," a grim fable whose shocking conclusion implies that modern man needs scapegoats just as his primitive ancestors did, was much discussed when it was first published in the *New Yorker* in 1948. It has since been reprinted frequently in short-story anthologies. In other stories also, collected in *The Lottery: or The Adventures of James Harris* (1949), she makes effective use of supernatural and other unusual materials usually placed in realistic, everyday settings, as she does also in such novels as *Hangsaman* (1951), *The Haunting of Hill House* (1952), and *We Have Always Lived in the Castle* (1962).

CARSON McCULLERS (1917–67)

In addition to three well-known and highly praised novels, *The Heart Is a Lonely Hunter* (1940), *Reflections in a Golden*

*Eye* (1941), and *The Member of the Wedding* (1946), Carson McCullers is the author of a notable group of shorter narratives collected in *The Ballad of the Sad Café*. The long title story is a powerfully moving, if extreme, example of "Southern Gothic," with such grotesque characters as a six-foot female bootlegger, a Heathcliff-like hero, and a humpbacked dwarf. The characters in the other stories are less freakish, but they too suffer the plight of misfits or outcasts in a world full of loneliness and lacking in love.

JAMES PURDY (1923– )

One of the most individual of contemporary fiction writers is James Purdy, the author of three novels, *Malcolm* (1959), *The Nephew* (1960), and *Cabot Wright Begins* (1964), and of two short-story collections, *Color of Darkness: Eleven Stories and a Novella* (1957) and *Children Is All* (1962). The recipient of much critical praise, but not as well known to the general reader as he deserves to be, Purdy portrays with dramatic force and often with shocking impact characters made grotesque by loneliness and the inability to communicate, and human relationships in which hatred, cruelty, and callous indifference are all too prevalent.

PHILIP ROTH (1933– )

The author of four novels, *Letting Go* (1962), *When She Was Good* (1967), and the bestselling *Portnoy's Complaint* (1969) and *The Breast* (1972), Philip Roth also wrote an outstanding volume of short fiction, *Goodbye, Columbus and Five Short Stories* (1959), his first book and winner of the National Book Award. Often compared with Bernard Malamud, Roth rivals that writer in making his Jewish characters come alive and in portraying their conflicts and involvements with humor, irony and penetrating insight. All Roth's shorter pieces are very good, and he is at his best in the title short novel, a portrait of a young Jew striving to know himself and discover his place in the world, and in "Defender of the Faith," a richly ironic and subtle account of the morally frustrating relationship between a Jewish sergeant and a goldbricking Jewish army recruit.

IRWIN SHAW (1913– )

Irwin Shaw is one of the most professional and prolific of contemporary fiction writers. He has been criticized for deliberately exploiting topical subjects—World War II, anti-Semitism, and McCarthyism—and for writing overcontrived and sentimentalized stories, as for example, his famous wartime story about anti-Semitism, "Act of Faith," which compels admiration for its superb technique and its rhetorical persuasiveness in appealing to the emotions but fails to convince the reader as fully of its truth as do two other memorable stories of soldiers, William Styron's "The Long March" and Philip Roth's "Defender of the Faith." But Shaw's stories, in addition to their technical artistry, give the reader, as he wanted them to do, much insight into what people thought and felt during the middle decades of this century. *Mixed Company: Collected Short Stories* (1950) contains stories from three earlier volumes: *Sailor Off the Bremen and Other Stories* (1939), *Welcome to the City and Other Stories* (1942), and *Act of Faith and Other Stories* (1946). A later collection is *Tip on a Dead Jockey* (1957). Shaw also wrote *The Young Lions* (1948), one of the best World War II novels by an American writer.

PETER TAYLOR (1917– )

A native of Tennessee, Peter Taylor writes with a Jamesian artistry and sensibility about a middle-class South whose traditional values, morals, and manners have succumbed to a modern materialism which has made for instability and disorder. He published a novel, *A Woman of Means* (1950), and four volumes of short stories: *A Long Fourth* (1948), *The Widows of Thornton* (1954), *Happy Families Are All Alike* (1959), and *Miss Leonora, When Last Seen* (1963). His subtle and restrained manner of narration is particularly effective in "A Wife of Nashville," one of his many stories dealing with family relationships and domestic dilemmas, and in the brilliant "Venus, Cupid, Folly and Time," the 1959 O. Henry Award first-prize winner, which is both an account of social change and a moving story on the theme of the loss of innocence. *The Collected Stories of Peter Taylor* (1969) contains twenty-one stories, five of them not previously collected.

JOHN UPDIKE (1932– )

Most of John Updike's critics have viewed his work with mixed feelings. They concede that he has real talent, but they are also made a little uncomfortable by what is usually called Updike's "facility." They admire his achievement in giving reality to his fictional town of Olinger (modeled on the Shillington, Pennsylvania, of Updike's boyhood) and for writing so well about childhood and adolescence, but they deplore his overconcern with a narrowly limited autobiographical world which produces reminiscent "musings" or informal essays rather than genuine stories, or stories of husband-wife relationships which are sometimes slight and trivial. They praise his painter's eye for detail and the poetic quality of his prose, but they find him at times too elegant a stylist for his subject and too obvious a symbolist. Updike has written some memorable stories, some weak ones, and a good many in between. His best stories have substance and such meaningful themes as the influence of the past, life's transiency, and the power of love. Very prolific, he has so far published five volumes of stories, most of them originally printed in the *New Yorker*: *The Same Door* (1959), *Pigeon Feathers* (1962), *The Music School* (1966), *Bech: A Book* (1970), and *Museums and Women and Other Stories* (1972). He is also the author of several novels, among them *Rabbit Run* (1960), *Couples* (1968), and *Rabbit Redux* (1971), as well as a volume of verse and one of nonfiction prose.

Writers not mentioned above who have published collections of stories especially deserving of the reader's attention are John Barth, *Lost in the Funhouse* (1968); Donald Barthelme, *Come Back, Dr. Caligari* (1964), *Unspeakable Practices, Unnatural Acts* (1968), *City Life* (1970) and *Sadness* (1972); Saul Bellow, *Mosby's Memoirs and Other Stories* (1968); Ray Bradbury, *The Martian Chronicles* (1950); John Bell Clayton, *The Strangers Were There: Selected Stories* (1957); Robert Coover, *Pricksongs & Descants* (1969); George P. Elliott, *Among the Dangs* (1961); Herbert Gold, *Love and Like* (1960); Andrew Lytle, *A Novel, a Novella, and Four Stories* (1958); Robie Macauley, *The End of Pity and Other Stories* (1957); William March, *Trial Balance: Collected Short Stories* (1945); Vladimir Nabokov, *Nine Stories* (1947) and *Nabokov's Dozen* (1958); Joyce Carol Oates, *By the North Gate* (1963), *Upon the Sweeping Flood* (1966), *The Wheel of Love* (1970), and *Marriages and Infidelities* (1972); Reynolds Price, *The Names and Faces of Heroes* (1963); Mark Schorer, *The State of Mind* (1947); Delmore Schwartz, *The World is a Wedding* (1948); Isaac Bashevis Singer, *Gimpel the Fool and Other Stories* (1957) and *The Spinoza of Market Street* (1961); Wallace Stegner, *The Women on the Wall* (1949) and *The City of the Living*

(1956); Jesse Stuart, *Clearing in the Sky and Other Stories* (1950); Ruth Suckow, *Some Others and Myself: Seven Stories and a Memoir* (1952); Robert Penn Warren, *The Circus in the Attic and Other Stories* (1948); Jessamyn West, *The Friendly Persuasion* (1945); William Carlos Williams, *Make Light of It: Collected Stories* (1950); Edmund Wilson, *Memoirs of Hecate County* (1946); Donald Windham, *The Warm Country* (1962); and Richard Yates, *Eleven Kinds of Loneliness* (1962).

# BIBLIOGRAPHY

## GENERAL REFERENCES

The following listings are confined to works which the reader who wishes to pursue further his study of the short story will find particularly useful. With a few exceptions, collections of short stories cited in the text are not again listed here. For additional bibliographical information see *Literary History of the United States*, ed. by Robert E. Spiller et al. (3d ed., New York, 1963), II, and *Bibliography Supplement II* (1972); Lewis Leary, *Articles on American Literature, 1900–1950* (Durham, N.C., 1954); *American Bibliographies*, published annually in *Publications of the Modern Language Association; Eight American Authors: A Review of Research and Criticism*, ed. by Floyd Stovall (New York, 1963) (includes Poe, Hawthorne, Melville, Mark Twain, and Henry James); *American Literary Scholarship: An Annual*, ed. by James Woodress and J. Albert Robbins (Durham, N.C., 1963–67); Jarvis Thurston et al., *Short Fiction Criticism: A Checklist of Interpretation Since 1925 of Stories and Novelettes (American, British, Continental) 1800–1958* (Denver, 1960); and Warren S. Walker, *Twentieth Century Short Story Explication* (Hamden, Conn., 1967), and *Supplement* (1970). See also Dorothy E. Cook, ed.,

*Short Story Index* (New York, 1953), and supplements.

Useful bibliographies are provided in various numbers of the University of Minnesota Pamphlets on American Writers, ed. by Leonard Unger and George T. Wright. In the 103 pamphlets published to date the following short-story writers are included: Hemingway, Faulkner, Henry James, Mark Twain, Wharton, Melville, Fitzgerald, Hawthorne, Irving, Porter, Farrell, Cather, Bierce, Conrad Aiken, Anderson, Warren, Lardner, Salinger, O'Connor, London, Gordon, Jewett, Welty, McCarthy, Crane, Caldwell, Updike, O'Hara, McCullers, Singer, Poe, Barth, Nabokov, Styron, and Dreiser.

### *Anthologies of American Short Stories*

There are many anthologies of short stories in print, some designed for the general reader but the most as texts for courses in literature. Most of them also include stories by both American and English writers and often by European writers as well. Although they contain other stories in addition to American ones, the Brooks and Warren, and Gordon and Tate volumes are listed below because they have been highly influential, the former especially so, in promoting the close reading and intensive analysis that now characterize much short-story study and criticism.

William Abrahams, ed. *The O. Henry Award Prize Stories* (New York, annual publication); William Abrahams, ed. *Fifty Years of the American Short Story* (sixty stories chosen from the O. Henry award volumes, 1919–70, New York, 1970); Cleanth Brooks and Robert Penn Warren, eds., *Understanding Fiction*, 2d ed. (New York, 1959); Whit Burnett and Hallie Burnett, *Story: The Fiction of the Forties* (New York, 1949); John Henrik Clarke, ed., *American Negro Short Stories* (New York, 1966); Eugene Current-Garcia and Walton R. Patrick, eds., *American Short Stories: 1820 to the Present* (Chicago, 1964); Charles A. Fenton, ed., *The Best Short Stories of World War II: An American Anthology* (New

York, 1957); Martha Foley and David Burnett, eds., *The Best American Short Stories* (Boston, annual publication); Herbert Gold and David L. Stevenson, eds., *Stories of Modern America* (New York, 1961); Caroline Gordon and Allen Tate, eds., *The House of Fiction: An Anthology of the Short Story with Commentary*, 2d ed., (New York, 1960); Harold U. Ribalow, ed., *Treasury of American Jewish Stories* (New York, 1958); *Short Stories from the New Yorker* (New York, 1940); *55 Stories from the New Yorker* (New York, 1949); *Stories from the New Yorker* (New York, 1960); Wallace and Mary Stegner, eds., *Great American Short Stories* (Dell paperback, New York, 1957).

*History and Criticism of the American Short Story*

A comprehensive but now outdated survey of the short story from Irving to O. Henry is found in Fred Lewis Pattee, *The Development of the American Short Story* (New York, 1923). A brief survey of twentieth-century writers is found in Ray B. West, *The Short Story in America, 1900–1950* (Chicago, 1952). Considerable attention is given to the short story in Arthur Hobson Quinn, *American Fiction* (New York, 1936); and in George D. Snell, *The Shapers of American Fiction* (New York, 1947). Some attention is given to American writers, along with those of other nationalities, in H. E. Bates, *The Modern Short Story* (London, 1941 and Boston, 1950); in Frank O'Connor, *The Lonely Voice: A Study of the Short Story* (Cleveland, 1963); and in T. O. Beachcroft, *The Modest Art: A Survey of the Short Story in English* (London, 1968). Brief accounts of the development of the American short story are given in *American Short Stories: 1820 to the Present*, ed. by Eugene Current-Garcia and Walton R. Patrick (Chicago, 1964); in *Great American Short Stories*, ed. by Wallace and Mary Stegner (Dell paperback, New York, 1957); and in Danforth Ross, *The American Short Story*, (University of Minnesota Pamphlet on American Writers, Minneapolis, 1961). Books dealing with limited periods are Austin McGiffert Wright, *The American Short Story in the*

*Twenties* (Chicago, 1961); and William Peden, *The American Short Story* (Boston, 1964), a survey of the period from 1940 to 1963.

### *Chapter 1*

Editions of Irving's collected writings are *The Works of Washington Irving*, 21 vols. (New York, 1860–61); 12 vols. (New York, 1881); 40 vols. (New York, 1897). Well-edited volumes of selections are *Washington Irving: Representative Selections*, ed. by Henry A. Pochmann (New York, 1934); and *Washington Irving: Selected Prose*, ed. by Stanley T. Williams (Rinehart Edition, New York, 1950).

Accounts of Irving's life and work are given in Stanley T. Williams, *The Life of Washington Irving*, 2 vols. (New York, 1935); Edward Wagenknecht, *Washington Irving: Moderation Displayed* (New York, 1962); and William L. Hedges, *Washington Irving: An American Study, 1802–1832* (Baltimore, 1965). See also Walter A. Reichart, *Washington Irving and Germany* (Ann Arbor, 1957); Henry A. Pochmann, "Irving's German Tour and its Influence on His Tales," *Publications of the Modern Language Association*, Vol. XLV (December, 1930), 1150–87; Pochmann, "Irving's German Sources in *The Sketch Book*," *Studies in Philology*, Vol. XXVII (July, 1930), 477–507; Louis LeFevre, "Paul Bunyan and Rip Van Winkle," *Yale Review*, Vol. XXXVI (Autumn, 1946), 66–76; Daniel G. Hoffman, "Irving's Use of American Folklore in 'The Legend of Sleepy Hollow,'" *Publications of the Modern Language Association*, Vol. LXVIII (June, 1953), 425–35; Terence Martin, "Rip, Ichabod and the American Imagination," *American Literature*, Vol. XXXI (May, 1959), 137–49; Philip Young, "Fallen from Time: The Mythic Rip Van Winkle," *Kenyon Review*, Vol. XXII (Autumn, 1960), 547–73.

### *Chapter 2*

Editions of Hawthorne's collected writings are *The Complete Works of Nathaniel Hawthorne*, 12 vols. (Riverside Edition, Boston,

1883); and *The Writings of Nathaniel Hawthorne*, 22 vols. (Old Manse Edition, Boston, 1900). Volumes of selections with useful editorial material are *Hawthorne: Representative Selections*, ed. by Austin Warren (New York, 1934); *The Complete Novels and Selected Tales of Nathaniel Hawthorne*, ed. by Norman Holmes Pearson (Modern Library, New York, 1937); *Hawthorne's Short Stories*, ed. by Newton Arvin (New York, 1946); *The Portable Hawthorne*, ed. by Malcolm Cowley (New York, 1948); *Nathaniel Hawthorne: Selected Tales and Sketches*, ed. by Hyatt H. Waggoner (Rinehart Edition, New York, 1950).

Of the many biographies of Hawthorne, Randall Stewart, *Nathaniel Hawthorne* (New Haven, 1948), comes closest to being definitive but contains relatively little criticism of Hawthorne's work. Notes and sources for many of the tales are to be found in *The American Notebooks of Nathaniel Hawthorne*, ed. by Randall Stewart (New Haven, 1932). Discussions of the short fiction appear in Newton Arvin, *Hawthorne* (Boston, 1929); Mark Van Doren, *Hawthorne* (New York, 1949); Edward Wagenknecht, *Nathaniel Hawthorne* (New York, 1961); and Hubert H. Hoeltje, *Inward Sky: The Mind and Heart of Nathaniel Hawthorne* (Durham, N.C., 1962). An early critical biography is Henry James, *Hawthorne* (New York, 1879).

For additional criticism see F. O. Matthiessen, *American Renaissance* (New York, 1941); Arlin Turner, *Nathaniel Hawthorne* (New York, 1961); Hyatt H. Waggoner, *Hawthorne* (Cambridge, Mass., 1963); Richard H. Fogle, *Hawthorne's Fiction: The Light and the Dark* (Norman, Okla., 1964); Terence Martin, *Nathaniel Hawthorne* (New York, 1965); Frederick C. Crews, *The Sins of the Fathers: Hawthorne's Psychological Themes* (New York, 1966). Essays by Q. D. Leavis, Yvor Winters, and others dealing with the short stories are included in *Hawthorne: A Collection of Critical Essays*, ed. by A. N. Kaul (New York, 1966).

All fifteen of the tales and sketches published by Melville, along with an additional unpublished tale, "The Two Temples," are included in *The Complete Stories of Herman Melville*, ed. by Jay

Leyda (New York, 1949). Volumes of selected tales are *Herman Melville, Selected Tales and Poems*, ed. by Richard Chase (Rinehart Edition, New York, 1950); *Herman Melville: Representative Selections*, ed. by Willard Thorp (New York, 1938); *The Portable Melville*, ed. by Jay Leyda (New York, 1952); *Billy Budd and Other Tales* (Signet paperback, New York, 1961); *Melville's Billy Budd*, ed. by F. Barron Freeman (Cambridge, Mass., 1948); *Billy Budd, Sailor: An Inside Narrative*, ed. by Harrison Hayford and Merton M. Sealts, Jr. (Chicago, 1962). Biographical and critical studies of value to the student of Melville's shorter fiction include F. O. Matthiessen, *American Renaissance* (New York, 1941); Richard Chase, *Herman Melville: A Critical Study* (New York, 1949); Newton Arvin, *Herman Melville* (New York, 1950); Leon Howard, *Herman Melville: A Biography* (Berkeley, Calif., 1951); Jay Leyda, *The Melville Log* (New York, 1951); Richard H. Fogle, *Melville's Shorter Tales* (Norman, Okla., 1960); James E. Miller, *A Reader's Guide to Herman Melville* (New York, 1962); Tyrus Hillway, *Herman Melville* (New York, 1963); Charles G. Hoffman, "The Shorter Fiction of Herman Melville," *South Atlantic Quarterly*, Vol. LII (July, 1953), 414–30. A few of the many studies of the more important stories are Leo Marx, "Melville's Parable of the Walls" [on "Bartleby"], *Sewanee Review*, Vol. XLI (Autumn, 1953), 602–27; Mordecai Marcus, "Melville's Bartleby as a Psychological Double," *College English*, Vol. XXIII (February, 1962), 365–68; Marvin Felheim, "Meaning and Structure in 'Bartleby,' " *College English*, Vol. XXIII (February, 1962), 369–76; Joseph Schiffman, "Melville's Final Stage, Irony: A Re-examination of Billy Budd Criticism," *American Literature*, Vol. XXII (May, 1950), 128–36; Karl E. Zink, "Herman Melville and the Forms: Irony and Social Criticism in *Billy Budd*, " *Accent*, Vol. XII (Summer, 1952), 132–39; and Wendell Glick, "Expediency and Absolute Morality in *Billy Budd*," *Publications of the Modern Language Association*, Vol. LXVIII (March, 1953), 103–10. For additional criticism of the shorter fiction see "Criticism of Herman Melville:

A Selected Checklist," *Modern Fiction Studies*, Vol. VIII (Autumn, 1962), 312–46.

*Chapter 3*

The comprehensive edition of Poe's writings is *The Complete Works of Edgar Allan Poe*, 17 vols., ed. by James A. Harrison, (New York, 1902). Two convenient collections of the tales are *The Complete Tales and Poems of Edgar Allan Poe* (Modern Library, New York, 1938); and *The Complete Poems and Stories of Edgar Allan Poe*; 2 vols., ed. by Arthur Hobson Quinn and Edward H. O'Neill (New York, 1946). Volumes of selections are *Edgar Allan Poe: Representative Selections*, ed. by Margaret Alterton and Hardin Craig (New York, 1935); *The Portable Edgar Allan Poe*, ed. by Philip Van Doren Stern (New York, 1945); *Edgar Allan Poe: Selected Prose and Poetry*, ed. by W. H. Auden (Rinehart Edition, New York, 1950); and *Selected Writings of Edgar Allan Poe*, ed. by Edward H. Davidson (Riverside Edition, New York, 1956). Arthur H. Quinn, *Edgar Allan Poe: A Critical Biography* (New York, 1941), is detailed and scholarly.

Other biographical and critical studies are George E. Woodberry, *The Life of Edgar Allan Poe* (Boston, 1909); Hervey Allen, *Israfel: The Life and Times of Edgar Allan Poe* (New York, 1926); Marie Bonaparte, *The Life and Works of Edgar Allan Poe: A Psycho-Analytic Interpretation* (London, 1949); Edward H. Davidson, *Poe: A Critical Study* (Cambridge, Mass., 1957); Vincent Buranelli, *Edgar Allan Poe* (New York, 1961); and Edward Wagenknecht, *Edgar Allan Poe: The Man Behind the Legend* (New York, 1963). See also *The Letters of Edgar Allan Poe*, ed. by John Ward Ostrom (Cambridge, Mass., 1948, rev. ed., 1966). Two extensive collections of critical essays are *The Recognition of Edgar Allan Poe*, ed. by Eric W. Carlson (Ann Arbor, Mich., 1966); and *Poe: A Collection of Critical Essays*, ed. by Robert Regan (New York, 1967). For discussions of Poe's detective stories see the essays by E. M. Wrong, Dorothy Sayers, W. H. Wright, and others in *The Art of the Mystery Story*, ed. by Howard Haycraft (New York, 1946); W. K. Wimsatt, "Poe and the Mystery of Mary Rogers," *Publications of the Modern Language Association*, Vol. LV (March, 1941), 230–48; John Walsh, *Poe the Detective: The Curious Circumstances Behind the Mystery of Marie Roget* (New Brunswick, N. J., 1968); and Joseph J. Moldenhauer, "Murder as a Fine Art: Basic Connections Between Poe's Aesthetics, Psychology, and Moral Vision," *Publications of the Modern Language Association*, Vol. LXXXIII (May, 1968), 284–97.

## *Chapter 4*

Collections of local-color stories are *American Local-Color Stories*, ed. by Harry R. Warfel and G. Harrison Orians (New York, 1941); and *The Local Colorists*, ed. by Claude M. Simpson (New York, 1960). See also "Regionalism and Local Color," in *Literary History of the United States*, ed. by Robert E. Spiller et al. (New York, 1948), III, 304–25; and Donald A. Dike, "Notes on Local Color and Its Relation to Realism," *College English*, Vol. XIV (November, 1952), 81–88.

Editions of Harte's collected works are *The Writings of Bret Harte*, 19 vols. (Boston, 1896–1914); and *The Works of Bret Harte*, 25 vols. (New York, 1914). Volumes of selections are *The Luck of Roaring Camp and Selected Stories and Poems*, ed. by George R. Stewart, Jr. (New York, 1928); *The Luck of Roaring Camp, California Tales and Poems* (Everyman's Library, London, 1929); *The Best Stories of Bret Harte* (Modern Library, New York, 1947); and *The Outcasts of Poker Flat and Other Tales* (Signet paperback, New York, 1961). The introduction to *Bret Harte: Representative Selections*, ed. by Joseph B. Harrison (New York, 1941), discusses Harte's stories in considerable detail. For biography see George R. Stewart, Jr., *Bret Harte: Argonaut and Exile* (Boston, 1931); and Richard O'Connor, *Bret Harte: A Biography* (Boston, 1966).

The definitive edition of Twain's collected works is *The Writings of Mark Twain*, 37 vols., ed. by Albert B. Paine (New York, 1922–25). There are sixty stories and sketches in *The Complete Short Stories of Mark Twain*, ed. by Charles Neider (Bantam paperback, New York, 1958). Other collections are *The Family Mark Twain* (New York, 1935); *The Mark Twain Omnibus*, ed. by Max J. Herzberg (New York, 1935); *The Portable Mark Twain*, ed. by Bernard De Voto (New York, 1946); *Mark Twain; Representative Selections*, ed. by Fred L. Pattee (New York, 1935).

The fullest discussions of the tales and sketches are to be found in Bernard De Voto, *Mark Twain's America* (Boston, 1932); and Gladys Carmen Bellamy, *Mark Twain as a Literary Artist* (Norman, Okla., 1950). See also *Mark Twain: A Collection of Critical Essays*, ed. by Henry Nash Smith (New York, 1963). Critical biographies include Edward Wagenknecht, *Mark Twain: The Man and His Work* (Norman, Okla., 1961); and J. DeLancey Ferguson, *Mark Twain: Man and Legend* (New York, 1943). The authorized life is Albert B. Paine, *Mark Twain, A Biography* (New York, 1912).

## *Chapter 5*

Harriet Beecher Stowe, Rose Terry Cooke, Sarah Orne Jewett, and Mary E. Wilkins Freeman are discussed in Babette May Levy, "Mutations in New England Local Color," *New England Quarterly*, Vol. XIX (September, 1946), 338–58. John R. Adams, *Harriet Beecher Stowe* (New York, 1963), is a critical study.

Representative collections of Rose Terry Cooke's stories are *Somebody's Neighbors* (1881) and *Huckleberries Gathered from New England Hills* (1891).

A substantial amount of Sarah Orne Jewett's work is included in *Stories and Tales*, 7 vols. (Boston, 1910). Another collection is *The Best Stories of Sarah Orne Jewett*, ed. by Willa Cather (Boston, 1925, reissued as *The Country of the Pointed Firs and Other Stories*, Garden City, N.Y., 1954). Studies of Miss Jewett's life and work are F. O. Matthiessen, *Sarah Orne Jewett* (Boston, 1929); and Richard Cary, *Sarah Orne Jewett* (New York, 1962).

See also Willa Cather's essay on Miss Jewett in *Not Under Forty* (New York, 1936), 76–95; and *Letters of Sarah Orne Jewett*, ed. by Annie Fields (Boston, 1911).

Volumes of Mary E. Wilkins Freeman's stories not mentioned in the text are *Young Lucretia and Other Stories* (1892); *Silence and Other Stories* (1898); and *The Love of Parson Lord and Other Stories* (1914). The only volume of selections is *The Best Stories of Mary E. Wilkins Freeman*, ed. by Henry W. Lanier (New York, 1927). For biography and criticism see Edward Foster, *Mary E. Wilkins Freeman* (New York, 1956); and Perry D. Westbrook, *Mary Wilkins Freeman* (New York, 1967).

Cable and the other southern writers mentioned in the text are discussed by Carlos Baker in Chapter 52 of *Literary History of the United States*, ed. by Robert E. Spiller et al., 3d ed., (New York, 1963), II, 843–61. *Old Creole Days* (Heritage Press Edition, New York, 1943), with an introduction by Edward L. Tinker and an essay on Cable's New Orleans by Lafcadio Hearn, contains the seven stories of the original edition plus "Madame Delphine" and "Père Raphaël." See also Edward L. Tinker, "Cable and the Creoles," *American Literature*, Vol. V (January, 1934), 313–26. Arlin W. Turner, *George W. Cable* (Durham, N.C., 1956), and Philip Butcher, *George W. Cable* (New York, 1962), are biographical and critical studies.

Grace Elizabeth King writes of the traditions and manners of the Creoles of New Orleans in *Memories of a Southern Woman of Letters* (New York, 1932), as well as in her stories. In addition to biography and criticism, Daniel S. Rankin, *Kate Chopin and Her Creole Stories* (Philadelphia, 1932), includes previously unpublished and uncollected stories.

Biographical and critical studies of other writers discussed in this chapter are John D. Kern, *Constance Fenimore Woolson* (Philadelphia, 1934); Rayburn S. Moore, *Constance F. Woolson* (New York, 1963); Edd Winfield Parks, *Charles Egbert Craddock* (*Mary Noailles Murfree*) (Chapel Hill, N.C., 1941); Theodore L. Gross, *Thomas Nelson Page* (New York, 1967); Grant C. Knight, *James*

*Lane Allen and the Genteel Tradition* (Chapel Hill, N.C., 1935); and William K. Bottorf, *James Lane Allen* (New York, 1964).

For criticism of Hamlin Garland's fiction see Walter F. Taylor, *The Economic Novel in America* (Chapel Hill, N.C., 1942), 148–83; Bernard I. Duffey, "Hamlin Garland's 'Decline' from Realism," *American Literature*, Vol. XXV (March, 1953), 69–74; and Jean Holloway, *Hamlin Garland* (Austin, Texas, 1960). Garland has a good deal to say about his stories in *A Son of the Middle Border* (1917). *Main-Travelled Roads*, ed. by Thomas A. Bledsoe (Rinehart Edition, New York, 1954), reproduces the text of the six stories of the first edition published in 1891.

## *Chapter 6*

For the stories of Hale and Aldrich see *The Works of Edward Everett Hale*, 10 vols. (Boston, 1898); and *The Writings of Thomas Bailey Aldrich*, 9 vols. (Boston, 1907). Biographical and critical studies of Aldrich are Ferris Greenslet, *The Life of Thomas Bailey Aldrich* (Boston, 1908); and Charles E. Samuels, *Thomas Bailey Aldrich* (New York, 1965).

Stockton's writings are collected in *The Novels and Stories of Frank R. Stockton*, 24 vols. (New York, 1899–1904). A biographical and critical study is Martin I. J. Griffin, *Frank R. Stockton* (Philadelphia, 1939).

Most of Bunner's prolific output of stories is included in *The Stories of H. C. Bunner: First and Second Series*, 4 vols. (New York, 1916). For biography and criticism see Gerard E. Jensen, *The Life and Letters of Henry Cuyler Bunner* (Durham, N.C., 1939); and Gabriel Leeb, "The United States Twist: Some Plot Revisions by Henry Cuyler Bunner," *American Literature*, Vol. IX (January, 1938), 431–41.

Editions of Fitz-James O'Brien's writings are *The Poems and Stories of Fitz-James O'Brien*, ed. by William Winter (Boston, 1881); *The Collected Stories of Fitz-James O'Brien*, ed. by Edward J. O'Brien (New York, 1925); and *The Diamond Lens and Other Stories* (New York, 1932).

Bierce's stories are reprinted in *The Collected Writings of Ambrose Bierce*, ed. by Clifton Fadiman (New York, 1946). Biographical and critical studies are Carey McWilliams, *Ambrose Bierce: A Biography* (New York, 1929); C. Hartley Grattan, *Bitter Bierce: A Mystery of American Letters* (New York, 1929); Napier Wilt, "Ambrose Bierce and the Civil War," *American Literature*, Vol. I (November, 1929) 260–85; Arthur M. Miller, "The Influence of Edgar Allan Poe on Ambrose Bierce," *American Literature*, Vol. IV (May, 1932), 130–50; Eric Solomon, "The Bitterness of Battle: Ambrose Bierce's War Fiction," *Midwest Quarterly*, Vol. V (Winter, 1964), 147–65; Richard O'Connor, *Ambrose Bierce: A Biography* (Boston, 1967); and M. E. Grenander, *Ambrose Bierce* (New York, 1971).

Collected editions of O. Henry include *Complete Writings of O. Henry*, 14 vols. (Garden City, N.Y., 1917); *O. Henry Biographical Edition*, 18 vols. (Garden City, N.Y., 1929); *The Complete Works of O. Henry*, 2 vols. (Garden City, N.Y., 1953). Selections of stories are *Selected Stories from O. Henry*, ed. by C. Alphonso Smith (Garden City, N.Y., 1922); *The Voice of the City and Other Stories by O. Henry*, ed. by Clifton Fadiman (New York, 1935); and *Best Short Stories of O. Henry*, ed. by Bennett Cerf and Van H. Cartmell (Modern Library, New York, 1945). Book-length studies of O. Henry are Hudson Long, *O. Henry: The Man and His Work* (Philadelphia, 1949); Gerald Langford, *Alias O. Henry* (New York, 1957); and Eugene Current-Garcia, *O. Henry* (New York, 1965). Earlier biographies are C. Alphonso Smith, *O. Henry Biography* (New York, 1916); and Robert H. Davis and Arthur B. Maurice, *The Caliph of Bagdad* (New York, 1931). See also "The Age of O. Henry," in Fred L. Pattee, *Side-lights on American Literature* (New York, 1922), 3–55.

## *Chapter 7*

There are two collected editions of James's fiction. *The Novels and Tales of Henry James*, 24 vols. (New York Edition, New York, 1907–1909), was assembled by James himself. To this edition were added the unfinished novels, *The Ivory Tower* and *The Sense of the Past* in 1917, and the *Letters* in 1920. *The Novels and Stories of Henry James* (London, 1921–23) contains the texts and prefaces of the New York Edition plus several novels and a considerable number of tales not included in that edition. The most comprehensive collection of James's shorter fiction is *The Complete Tales of Henry James*, 12 vols., ed. by Leon Edel (Philadelphia, 1962–65). Volumes of selected stories, most of them with useful commentary, are *The Short Stories of Henry James*, ed. by Clifton Fadiman (New York, 1945); *Henry James: Selected Short Stories*, ed. by Quentin Anderson (Rinehart Edition, New York, 1950); *The Great Short Novels of Henry James*, ed. by Philip Rahv (New York, 1944); *Eight Uncollected Tales of Henry James*, ed. by Edna Kenton (New Brunswick, N.J., 1950); *Henry James: Stories of Writers and Artists*, ed. by F. O. Matthiessen (New York, 1944); *The Ghostly Tales of Henry James*, ed. by Leon Edel (New Brunswick, N.J., 1948); *Henry James: Representative Selections*, ed. by Lyon N. Richardson (Cincinnati, 1941); *The Portable Henry James*, ed. by Morton Dauwen Zabel (New York, 1951); *Henry James: Fifteen Short Stories*, ed. by Morton Dauwen Zabel (Bantam paperback, New York, 1961). An indispensable book for the student of James's fiction is *The Notebooks of Henry James*, ed. by F. O. Matthiessen and Kenneth B. Murdock (New York, 1947). For biography and criticism see F. W. Dupee, *Henry James* (New York, 1951); and Leon Edel, *Henry James*, 5 vols. (New York, 1953–72). General studies of James's fiction are Joseph Warren Beach, *The Method of Henry James* (New Haven, 1918); Elizabeth Stevenson, *The Crooked Corridor: A Study of Henry James* (New York, 1949); Bruce R. McElderry, Jr., *Henry James* (New York, 1965); S. Gorley Putt, *Henry James: A Reader's Guide* (Ithaca, N.Y., 1966). James's writings up to 1881 are discussed in detail in Cornelia P. Kelley, *The Early Development of Henry James* (Urbana, Ill., 1930). F. O. Matthiessen, *Henry James: The Major Phase* (New York, 1944), is a study of the novels and stories of the period 1895–1910. A collection of essays on various aspects of James is *The Question of Henry James*, ed. by F. W. Dupee (New York, 1945). For a comprehensive listing of critical articles on James's individual stories, which are so numerous that it is impossible to give an adequate indication of the best of them here, see Maurice Beebe and William T. Stafford, "Criticism of Henry James: A Selected Checklist," *Modern Fiction Studies*, Vol. XII (Spring, 1966), 117–77.

### *Chapter 8*

*The Works of Stephen Crane*, ed. by Wilson Follett, 12 vols. (New York, 1925–27), has introductions by Amy Lowell, Willa Cather, Sherwood Anderson, and others. Volumes of stories in *The Works of Stephen Crane*, ed. by Fredson Bowers (Charlottesville, Va.), all with excellent scholarly introductions, are: Vol. 1, *Bowery Tales* (1967); Vol. 5, *Tales of Adventure* (1970); Vol. 6, *Tales of War* (1970); Vol. 7, *Tales of Whilomville* (1969); Vol. 8, *Tales, Sketches and Reports* (1973). A single-volume collected edition is *The Complete Short Stories and Sketches of Stephen Crane*, ed. by Thomas A. Gullason (New York, 1963). *Twenty Stories by Stephen Crane*, ed. by Carl Van Doren (New York, 1940), contains most of Crane's best stories. Other volumes containing selected stories are *Men, Women and Boats* (New York, 1921); *Stephen Crane: An Omnibus*, ed. by Robert W. Stallman (New York, 1952); *Stephen Crane: The Red Badge of Courage and Selected Prose and Poetry*, ed. by William M. Gibson (Rinehart Edition, New York, 1956). *Sullivan County Sketches*, ed. by Melvin Schoberlin (Syracuse, N.Y., 1949), is a collection of early writings. Biographical and critical studies are Thomas Beer, *Stephen Crane: A Study in American Letters* (New York, 1923); John Berryman, *Stephen Crane* (New York, 1950); Edwin H. Cady, *Stephen Crane* (New York, 1962); Robert W. Stallman, *Stephen Crane* (New York, 1968); and *Stephen Crane's Career: Perspectives and Evaluations*, ed. by Thomas A. Gullason (New York, 1971). For additional criticism see Maurice Beebe and Thomas A. Gullason, "Criticism of Stephen Crane:

*Eros* (1928), and *Among the Lost People* (1934). "Mr. Arcularis" and Aiken's play of the same title based on the story are discussed in Rufus A. Blanchard, "Metamorphosis of a Dream," *Sewanee Review*, Vol. LXV (Oct.–Dec., 1957), 694–702. See also James W. Tuttleton, "Aiken's 'Mr. Arcularis': Psychic Regression and the Death Instinct," *American Imago*, Vol. XX (Winter, 1963), 295–314. Frederick J. Hoffman, *Conrad Aiken* (New York, 1962), is a critical study.

*Selected Works of Stephen Vincent Benét*, New York, 1943, contains twenty-two stories. Other collections are *Twenty-five Short Stories of Stephen Vincent Benét* (Garden City, N.Y., 1943), and *The Last Circle* (New York, 1946). For biography and criticism see Charles A. Fenton, *Stephen Vincent Benét* (New Haven, 1958), and Parry Stroud, *Stephen Vincent Benét* (New York, 1962).

## *Chapter 11*

Several of Hemingway's stories published in magazines have not been collected in book form. They include "The Denunciation" (*Esquire*, November, 1938), "The Butterfly and the Tank" (*Esquire*, December, 1938), "Night After Battle" (*Esquire*, February, 1939), "Nobody Ever Dies" (*Cosmopolitan*, March, 1939), "Under the Ridge (*Cosmopolitan*, October, 1939), and two stories in the *Atlantic Monthly* (November, 1957): "Man of the World," and "Get a Seeing Eye Dog." Also, eight of the twenty-four stories about Hemingway's semiautobiographical hero in *The Nick Adams Stories* (New York, 1972) were previously unpublished.

Critical studies are Carlos Baker, *Hemingway: The Writer as Artist* (Princeton, N.J., 1952); Philip Young, *Ernest Hemingway* (New York, 1952, rev. ed., 1966); and Earl Rovit, *Ernest Hemingway* (New York, 1963). See also C. A. Fenton, *The Apprenticeship of Ernest Hemingway* (New York, 1954). The authorized biography is Carlos Baker, *Ernest Hemingway: A Life Story* (New York, 1969). Collections of critical essays are *Ernest Hemingway: The Man and His Work*, ed. by John K. M. McCaffery (Cleveland and New York, 1950); and *Hemingway and His Critics*, ed. by



A Selected Checklist," *Modern Fiction Studies*, Vol. V (Autumn, 1959), 282–91.

Volumes of selected stories by Jack London are *The Call of the Wild and Other Stories*, ed. by Frank L. Mott (New York, 1935); *Best Short Stories of Jack London* (Garden City, N.Y., 1945); *Jack London's Tales of Adventure*, ed. by Irving Shepard (Garden City, N.Y., 1956); *The Call of the Wild and Selected Stories*, ed. by Franklin Walker, Signet paperback (New York, 1960). For biography and criticism see Irving Stone, *Sailor on Horseback: The Biography of Jack London* (New York, 1947); Richard O'Connor, *Jack London: A Biography* (New York, 1964); and Franklin Walker, *Jack London and the Klondike* (San Marino, Calif., 1966).

Collections of Edith Wharton's stories are *An Edith Wharton Treasury*, ed. by Arthur Hobson Quinn (New York, 1950); *The Best Short Stories of Edith Wharton*, ed. by Wayne Andrews (New York, 1958); *The Edith Wharton Reader*, ed. by Louis Auchincloss (New York, 1965); and *The Collected Short Stories of Edith Wharton*, ed. by R. W. B. Lewis, 2 vols. (New York, 1968). Mrs. Wharton published *The Writing of Fiction* in 1925 and the autobiographical *A Backward Glance* in 1934, both of which provide insight into her aims and methods in writing fiction. Biographical and critical studies are Percy Lubbock, *Portrait of Edith Wharton* (New York, 1947), and Blake Nevius, *Edith Wharton: A Study of Her Fiction*, (Berkeley and Los Angeles, 1953). See also Millicent Bell, "Edith Wharton and Henry James: The Literary Relation," *Publications of the Modern Language Association*, Vol. LXXIV (December, 1959), 619–37.

Willa Cather's fiction is collected in *The Novels and Stories of Willa Cather*, 13 vols. (Boston, 1937–41). For her early writings see Willa Cather, *Collected Short Fiction, 1892–1912*, ed. by Mildred R. Bennett (Lincoln, Neb. 1965). Critical studies are David Daiches, *Willa Cather: A Critical Introduction* (Ithaca, N.Y., 1951); and E. K. Brown and Leon Edel, *Willa Cather: A Critical Biography* (New York, 1953). See also Mildred R. Bennett, *The*

*World of Willa Cather* (New York, 1951); Elizabeth Shepley Sergeant, *Willa Cather: A Memoir* (Philadelphia and New York, 1953); and Edward A. and Lillian D. Bloom, *Willa Cather's Gift of Sympathy* (Carbondale, Ill., 1962).

*The Best Short Stories of Theodore Dreiser,* ed. by Howard Fast (Cleveland, 1947), contains fourteen stories, most of them taken from *Free and Other Stories* (1918). Critical biographies are Robert H. Elias, *Theodore Dreiser: Apostle of Nature* (New York, 1949); F. O. Matthiessen, *Theodore Dreiser* (New York, 1951); Philip Gerber, *Theodore Dreiser* (New York, 1964); and W. A. Swanberg, *Dreiser* (New York, 1965). A collection of critical essays and other writings concerning Dreiser is *The Stature of Theodore Dreiser,* ed. by Alfred Kazin and Charles Shapiro (Bloomington, Ind., 1955).

*Chapter 9*

The best edition of Anderson's *Winesburg, Ohio* is that edited by Malcolm Cowley (New York, 1960). Selected *Winesburg* pieces and other stories are included in *The Sherwood Anderson Reader,* ed. by Paul Rosenfeld (Boston, 1947); and in *The Portable Sherwood Anderson,* ed. by Horace Gregory (New York, 1949). Biographical and critical studies are Irving Howe, *Sherwood Anderson* (New York, 1951); James Schevill, *Sherwood Anderson* (Denver, 1951); Rex Burbank, *Sherwood Anderson* (New York, 1964); and David D. Anderson, *Sherwood Anderson* (New York, 1967). Critical and scholarly studies include Lionel Trilling, "Sherwood Anderson," *Kenyon Review,* Vol. III (Summer, 1941), 293–302 (reprinted in Trilling's *The Liberal Imagination* [New York, 1950], 22–33); Frederick J. Hoffman, *Freudianism and the Literary Mind* (Baton Rouge, La., 1945), 230–55; William Phillips, "How Sherwood Anderson Wrote *Winesburg, Ohio,*" *American Literature,* Vol. XXIII (March, 1951), 7–30; Jon S. Lawry, " 'Death in the Woods' and the Artist's Self in Sherwood Anderson," *Publication of the Modern Language Association,* Vol. LXXIV (June, 1959), 306–11. For additional criticism see *The Achievement of Sherwood Anderson,* ed. by Ray Lewis White (Chapel Hill, N. C., 1966). See also *The Letters of Sherwood Anderson,* ed. by Howard Mumford Jones and Walter B. Rideout (Boston, 1953).



*Chapter 10*

*The Best Short Stories of Wilbur Daniel Steele* (New York, 1946), is a representative selection of twenty-four stories. For biography and criticism see Martin Bucco, *Wilbur Daniel Steele* (New York, 1972).

Most of Ring Lardner's published stories are included in *Round Up: The Stories of Ring W. Lardner* (New York, 1929). Critical and biographical studies are Donald Elder, *Ring Lardner* (New York, 1956); and Walton R. Patrick, *Ring Lardner* (New York, 1963).

F. Scott Fitzgerald published four volumes of short stories: *Flappers and Philosophers* (1921), *Tales of the Jazz Age* (1922), *All the Sad Young Men* (1926), *Taps at Reveille* (1935). *The Stories of F. Scott Fitzgerald,* ed. by Malcolm Cowley (New York, 1953), contains twenty-eight stories, nine of them previously uncollected. Fourteen other uncollected stories are included in *Afternoon of an Author,* ed. by Arthur Mizener (New York, 1957). Biographical and critical studies are Arthur Mizener, *The Far Side of Paradise* (Boston, 1951); *F. Scott Fitzgerald: The Man and His Work,* ed. by Alfred Kazin (Cleveland and New York, 1951); Andrew Turnbull, *Scott Fitzgerald* (New York, 1962); Kenneth E. Eble, *F. Scott Fitzgerald* (New York, 1963); James E. Miller, *F. Scott Fitzgerald: His Art and His Technique* (New York, 1964); Henry Dan Piper, *F. Scott Fitzgerald: A Critical Portrait* (New York, 1965). For additional critical studies see Maurice Beebe and Jackson R. Bryer. "Criticism of F. Scott Fitzgerald: A Selected Checklist," *Modern Fiction Studies,* Vol. VII (Spring, 1961), 82–94.

*The Short Stories of Conrad Aiken,* New York, 1950, contains twenty-nine stories, most of which are reprinted from three earlier collections: *Bring! Bring! and Other Stories* (1925), *Costumes by*

Carlos Baker (New York, 1961). There is a full listing of critical studies in Maurice Beebe and John Feaster, "Criticism of Ernest Hemingway: A Selected Checklist," *Modern Fiction Studies,* Vol. XIV (Autumn, 1968), 337–69. Information concerning the publication of the short stories is given in Lee Samuels, *A Hemingway Checklist* (New York, 1951).

## *Chapter 12*

The history of the publication of Faulkner's short stories in magazines and in book form, a somewhat complicated one, is given in James B. Meriwether, *The Literary Career of William Faulkner: A Bibliographical Study* (Princeton, N.J., 1961). A collection of Faulkner's stories not mentioned in the text is *Big Woods* (New York, 1955). It includes, in addition to three previously collected hunting stories ("The Bear," "The Old People," "A Bear Hunt"), "Race at Morning," originally published in the *Saturday Evening Post,* (March 5, 1955). Not collected is "By the People," *Mademoiselle* (October, 1955). A volume of early writings is William Faulkner, *New Orleans Sketches,* ed. by Carvel Collins (New York, 1968). *The Portable Faulkner,* ed. by Malcolm Cowley (New York, 1954), includes selected stories. For criticism see Irving Howe, *William Faulkner: A Critical Study* (New York, 1952); William Van O'Connor, *The Tangled Fire of William Faulkner* (Minneapolis, 1954); and Hyatt H. Waggoner, *William Faulkner* (Lexington, Ky., 1959). See also *William Faulkner: Two Decades of Criticism,* ed. by F. J. Hoffman and Olga Vickery (East Lansing, Mich., 1951, rev. ed. 1960). For additional critical studies see Maurice Beebe, "Criticism of William Faulkner: A Selected Checklist," *Modern Fiction Studies,* Vol. XIII (Spring, 1967), 115–61.

## *Chapter 13*

Proletarian writers and various aspects of proletarian literature are discussed by Leslie Fiedler, Irving Howe, and others in *Proletarian Writers of the Thirties,* ed. by David Madden (Carbondale,

Ill., 1968). A representative volume of proletarian short stories is Albert Maltz's *The Way Things Are* (New York, 1938).

Two comprehensive collections of Caldwell's short fiction are *Jackpot: The Short Stories of Erskine Caldwell* (New York, 1948); and *The Complete Stories of Erskine Caldwell* (New York, 1953). Caldwell discusses his fiction and his writing methods in *Call It Experience: The Years of Learning How to Write* (New York, 1951).

An extensive collection of Farrell's earlier and most characteristic stories is *The Short Stories of James T. Farrell* (New York, 1937). Farrell's views on literature and society are expressed in *A Note on Literary Criticism* (New York, 1936); and in *Literature and Morality* (New York, 1947). Edgar M. Branch, *James T. Farrell* (New York, 1971) is a critical study.

Critical studies of Steinbeck are Peter Lisca, *The Wide World of John Steinbeck* (New Brunswick, N.J., 1958); Warren French, *John Steinbeck* (New York, 1961); and Joseph Fontenrose, *John Steinbeck* (New York, 1963). For additional criticism see Maurice Beebe and Jackson R. Bryer, "Criticism of John Steinbeck: A Selected Checklist," *Modern Fiction Studies*, Vol. XI (Spring, 1965), 90–103.

Anthologies of Saroyan's short fiction are *The Saroyan Special* (New York, 1948), an omnibus collection of ninety-two stories, and *The William Saroyan Reader* (New York, 1958). For criticism see Howard R. Floan, *William Saroyan* (New York, 1966).

Critical studies of John O'Hara are E. Russell Carson, *The Fiction of John O'Hara* (Pittsburgh, 1961); and Sheldon Grebstein, *John O'Hara* (New York, 1966). A posthumous volume of thirty-four previously uncollected stories is *The Time Element and Other Stories* (New York, 1972).

*Here Lies: The Collected Stories of Dorothy Parker* (New York, 1939), reprints the stories in two earlier volumes. *Laments for the Living* (New York, 1930), and *After Such Pleasures* (New York, 1933). For criticism see Mark Van Doren, "Dorothy Parker," *English Journal*, Vol. XXIII (September, 1934), 535–43.

There is also some discussion of her stories in John Keats, *You Might As Well Live* (New York, 1970).

Most of James Thurber's pieces which can be considered short stories, as distinguished from his humorous essays, autobiographical sketches, and other literary journalism, are to be found in *The Middle-Aged Man on the Flying Trapeze* (New York, 1935); *My World and Welcome to It* (New York, 1942); *The Thurber Carnival* (New York, 1945); and *Thurber Country* (New York, 1953). For criticism see Robert H. Elias, "James Thurber: The Primitive, the Innocent, and the Individual," *American Scholar*, Vol. XXVII (Summer, 1958), 355–63; and Robert E. Morsburger, *James Thurber* (New York, 1964).

Kay Boyle's *Thirty Stories* (New York, 1946), contains stories from several earlier collections. Discussions of her work are Struthers Burt, "The Mature Craft of Kay Boyle," *Saturday Review of Literature*, Vol. XXIX (November 30, 1946), 11, and Richard C. Carpenter, "Kay Boyle," *College English*, Vol. XV (November, 1953), 81–87.

*Chapter 14*

The fullest critical studies are Harry John Mooney, *The Fiction and Criticism of Katherine Anne Porter* (Pittsburgh, 1962); and George Hendrick, *Katherine Anne Porter* (New York, 1965). For additional criticism see Robert Penn Warren, "Irony with a Center; Katherine Anne Porter," in *Selected Essays*, (New York, 1958); Sarah Youngblood, "Structure and Imagery in Katherine Anne Porter's 'Pale Horse, Pale Rider,'" *Modern Fiction Studies*, Vol. V (Winter, 1959), 344–52; Marjorie Ryan, "*Dubliners* and the Stories of Katherine Anne Porter," *American Literature*, Vol. XXXI (January, 1960), 464–73; James William Johnson, "Another Look at Katherine Anne Porter," *The Virginia Quarterly Review*, Vol. XXXVI (Autumn, 1960), 598–613; Charles A. Allen, "The Nouvelles of Katherine Anne Porter," *University of Kansas City Review*, Vol. XXIX (December, 1962), 87–93; Daniel Curley, "Katherine Anne Porter: The Larger Plan," *Kenyon Review*,

Vol. XXVI (Autumn, 1963), 671–95; and *Katherine Anne Porter: A Critical Symposium,* ed. by Lodwick Hartley and George Core (Athens, Ga., 1969). The origins of "Noon Wine" are described by Miss Porter in " 'Noon Wine': The Sources," *Yale Review,* Vol. XLVI (Autumn, 1956), 22–39 (reprinted in Cleanth Brooks and Robert Penn Warren, *Understanding Fiction,* 2d ed. (New York, 1959), 610–20.

*Chapter 15*

Critical studies of Eudora Welty include Robert Penn Warren, "Love and Separateness in Eudora Welty," *Kenyon Review,* Vol. VI (Spring, 1944), 246–59 (reprinted in *Selected Essays* [New York, 1958], 156–69); Eunice Glenn, "Fantasy in the Fiction of Eudora Welty," *A Southern Vanguard,* ed. by Allen Tate (New York, 1947, reprinted in *Critiques and Essays on Modern Fiction, 1920–1951,* ed. by John Aldridge [New York, 1952], 506–17); Granville Hicks, "Eudora Welty," *College English,* Vol. XIV (November, 1952), 69–76; Robert Daniel, "The World of Eudora Welty," *Hopkins Review,* Vol. VII (Winter, 1953), 49–58 (reprinted in *Southern Renascence,* ed. by Louis D. Rubin, Jr., and Robert D. Jacobs [Baltimore, 1953], 306–15); Ruth M. Vande Kieft, *Eudora Welty* (New York, 1962); Alun R. Jones, "The World of Love: The Fiction of Eudora Welty," in *The Creative Present,* ed. by Nona Balakian and Charles Simmons (New York, 1963), 175–92.

Miss Welty has discussed her theory and practice of fiction in several essays: "How I Write," *Virginia Quarterly Review,* Vol. XXXI (Spring, 1955), 240–51 (reprinted in Cleanth Brooks and Robert Penn Warren, *Understanding Fiction,* 2d ed. [New York, 1959], 545–53); *Short Stories* (New York, 1959, an expanded version of "The Reading and Writing of Short Stories," *Atlantic Monthly,* Vol. CLXXXIII [February–March, 1949], 54–58 and 46–49); "Must the Novelist Crusade?" *Atlantic Monthly,* Vol. CCXVI (October, 1965), 104–108; "Eye of the Story," *Yale Review,* Vol. LV (December, 1965), 265–74. Recently published uncollected stories, all of which appeared in the *New Yorker,* are "Where Is the Voice Coming from?" (July 6, 1963), "The Demonstrators," (November 26, 1966), and "The Optimist's Daughter" (March 15, 1969).

Mary McCarthy's fiction is discussed in Barbara McKenzie, *Mary McCarthy* (New York, 1966); and in Doris Grumbach, *The Company She Kept* (New York, 1967).

*The Collected Stories of Jean Stafford* (New York, 1969) contains thirty stories. Critical studies are Louis Auchincloss, "Jean Stafford," *Pioneers and Caretakers* (Minneapolis, 1965), 152–60; Ihab H. Hassan, "Jean Stafford: The Expense of Style and the Scope of Sensibility," *Western Review,* Vol. XIX (Spring, 1955), 185–203; and Olga W. Vickery, "Jean Stafford and the Ironic Vision," *South Atlantic Quarterly,* Vol. LXI (Autumn, 1962), 484–91. Miss Stafford comments on her work in "Truth in Fiction," *Library Journal,* Vol. LXXXXI (October 1, 1966), 4557–65.

The most detailed critical discussion of J. F. Powers is John V. Hagopian, *J. F. Powers* (New York, 1968). Other critical studies are Naomi Lebowitz, "The Stories of J. F. Powers: The Sign of the Contradiction," *Kenyon Review,* Vol. XX (Summer, 1958), 494–99, and Robert G. Twombly, "Hubris, Health and Holiness: The Despair of J. F. Powers," *Seven Contemporary Authors,* ed. by Thomas B. Whitbread (Austin, 1966), 143–62.

Warren French, *J. D. Salinger* (New York, 1963), is a comprehensive critical study. Many of the numerous essays on Salinger can be found in *Salinger: A Critical and Personal Portrait,* ed. by Henry Anatole Grunwald (New York, 1962); *Studies in J. D. Salinger,* ed. by Marvin Laser and Norman Fruman (New York, 1963); and *J. D. Salinger and the Critics,* ed. by William F. Belcher and James W. Lee (Belmont, Calif., 1962). See also Maurice Beebe and Jennifer Sperry, "Criticism of J. D. Salinger: A Selected Checklist," *Modern Fiction Studies,* Vol. XII (Autumn, 1966), 377–90.

Recently published uncollected stories of Bernard Malamud are "Man in the Drawer" (*Atlantic Monthly,* April, 1968), named the 1968 O. Henry Memorial Award first-prize story, and "An

Exorcism" (*Harper's*, December, 1968). For criticism see Earl H. Rovit, "Bernard Malamud and the Jewish Literary Tradition," *Critique*, Vol. XII (Winter–Spring, 1960), 3–10; Granville Hicks, "Generations of the Fifties: Malamud, Gold, and Updike," in *The Creative Present*, ed. by Nona Balakian and Charles Simmons (New York, 1963), 217–37; Sidney Richman, *Bernard Malamud* (New York, 1966); Leslie A. Field and Joyce W. Field, eds., *Bernard Malamud and the Critics* (New York, 1970).

Flannery O'Connor, *Mystery and Manners*, ed. by Sally and Robert Fitzgerald (New York, 1969), is a posthumous collection of articles and lectures on writing. For discussions of Miss O'Connor's stories see Robert Fitzgerald, "The Countryside and the True Country," *Sewanee Review*, Vol. LXX (Summer, 1962), 380–94; Robert Drake, *Flannery O'Connor: A Critical Essay* (Grand Rapids, Mich., 1966); Ruth Vande Kieft, "Judgment in the Fiction of Flannery O'Connor," *Sewanee Review*, Vol. LXXVI (April–June, 1968), 337–56; Thomas M. Carlson, "Flannery O'Connor: The Manichaean Dilemma," *Sewanee Review*, Vol. LXXVII (Spring, 1969), 254–76; and Josephine Hendin, *The World of Flannery O'Connor* (Bloomington, Ind., 1970).

# INDEX

Note: Titles of books and stories are listed alphabetically under the names of the authors.